CAMBRIDGE LIBRARY COLLECTION

Books of enduring scholarly value

Physical Sciences

From ancient times, humans have tried to understand the workings of
the world around them. The roots of modern physical science go back to
the very earliest mechanical devices such as levers and rollers, the mixing
of paints and dyes, and the importance of the heavenly bodies in early
religious observance and navigation. The physical sciences as we know them
today began to emerge as independent academic subjects during the early
modern period, in the work of Newton and other 'natural philosophers',
and numerous sub-disciplines developed during the centuries that followed.
This part of the Cambridge Library Collection is devoted to landmark
publications in this area which will be of interest to historians of science
concerned with individual scientists, particular discoveries, and advances in
scientific method, or with the establishment and development of scientific
institutions around the world.

James Nasmyth, Engineer

This autobiography was first published in 1883, and recounts the life of the
Scottish scientist and inventor James Nasmyth (1808–90), who was arguably
the last of the early pioneers of the machine tool industry, most famously
remembered for his invention of the steam hammer. He also produced and
manufactured several other important machine tools, including a hydraulic
press which used water pressure to force tight-fitting machine parts together.
All of these machines became popular in manufacturing, and all are still
in use today in modified forms. Nasmyth retired from business in 1856 at
the age of just 48, and pursued his various hobbies including astronomy; he
was co-author of *The Moon: Considered as a Planet, a World, and a Satellite*
(1874) with James Carpenter. This autobiography follows a chronological
order, and a list of Nasmyth's inventions is given at the end of the book.

Cambridge University Press has long been a pioneer in the reissuing of out-of-print titles from its own backlist, producing digital reprints of books that are still sought after by scholars and students but could not be reprinted economically using traditional technology. The Cambridge Library Collection extends this activity to a wider range of books which are still of importance to researchers and professionals, either for the source material they contain, or as landmarks in the history of their academic discipline.

Drawing from the world-renowned collections in the Cambridge University Library, and guided by the advice of experts in each subject area, Cambridge University Press is using state-of-the-art scanning machines in its own Printing House to capture the content of each book selected for inclusion. The files are processed to give a consistently clear, crisp image, and the books finished to the high quality standard for which the Press is recognised around the world. The latest print-on-demand technology ensures that the books will remain available indefinitely, and that orders for single or multiple copies can quickly be supplied.

The Cambridge Library Collection will bring back to life books of enduring scholarly value (including out-of-copyright works originally issued by other publishers) across a wide range of disciplines in the humanities and social sciences and in science and technology.

James Nasmyth, Engineer

An Autobiography

James Nasmyth
Edited by Samuel Smiles

CAMBRIDGE UNIVERSITY PRESS

Cambridge, New York, Melbourne, Madrid, Cape Town, Singapore,
São Paolo, Delhi, Dubai, Tokyo

Published in the United States of America by Cambridge University Press, New York

www.cambridge.org
Information on this title: www.cambridge.org/9781108014465

© in this compilation Cambridge University Press 2010

This edition first published 1883
This digitally printed version 2010

ISBN 978-1-108-01446-5 Paperback

JAMES NASMYTH

JAMES NASMYTH

ENGINEER

AN AUTOBIOGRAPHY

EDITED BY

SAMUEL SMILES, LL.D.
AUTHOR OF 'LIVES OF THE ENGINEERS'

WITH A PORTRAIT BY GEORGE REID, R.S.A., ETCHED BY PAUL RAJON,
AND NUMEROUS ILLUSTRATIONS

LONDON
JOHN MURRAY, ALBEMARLE STREET
1883

PREFACE.

I HAVE had much pleasure in editing the following Memoir of my friend Mr. Nasmyth. Some twenty years since (in April 1863), when I applied to him for information respecting his mechanical inventions, he replied : " My life presents no striking or remarkable incidents, and would, I fear, prove but a tame narrative. The sphere to which my endeavours have been confined has been of a comparatively quiet order ; but, vanity apart, I hope I have been able to leave a few marks of my existence behind me in the shape of useful contrivances, which are in many ways helping on great works of industry."

Mr. Nasmyth, nevertheless, kindly furnished me with information respecting himself, as well as his former master and instructor, Henry Maudsley of London, for the purpose of being inserted in *Industrial Biography, or Ironworkers and Toolmakers*, which was published at the end of 1863. He was of opinion that the outline of his life there presented was sufficiently descriptive of his career as a mechanic and in-ventor.

During the years that have elapsed since then, Mr. Nasmyth has been prevailed upon by some of his friends—more especially by Sir John Anderson, late of Woolwich

Arsenal—to note down the reminiscences of his life, with
an account of his inventions, and to publish them for the
benefit of others.   He has accordingly spent some of his
well-earned leisure during the last two years in writing out
his numerous recollections.   Having consulted me on the
subject, I recommended that they should be published in the
form of an Autobiography, and he has willingly given his
consent.

Mr. Nasmyth has furnished me with abundant notes of
his busy life, and he has requested me, in preparing them for
publication, to "make use of the pruning-knife."   I hope,
however, that in editing the book I have not omitted anything
that is likely to be interesting or instructive.   I must add
that everything has been submitted to his correction and
received his final approval.

The narrative abundantly illustrates Mr. Nasmyth's own
definition of Engineering; namely, *common sense applied to
the use of materials*.   In his case, common sense has been
more especially applied to facilitating and perfecting work by
means of Machine Tools.   Civilisation began with tools ; and
every step in advance has been accomplished through their
improvement.   Handicraft labour, in bone, stone, or wood,
was the first stage in the development of man's power ; and
tools or machines, in iron or steel, are the last and most
efficient method of economising it, and enabling him to intelligently direct the active and inert forces of nature.

It will be observed that Mr. Nasmyth, on his first start
in life, owed much to the influence of his father, who was
not only an admirable artist —"the founder," as Sir David
Wilkie termed him, " of the landscape painting school of
Scotland "— but an excellent mechanic.   His " bow-and-
string" roofs and bridges show his original merits as a

designer ; and are sufficient to establish his ability as a
mechanical engineer.   Indeed, one of Mr. Nasmyth's principal
objects, in preparing the notes of the following work, has been
to introduce a Memorial to the memory of his father, to
whom he owed so much, and to whom he was so greatly
attached through life.   Hence the numerous references to
him, and the illustrations from his works of art, of architecture, as well as of mechanics, given in the early part of the
book.

I might point out that Mr. Nasmyth's narrative has a
strong bearing upon popular education ; not only as regards economical use of time, careful observation, close
attention to details, but as respects the uses of Drawing.
The observations which he makes as to the accurate knowledge
of this art are very important.   In this matter he concurs
with Mr. Herbert Spencer in his work on *Education*.   " It is
very strange," Mr. Nasmyth said some years ago, " that
amidst all our vaunted improvements in education, the
faculty of comparison by sight, or what may be commonly
called *the correctness of eye*, has been so little attended to."
He accordingly urges the teaching of rudimentary drawing in
all public schools.   "Drawing is," he says, "the Education of
the Eye.   It is more interesting than words.   It is graphic
language."

The illustrations given in the course of the following
book will serve to show his own mastery of drawing —
whether as respects Mechanical details, the Moon's surface, or
the fairy-land of Landscape.   It is perhaps not saying too
much to aver that had he not devoted his business life to
Mechanics, he would, like his father, his brother Patrick, and
his sisters, have taken a high position as an artist.   In the
following Memoir we have only been able to introduce a

few specimens of his drawings; but "The Fairies," "The Antiquary," and others, will give the reader a good idea of Mr. Nasmyth's artistic ability.

Since his retirement from business life, at the age of forty-eight, Mr. Nasmyth's principal pursuit has been Astronomy. His Monograph on "The Moon," published in 1874, exhibits his ardent and philosophic love for science in one of its sublimest aspects. His splendid astronomical instruments, for the most part made entirely by his own hands, have enabled him to detect the "willow leaf-shaped" objects which form the structural element of the Sun's luminous surface. The discovery was shortly after verified by Sir John Herschel and other astronomers, and is now a received fact in astronomical science.

A Chronological List of some of Mr. Nasmyth's contrivances and inventions is given at the end of the volume, which shows, so far, what he has been enabled to accomplish during his mechanical career. These begin at a very early age, and were continued for about thirty years of a busy and active life. Very few of them were patented ; many of them, though widely adopted, are unacknowledged as his invention. They, nevertheless, did much to advance the mechanical arts, and still continue to do excellent service in the engineering world.

The chapter relating to the origin of the Cuneiform Character, and of the Pyramid or Sun-worship in its relation to Egyptian Architecture, is placed at the end, so as not to interrupt the personal narrative. That chapter, it is believed, will be found very interesting, illustrated, as it is, by Mr. Nasmyth's drawings.   S. S.

LONDON, *January* 1883.

CONTENTS.

CHAPTER I.

MY ANCESTRY.

CHAPTER II.

ALEXANDER NASMYTH.

CHAPTER III.

AN ARTIST'S FAMILY.

CHAPTER IV.

MY EARLY YEARS.

CHAPTER V.

MY SCHOOL-DAYS.

CHAPTER VI.

MECHANICAL BEGINNINGS.

CHAPTER VII.

HENRY MAUDSLEY, LONDON.

CHAPTER XII.

FREE TRADE IN ABILITY—THE STRIKE—DEATH OF MY FATHER.

CHAPTER XIII.

MY MARRIAGE—THE STEAM HAMMER.

CHAPTER XIV.

TRAVELS IN FRANCE AND ITALY.

CHAPTER XV.

STEAM HAMMER PILE DRIVER.

ILLUSTRATIONS.

ILLUSTRATIONS.

EDINBURGH CASTLE FROM THE VENNEL, GRASSMARKET. BY ALEXANDER NASMYTH.

AUTOBIOGRAPHY

CHAPTER I.

MY ANCESTRY.

OUR history begins long before we are born. We represent the hereditary influences of our race, and our ancestors virtually live in us. The sentiment of ancestry seems to be inherent in human nature, especially in the more civilised races. At all events, we cannot help having a due regard for our forefathers. Our curiosity is stimulated by their immediate or indirect influence upon ourselves. It may be a generous enthusiasm, or, as some might say, a harmless vanity, to take pride in the honour of their name. The gifts of nature, however, are more valuable than those of fortune; and no line of ancestry, however honourable, can absolve us from the duty of diligent application and perseverance, or from the practice of the virtues of self-control and self-help.

Sir Bernard Burke, in his *Peerage and Baronetage*, gives a faithful account of the ancestors from whom I am lineally descended.[1] " The family of Naesmyth," he says, " is one of remote antiquity in Tweeddale, and has possessed lands there since the 13th century." They fought in the wars of Bruce and Baliol, which ended in the independence of Scotland.

---

[1] Sir B. Burke's *Peerage and Baronetage.* Ed. 1879. Pp. 885-6.

The following is the family legend of the origin of the name of Naesmyth :—

In the troublous times which prevailed in Scotland before the union of the Crowns, the feuds between the King and the Barons were almost constant. In the reign of James III. the House of Douglas was the most prominent and ambitious. The Earl not only resisted his liege lord, but entered into a combination with the King of England, from whom he received a pension. He was declared a rebel, and his estates were confiscated. He determined to resist the royal power, and crossed the Border with his followers. He was met by the Earl of Angus, the Maxwells, the Johnstons, and the Scotts. In one of the engagements which ensued the Douglases appeared to have gained the day, when an ancestor of the Naesmyths, who fought under the royal standard, took refuge in the smithy of a neighbouring village. The smith offered him protection, disguised him as a hammerman, with a leather apron in front, and asked him to lend a hand at his work.

While thus engaged a party of the Douglas partisans entered the smithy. They looked with suspicion on the disguised hammerman, who, in his agitation, struck a false blow with the sledge hammer, which broke the shaft in two. Upon this, one of the pursuers rushed at him, calling out, " Ye're *nae smyth !*" The stalwart hammerman turned upon his assailant, and, wrenching a dagger from him, speedily overpowered him. The smith himself, armed with a big hammer, effectually aided in overpowering and driving out the Douglas men. A party of the royal forces made their appearance, when Naesmyth rallied them, led them against the rebels, and converted what had been a temporary defeat into a victory. A grant of lands was bestowed upon him for his service. His armorial bearings consisted of a hand dexter with a dagger, between two broken hammer-shafts, and there they remain to this day. The motto was, *Non arte sed marte,*

" Not by art but by war."　In my time I have reversed the motto (*Non marte sed arte*); and instead of the broken hammer-shafts, I have adopted, not as my " arms " but as a device, the most potent form of mechanical art—the Steam Hammer.

Sir Michael Naesmyth, Chamberlain of the Archbishop of St. Andrews, obtained the lands of Posso and Glenarth in

ORIGIN OF THE NAME.　BY JAMES NASMYTH.

1544, by right of his wife, Elizabeth, daughter and heiress of John Baird of Posso.　The Bairds have ever been a loyal and gallant family.　Sir Gilbert, father of John Baird, fell at Flodden in 1513, in defence of his king.　The royal eyrie of Posso Crag is on the family estate; and the Lure worn by Queen Mary, and presented by her son James VI. to James Naesmyth, the Royal Falconer, is still preserved as a family heirloom.

During the intestine troubles in Scotland, in the reign
of Mary, Sir Michael Naesmyth espoused the cause of the
unfortunate Queen.   He fought under her banner at Lang-
side in 1568.   He was banished, and his estates were seized
by the Regent Moray.   But after the restoration of peace,
the Naesmyths regained their property.   Sir Michael died
at an advanced age.

He had many sons.   The eldest, James, married Joana,
daughter of William Veitch or Le Veitch of Dawick.   By
this marriage the lands of Dawick came into the family.   He
predeceased his father, and was succeeded by his son James,
the Royal Falconer above referred to.   Sir Michael's second
son, John, was chief chirurgeon to James VI. of Scotland,
afterwards James I. of England, and to Henry, Prince of
Wales.   He died in London in 1613, and in his testament
he leaves " his hert to his young master, the Prince's grace."
Charles I., in his instructions to the President of the Court
of Session, enjoins " that you take special notice of the
children of John Naesmyth, so often recommended by our late
dear father and us."   Two of Sir Michael's other sons were
killed at Edinburgh in 1588, in a deadly feud between the
Scotts and the Naesmyths.   In those days a sort of Corsican
vendetta was carried on between families from one generation
to another.

Sir Michael Naesmyth, son of the Royal Falconer, succeeded to the property.   His eldest son James was appointed
to serve in Claverhouse's troop of horse in 1684.   Among
the other notable members of the family was James Naesmyth, a very clever lawyer.   He was supposed to be so
deep that he was generally known as the " Deil o' Dawyk."
His eldest son was long a member of Parliament for the
county of Peebles ; he was, besides, a famous botanist, having
studied under Linnæus.   Among the inter-marriages of the
family were those with the Bruces of Lethen, the Stewarts
of Traquhair, the Murrays of Stanhope, the Pringles of Clif-

ton, the Murrays of Philiphaugh, the Keiths (of the Earl
Marischal's family), the Andersons of St. Germains, the
Marjoribanks of Lees, and others.

In the fourteenth century a branch of the Naesmyths of
Posso settled at Netherton, near Hamilton. They bought
an estate and built a residence. The lands adjoined part of
the Duke of Hamilton's estate, and the house was not far
from the palace. There the Naesmyths remained until the
reign of Charles II. The King, or his advisers, determined
to introduce Episcopacy, or, as some thought, Roman Catho-
licism, into the country, and to enforce it at the point of the
sword.

The Naesmyths had always been loyal until now. But
to be cleft by sword and pricked by spear into a religion
which they disbelieved, was utterly hateful to the Netherton
Naesmyths. Being Presbyterians, they held to their own
faith. They were prevented from using their churches,
and they accordingly met on the moors, or in unfrequented
places for worship.[1] The dissenting Presbyterians assumed
the name of Covenanters. Hamilton was almost the centre
of the movement. The Covenanters met, and the King's
forces were ordered to disperse them. Hence the inter-
necine war that followed. There were Naesmyths on both
sides—Naesmyths for the King, and Naesmyths for the
Covenant.

In an early engagement at Drumclog, the Covenanters
were victorious. They beat back Claverhouse and his
dragoons. A general rising took place in the West
Country. About 6000 men assembled at Hamilton,
mostly raw and undisciplined countrymen. The King's
forces assembled to meet them,—10,000 well-disciplined

---

[1] In the reign of James II. of England and James VII. of Scotland a law
was enacted, " that whoever should preach in a conventicle under a roof, or
should attend, either as a preacher or as a hearer, a conventicle in the open
air, *should be punished with death and confiscation of property.*"

troops, with a complete train of field artillery. What chance had the Covenanters against such a force ? Nevertheless, they met at Bothwell Bridge, a few miles west of Hamilton.

It is unnecessary to describe the action.[1] The Covenanters, notwithstanding their inferior force, resisted the cannonade and musketry of the enemy with great courage. They defended the bridge until their ammunition failed. When the English Guards and the artillery crossed the bridge, the battle was lost. The Covenanters gave way, and fled in all directions ; Claverhouse, burning with revenge for his defeat at Drumclog, made a terrible slaughter of the unresisting fugitives. One of my ancestors brought from the battlefield the remnant of the standard ; a formidable musquet—" Gun Bothwell " we afterwards called it ; an Andrea Ferrara ; and a powder-horn. I still preserve these remnants of the civil war.

My ancestor was condemned to death in his absence, and his property at Netherton was confiscated. What became of him during the remainder of Charles II.'s reign, and the reign of that still greater tormentor, James II., I do not know. He was probably, like many others, wandering about from place to place, hiding "in wildernesses or caves, destitute, afflicted, and tormented." The arrival of William III. restored religious liberty to the country, and Scotland was again left in comparative peace.

My ancestor took refuge in Edinburgh, but he never recovered his property at Netherton. The Duke of Hamilton, one of the trimmers of the time, had long coveted the possession of the lands, as Ahab had coveted Naboth's vineyard. He took advantage of the conscription of the men engaged in the Bothwell Brig conflict, and had the lands

---

[1] See the account of a Covenanting Officer in the Appendix to the *Scots Worthies*. See also Sir Walter Scott's *Old Mortality*, where the battle of Bothwell Brig is described.

forfeited in his favour.   I remember my father telling me
that, on one occasion when he visited the Duke of Hamilton
in reference to some improvement of the grounds adjoining
the palace, he pointed out to the Duke the ruined remains
of the old residence of the Naesmyths. As the first French
Revolution was then in full progress, when ideas of society
and property seemed to have lost their bearings, the Duke
good-humouredly observed, " Well, well, Naesmyth, there's no
saying but what, some of these days, your ancestors' lands
may come into your possession again! "

Before I quit the persecutions of "the good old times,"
I must refer to the burning of witches.   One of my ancient
kinswomen, Elspeth Naesmyth, who lived at Hamilton, was
denounced as a witch.   The chief evidence brought against
her was that she kept four black cats, and read her Bible
with two pairs of spectacles! a practice which shows that
she possessed the spirit of an experimental philosopher.   In
doing this she adopted a mode of supplementing the power
of spectacles in restoring the receding power of the eyes.
She was in all respects scientifically correct.   She increased
the magnifying power of the glasses ; a practice which is
preferable to single glasses of the same power, and which I
myself often follow.   Notwithstanding this improved method
of reading her Bible, and her four black cats, she was condemned to be burnt alive !   She was about the last victim
in Scotland to the disgraceful superstition of witchcraft.

The Naesmyths of Netherton having lost their ancestral
property, had to begin the world again.   They had to begin
at the beginning.   But they had plenty of pluck and
energy.   I go back to my great-great-grandfather, Michael
Naesmyth, who was born in 1652.   He occupied a house
in the Grassmarket, Edinburgh, which was afterwards
rebuilt, in 1696.   His business was that of a builder and
architect.   His chief employment was in designing and
erecting new mansions, principally for the landed gentry

and nobility.   Their old castellated houses or towers were
found too dark and dreary for modern uses.   The draw-
bridges were taken down, and the moats were filled up.
Sometimes they built the new mansions as an addition to
the old.   But oftener they left the old castles to go to ruin;

MICHAEL NAESMYTH'S HOUSE, GRASSMARKET.[1]

or, what was worse, they made use of the stone and other
materials of the old romantic buildings for the construction
of their new residences.

Michael Naesmyth acquired a high reputation for the
substantiality of his work.   His masonry was excellent, as
well as his woodwork.   The greater part of the latter was

[1] The lower house, at the right hand corner of the engraving, with the
three projecting gable ends, is the house in question.

executed in his own workshops at the back of his house in
the Grassmarket.   His large yard was situated between the
back of the house and the high wall that bounded the Greyfriars Churchyard, to the east of the flight of steps which
forms the main approach to George Heriot's Hospital.

INVERSNAID FORT.   AFTER A DRAWING BY ALEXANDER NASMYTH.

The last work that Michael Naesmyth was engaged in
cost him his life.   He had contracted with the Government
to build a fort at Inversnaid, at the northern end of Loch
Lomond.   It was intended to guard the Lowlands, and keep
Rob Roy and his caterans within the Highland Border.   A
promise was given by the Government that during the progress of the work a suitable force of soldiers should be quartered
close at hand to protect the builder and his workmen.

Notwithstanding many whispered warnings as to the danger of undertaking such a hazardous work, Michael Naesmyth and his men encamped upon the spot, though without the protection of the Government force. Having erected a temporary residence for their accommodation, he proceeded with the building of the fort. The work was well advanced by the end of 1703, although the Government had treated all Naesmyth's appeals for protection with evasion or contempt.

Winter set in with its usual force in those northern regions. One dark and snowy night, when Michael and his men had retired to rest, a loud knocking was heard at the door. "Who's there?" asked Michael. A man outside replied, "A benighted traveller overtaken by the storm." He proceeded to implore help, and begged for God's sake that he might have shelter for the night. Naesmyth, in the full belief that the traveller's tale was true, unbolted and unbarred the door, when in rushed Rob Roy and his desperate gang. The men, with the dirks of the Macgregors at their throats, begged hard for their lives. This was granted on condition that they should instantly depart, and take an oath that they should never venture within the Highland border again.

Michael Naesmyth and his men had no alternative but to submit, and they at once left the bothy with such scanty clothing as the Macgregors would permit them to carry away. They were marched under an armed escort through the snowstorm to the Highland border, and were there left with the murderous threat that, if they ever returned to the fort, certain death would meet them.[1]

Poor Michael never recovered from the cold which he

---

[1] Another attempt was made to build the fort at Inversnaid. But Rob Roy again surprised the small party of soldiers who were in charge. They were disarmed and sent about their business. Finally, the fort was rebuilt, and placed under command of Captain (afterwards General) Wolfe. When peace fell upon the Highlands and Rob Roy's country became the scene of picnics, the fort was abandoned and allowed to go to ruin.

caught during his forced retreat from Inversnaid. The effects of this, together with the loss and distress of mind which he experienced from the Government's refusal to pay for his work—notwithstanding their promise to protect him and his workmen from the Highland freebooters—so preyed upon his mind that he was never again able to devote himself to business. One evening, whilst sitting at his fireside with his grandchild on his knee, a death-like faintness came over him; he set the child down carefully by the side of his chair, and then fell forward dead on his own hearthstone.

Thus ended the life of Michael Naesmyth in 1705, at the age of fifty-three. He was buried by the side of his ancestors in the old family tomb in the Greyfriars Churchyard.

This old tomb, dated 1614, though much defaced, is one of the most remarkable of the many which surround the walls of that ancient and memorable burying-place. Greyfriars Churchyard is one of the most interesting places in Edinburgh. The National Covenant was signed there by the Protestant nobles and gentry of Scotland in 1638. The prisoners taken at the battle of Bothwell Brig were shut up there in 1679, and, after enduring great privations, a portion of the survivors were sent off to Barbadoes.

When I first saw the tombstone, an ash tree was growing out of the top of the main body of it, though that has since been removed. In growing, the roots had pushed out the centre stone, which has not been replaced. The tablet over it contains the arms of the family, the broken hammer-shafts, and the motto " *Non arte sed marte.* " There are the remains of a very impressive figure, apparently rising from her cerements. The body and extremities remain, but the head has been broken away. There is also a remarkable motto on the tablet above the tombstone—" *Ars mihi vim contra Fortunæ;* " which I take to be, " Art is my strength in contending against Fortune,"—a motto which is appropriate to my ancestors as well as to myself.

The business was afterwards carried on by Michael's son, my great-grandfather.   He was twenty-seven years old at the time of his father's death, and lived to the age of seventy-three.   He was a man of much ability and of large experience.   One of his great advantages in carrying on his

THE NAESMYTH TOMB IN GREYFRIARS CHURCHYARD.

business was the support of a staff of able and trustworthy foremen and workmen.   The times were very different then from what they are now.   Masters and men lived together in mutual harmony.   There was a kind of loyal family attachment among them, which extended through

many generations. Workmen had neither the desire nor
the means for shifting about from place to place. On the
contrary, they settled down with their wives and families in
houses of their own, close to the workshops of their employers. Work was found for them in the dull seasons
when trade was slack, and in summer they sometimes removed to jobs at a distance from headquarters. Much of
this feeling of attachment and loyalty between workmen
and their employers has now expired. Men rapidly remove
from place to place. Character is of little consequence.
The mutual feeling of goodwill and zealous attention to
work seems to have passed away. Sudden change, scamp-
ing, and shoddy have taken their place.

My grandfather, Michael Naesmyth, succeeded to the
business in 1751. He more than maintained the reputation of his predecessors. The collection of first-class works
on architecture which he possessed, such as the folio editions
of Vitruvius and Palladio, which were at that time both rare
and dear, showed the regard he had for impressing into his
designs the best standards of taste. The buildings he
designed and erected for the Scotch nobility and gentry were
well arranged, carefully executed, and thoroughly substantial.
He was also a large builder in Edinburgh. Amongst the
houses he erected in the Old Town were the principal
number of those in George Square. In one of these, No.
25, Sir Walter Scott spent his boyhood and youth. They
still exist, and exhibit the care which he took in the elegance
and substantiality of his works.

I remember my father pointing out to me the extreme
care and attention with which he finished his buildings.
He inserted small fragments of basalt into the mortar of the
external joints of the stones, at close and regular distances,
in order to protect the mortar from the adverse action of
the weather. And to this day they give proof of their
efficiency; the basalt protects the joints, and at the same time

gives a neat and pleasing effect to what would otherwise
have been the merely monotonous line of mason-work.

A great change was about to take place in the residences
of the principal people of Edinburgh. The cry was for
more light and more air. The extension of the city to the
south and west was not sufficient. There was a great
plateau of ground on the north side of the city, beyond the
North Loch. But it was very difficult to reach; being
alike steep on both sides of the Loch. At length, in 1767,
an Act was obtained to extend the royalty of the city over
the northern fields, and powers were obtained to erect a
bridge to connect them with the Old Town.

The magistrates had the greatest difficulty in inducing
the inhabitants to build dwellings on the northern side of
the city. A premium was offered to the person who should
build the first house; and £20 was awarded to Mr. John
Young on account of a mansion erected by him close to
George Street. Exemption from burghal taxes was also
granted to a gentleman who built the first house in Princes
Street. My grandfather built the first house in the south-
west corner of St. Andrew Square, for the occupation of
David Hume the historian, as well as the two most
important houses in the centre of the north side of the same
square. One of these last was occupied by the venerable
Dr. Hamilton, a very conspicuous character in Edinburgh.
He continued to wear the cocked hat, the powdered pigtail,
tights, and large shoe buckles, for about sixty years after the
costume had become obsolete. All these houses are still in
perfect condition, after resisting the ordinary tear and wear
of upwards of a hundred and ten northern winters. The
opposition to building houses across the North Loch soon
ceased; and the New Town arose, growing from day to day,
until Edinburgh became one of the most handsome and
picturesque cities in Europe.

There is one other thing that I must again refer to—

namely, the highly-finished character of my grandfather's work. Nothing merely moderate would do. The work must be of the very best. He took special pride in the sound quality of the woodwork and its careful workmanship. He chose the best Dantzic timber because of its being of purer grain and freer from knots than other wood. In those days the lower part of the walls of the apartments were wainscoted—that is, covered by timber framed in large panels. They were from three to four feet wide, and from six to eight feet high. To fit these in properly required the most careful joiner-work.

It was always a holiday treat to my father, when a boy, to be permitted to go down to Leith to see the ships discharge their cargoes of timber. My grandfather had a wood-yard at Leith, where the timber selected by him was piled up to be seasoned and shrunk, before being worked into its various appropriate uses. He was particularly careful in his selection of boards or stripes for floors, which must be perfectly level, so as to avoid the destruction of the carpets placed over them. The hanging of his doors was a matter that he took great pride in—so as to prevent any uneasy action in opening or closing. His own chamber doors were so well hung that they were capable of being opened and closed by the slight puff of a hand-bellows.

The excellence of my grandfather's workmanship was a thing that my own father always impressed upon me when a boy. It stimulated in me the desire to aim at excellence in everything that I undertook; and in all practical matters to arrive at the highest degree of good workmanship. I believe that these early lessons had a great influence upon my future career.

I have little to record of my grandmother. From all accounts, she was everything that a wife and mother should be. My father often referred to her as an example of the affection and love of a wife to her husband, and of a mother to her children. The only relic I possess of her handiwork is a sampler, dated 1743, the needlework of which is so delicate and neat, that to me it seems to excel everything of the kind that I have seen.

I am fain to think that her delicate manipulation in some respects descended to her grandchildren, as all of them have been more or less distinguished for the delicate use of their fingers—which has so much to do with the effective transmission of the artistic faculty into visible forms. The power of transmitting to paper or canvas the artistic conceptions of the brain through the fingers, and out at the end of the needle, the pencil, the pen, or brush, or even the modelling tool or chisel, is that which, in practical fact, constitutes the true artist.

This may appear a digression; though I cannot look at my grandmother's sampler without thinking that she had much to do with originating the Naesmyth love of the Fine Arts, and their hereditary adroitness in the practice of landscape and portrait painting, and other branches of the profession.

My grandfather died in 1803, at the age of eighty-four, and was buried by his father's side in the Naesmyth ancestral tomb in Greyfriars Churchyard. His wife, Mary Anderson, who died before him, was buried in the same place.

Michael Naesmyth left two sons—Michael and Alexander. The eldest was born in 1754. It was intended that he should have succeeded to the business; and, indeed, as soon as he reached manhood he was his father's right-hand man. He was a skilful workman, especially in the finer parts of joiner-work. He was also an excellent accountant and bookkeeper. But having acquired a taste for reading books about voyages and travels, of which his father's library was well supplied, his mind became disturbed, and he determined to see something of the world. He was encouraged by one of his old companions, who had been to sea, and realised some substantial results by his voyages to foreign parts. Accordingly Michael, notwithstanding the earnest remonstrances of his father, accompanied his friend on the next occasion when he went to sea.

After several voyages to the West Indies and other parts of the world, which both gratified and stimulated his natural taste for adventures, and also proved financially successful, his trading ventures at last met with a sad reverse, and he resolved to abandon commerce, and enter the service of the Royal Navy. He was made purser, and in this position he entered into a new series of adventures. He was present at many naval engagements. But he lost neither life nor limb. At last he was pensioned, and became a resident at Greenwich Hospital. He furnished the rooms that were granted him with all manner of curiosities, which his roving naval life had enabled him to collect. His original skill as a worker in wood came to life again. The taste of the workman and the handiness of the seaman enabled him to furnish his rooms at the Hospital in a most quaint and amusing manner.

My father had a most affectionate regard for Michael, and always spent some days with him when he had occasion to visit London. One bright summer day they went to have a stroll together on Blackheath ; and while my uncle was enjoying a nap on a grassy knoll, my father made a sketch of him, which I still preserve. Being of a most cheerful disposition, and having a great knack of detailing the incidents of his adventurous life, he became a great favourite with the resident officers of the Hospital ; and was always regarded by them as real good company. He ended his days there in peace and comfort, in 1819, at the age of sixty-four.

CHAPTER II.

My father, Alexander Nasmyth, was the second son of Michael Nasmyth. He was born in his father's house in the Grassmarket on the 9th of September 1758. The Grassmarket was then a lively place. On certain days of the week it was busy with sheep and cattle fairs. It was the centre of Edinburgh traffic. Most of the inns were situated there, or in the street leading up to the Greyfriars' Church gate.

The view from my grandfather's house was very grand. Standing up, right opposite, was the steep Castle rock, with its crown buildings and circular battery towering high overhead. They seemed almost to hang over the verge of the rock. The houses on the opposite side of the Grassmarket were crowded under the esplanade of the Castle Hill.

There was an inn opposite the house where my father was born, from which the first coach started from Edinburgh to Newcastle. The public notice stated that " The Coach would set out from the Grass Market ilka Tuesday at Twa o'clock in the day, GOD WULLIN', but *whether or no* on Wednesday." The " whether or no " was meant, I presume, as a precaution to passengers, in case all the places on the coach might not be taken on Wednesday.

The Grassmarket was also the place for public executions.

The gibbet stone was at the east end of the Market. It consisted of a mass of solid sandstone, with a quadrangular hole in the middle, which served as a socket for the gallows. Most of the Covenanters who were executed for conscience' sake in the reigns of Charles II. and James II. breathed their last at this spot. The Porteous mob, in 1736, had its culmination here. When Captain Porteous was dragged out of the Tolbooth in the High Street, and hurried down the West Bow, the gallows was not in its place; but the leaders of the mob hanged him from a dyer's pole, nearly

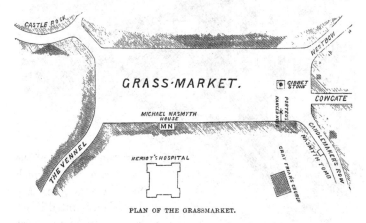

PLAN OF THE GRASSMARKET.

opposite the gallows stone, on the south side of the street, not far from my grandfather's door.[1]

I have not much to say about my father's education. For the most part, he was his own schoolmaster. I have heard him say that his mother taught him his A B C; and that he afterwards learned to read at Mammy Smith's. This old lady kept a school for boys and girls at the top of a house in the Grassmarket. There my father was taught to read his Bible, and to learn his Carritch.[2]

---

[1] See *Heart of Midlothian*. [2] *The Shorter Catechism*.

As it was only the bigger boys who could read the Bible, the strongest of them consummated the feat by climbing up the Castle rock, and reaching what they called "The Bibler's Seat." It must have been a break-neck adventure to get up to the place. The seat was almost immediately under

the window of the room in which James VI. was born. My father often pointed it out to me as one of the most dangerous bits of climbing in which he had been engaged in his younger years. The annexed illustration is from his own slight sepia drawing; the Bibler's Seat is marked †.

THE BIBLER'S SEAT.

Not so daring, but much more mischievous, was a trick which he played with some of his companions on the tops of the houses on the north side of the Grassmarket. The boys took a barrel to the Castlehill, filled it with small stones, and then shot it down towards the roofs of the houses in the Grassmarket. The barrel leapt from rock to rock, burst, and scattered a shower of stones far and wide. The fun was to see the "boddies" look out of their garret windows with their lighted lamps or candles, peer into the dark, and try to see what was the cause of the mischief. Sir David Baird, the hero of Seringapatam, played a trick of the same kind before he went to India.

Among my father's favourite companions were the two sons of Dr. John Erskine, minister of Old Greyfriars, in conjunction with the equally celebrated Dr. Robertson. Dr. Erskine was a man of great influence in his day, well known for his literary and theological works, as well as for his piety and practical benevolence.[1] On one occasion, when my father was at play with the boys, one of them threw a stone, which smashed a neighbour's window. A servant of the house ran out, and seeing the culprit, called out, " Very weel, Maister Erskine, I'll tell yeer faither wha broke the windae!" On which the boy, to throw her off the scent, said to his brother loudly, " Eh, keist! she thinks we're the boddy Erskine's sons."

The boddy Erskine! Who ever heard of such an ir-reverent nickname applied to that good and great man ? " The laddies couldna be his sons," thought the woman. She made no further inquiry, and the boys escaped scot free. The culprit afterwards entered the service of the East India Company. " The boy was father to the man." He acquired great reputation at the siege of Seringapatam, where he led the forlorn hope. Erskine was promoted, until in course of time he returned to his native city a full-blown general.

To return to my father's education. After he left " Mammy Smith's," he went for a short time to the original

---

[1] Dr. Erskine is well described by Scott in *Guy Mannering*, on the occasion when Pleydell and Mannering went to hear him preach a famous sermon.

High School. It was an old establishment, founded by
James VI. before he succeeded to the English throne. It
was afterwards demolished to make room for the University
buildings; and the new High School was erected a little
below the old Royal Infirmary. After leaving the High
School, Alexander Nasmyth was taught by his father, first
arithmetic and mensuration, next geometry and mathematics,
so far as the first three books of Euclid were concerned.
After that, his own innate skill, ability, and industry
enabled him to complete the rest of his education.

At a very early period my father exhibited a decided
natural taste for art. He used his pencil freely in sketching
from nature; and in course of time he showed equal skill in
the use of oil colour. At his own earnest request he was
bound apprentice to Mr. Crighton, then the chief coachbuilder in Edinburgh. He was employed in that special
department where artistic taste was necessary—that is, in
decorating the panels of the highest class of carriages, and
painting upon them coats of arms, with their crests and
supporters. He took great pleasure in this kind of work.
It introduced him to the practical details of heraldry, and
he made great progress in his business.

But, still further to improve himself in the art of drawing, my father devoted his evenings to attending the
Edinburgh Drawing Academy. This institution, termed
"The Trustees' Academy of Fine Art," had been formed and
supported by the funds arising from the estates confiscated
after the rebellions of 1715 and 1745. Part of these funds
was set apart by the Government for the encouragement
of drawing, and also for the establishment of the arts of
linen weaving, carpet manufacture, and other industrial
occupations.

These arts had been introduced into Scotland by the
French Protestants, who had been persecuted for conscience'
sake out of their own country, and settled in England,

Ireland, and Scotland, for the prosecution of their industrial callings. The Corporation was anxious to afford an asylum for these skilled and able workmen. The emigrants settled down with their families, and pursued their occupations of damask, linen, and carpet weaving. They were also required to take in Scotch apprentices, to teach them the various branches of their trade. The Magistrates caused cottages and workshops to be erected on a piece of unoccupied land near Edinburgh, where the street appropriately called Picardy Place now stands—the greater number of the weavers having come from Picardy in France.

In connection with the establishment of these industrial artisans, it was necessary to teach the young Scotch apprentices drawing, for the purpose of designing new patterns suitable for the market. Hence the establishment by the Trustees of the Forfeited Estate Funds of " The Academy of Fine Art." From the designing of patterns, the institution advanced to the improvement of the fine arts generally. Young men who had given proofs of their natural taste for drawing were invited to enter the school and participate in its benefits.

At the time that my father was apprenticed to the coach painter, the Trustees' Academy was managed by Alexander Runciman. He had originally been a house painter, from which business he proceeded to landscape painting. " Other artists," said one who knew him, " talked meat and drink ; but Runciman talked landscape." He went to Rome and studied art there. He returned to Edinburgh, and devoted himself to historical painting. He was also promoted to the office of master of the Trustees' Academy. When my father called upon him with his drawings from nature, Runciman found them so satisfactory that he was at once admitted as a student. After his admission he began to study with intent eagerness. The young men who had been occupied by their business during the day could only attend in the

evening.   And so the evenings were fixed for studying
drawing and design.   The Trustees' Academy made its mark
upon the art of Scotland: it turned out many artists of
great note—such as Raeburn, Wilkie, my father, and many
more.

At the time when my father entered as a student, the
stock of casts from the antique, and the number of drawings
from the old masters, were very small.   So much so that
Runciman was under the necessity of setting the students to
copy them again and again.   This became rather irksome
to the more ardent pupils.   My father had completed his
sixth copy of a fine chalk drawing of " The Laocoon."   It
was then set for him to copy again.   He begged Mr. Runciman for another subject.   The quick-tempered man at
once said, " I'll give you another subject."   And turning the
group of the Laocoon upside down, he added, " Now, then,
copy that!"   The patient youth set to work, and in a few
evenings completed a perfect copy.   It was a most severe
test; but Runciman was so proud of the skill of his pupil
that he had the drawing mounted and framed, with a note
of the circumstances under which it had been produced.   It
continued to be hung there for many years, and the story of
its achievement became traditional in the school.

During all this time my father continued in the employment of Crighton the carriage builder.   He improved in his
painting day by day.   But at length an important change
took place in his career.   Allan Ramsay, son of the author
of *The Gentle Shepherd,* and then court painter to George
III., called upon his old friend Crighton one day, to look
over his works.   There he found young Nasmyth painting
a coat of arms on the panel of a carriage.   He was so much
struck with the lad's artistic workmanship—for he was then
only sixteen—that he formed a strong desire to take him
into his service.   After much persuasion, backed by the offer
of a considerable sum of money, the coachbuilder was at

length induced to transfer my father's indentures to Allan
Ramsay.

It was, of course, a great delight to my father to be removed to London under such favourable auspices. Ramsay
had a large connection as a portrait painter. His object in
employing my father was that the latter should assist him
in the execution of the subordinate parts, or dress portions,
of portraits for the court, or of diplomatic personages. No
more favourable opportunity for advancement could have
presented itself. But it was entirely due to my father's
perseverance and advancing skill as an artist—the result of
his steady application and labour.

Ramsay was possessed of a very fine collection of drawings by the old masters, all of which were free for my father
to study. Ramsay was exceedingly kind to his young pupil.
He was present at all the discussions in the studio, even
when the sitters were present. Fellow-artists visited Ramsay from time to time. Among them was his intimate friend
Philip Renegal—an agreeable companion, and an excellent
artist. Renegal was one day so much struck with my father's
earnestness in filling up some work, that he then and there
got up a canvas and made a capital sketch-portrait of him
in oil. It only came into my father's possession some years
after Ramsay's death, and is now in my possession.

Among the many amusing recollections of my father's
life in London, there is one that I cannot resist narrating,
because it shows his *faculty of resourcefulness*—a faculty
which served him very usefully during his course through
life. He had made arrangements with a sweetheart to take
her to Ranelagh, one of the most fashionable places of public
amusement in London. Everybody went in full dress, and
the bucks and swells wore long striped silk stockings. My
father, on searching, found that he had only one pair of silk
stockings left. He washed them himself in his lodging-room,
and hung them up before the fire to dry. When he went to

look at them, they were so singed and burned that he could not put them on.   They were totally useless.   In this sad dilemma his resourcefulness came to his aid.   The happy idea occurred to him of painting his legs so as to resemble stockings.   He went to his water-colour box, and dexterously painted them with black and white stripes.   When

ALEXANDER NASMYTH.   AFTER RENEGAL'S PORTRAIT.

the paint dried, which it soon did, he completed his toilet, met his sweetheart, and went to Ranelagh.   No one observed the difference, except, indeed, that he was complimented on the perfection of the fit, and was asked " where he bought his stockings ?"   Of course he evaded all such questions, and left the gardens without any one discovering his artistic trick.

My father remained in Allan Ramsay's service until the end of 1778, when he returned to Edinburgh to practise on his own behalf the profession of portrait painter. He took with him the kindest good-wishes of his master, whose friendship he retained to the end of Ramsay's life. The artistic style of my father's portraits, and the excellent likenesses of his sitters, soon obtained for him ample employment. His portraits were for the most part full-lengths, but of a small or cabinet size. They generally consisted of family groups, with the figures about twelve to fourteen inches high. The groups were generally treated and arranged as if the personages were engaged in conversation with their children; and sometimes a favourite servant was introduced, so as to remove any formal aspect in the composition of the picture. In order to enliven the background, some favourite view from the garden or grounds, or a landscape, was given; which was painted with as much care as if it was the main feature of the picture. Many of these paintings are still to be found in the houses of the gentry in Scotland. Good examples of his art are to be seen at Minto House, the seat of the Earl of Minto, and at Dalmeny Park, the seat of the Earl of Rosebery.

Among my father's early employers was Patrick Miller, Esq., of Dalswinton, in Dumfriesshire. He painted Mr. Miller's portrait as well as those of several members of his family. This intercourse eventually led to the establishment of a very warm personal friendship between them. Miller had made a large fortune in Edinburgh as a banker; and after he had partially retired from business, he devoted much of his spare time to useful purposes. He was a man of great energy of character, and was never idle. At first he applied himself to the improvement of agriculture, which he did with great success on his estate of Dalswinton. Being one of the largest shareholders in the Carron Ironworks near Stirling, he also devoted much of his time to the improve-

ment of the guns of the Royal Navy.   He was the inventor
of that famous gun the Carronade.   The handiness of these
short and effective guns, which were capable of being loaded
and fired nearly twice as quickly as the long small-bore
guns, gave England the victory in many a naval battle, where
the firing was close and quick, yardarm to yardarm.

But Mr. Miller's greatest claim to fame arises from his
endeavours to introduce steam-power as an agent in the
propulsion of ships at sea.   Mr. Clerk of Eldin had already
invented the system of " breaking the line " in naval engagements—a system that was first practised with complete
success by Lord Rodney in his engagement off Martinico in
1780.   The subject interested Mr. Miller so much that he
set himself to work to contrive some mechanical method by
which ships of war might be set in motion, independently of
wind, tide, or calms, so that Clerk's system of breaking the
line might be carried into effect under all circumstances.

It was about this time that my father was often with
Miller; and the mechanical devices by means of which the
method of breaking the line could be best accomplished was
the subject of many of their conversations.   Miller found
that my father's taste for mechanical contrivances, and also
his ready skill as a draughtsman, were likely to be of much
use to him, and he constantly visited the studio.   My
father reduced Miller's ideas to a definite form, and prepared a series of drawings, which were afterwards engraved
and published.   Miller's favourite design was, to divide the
vessel into twin or triple hulls, with paddles between them,
to be worked by the crew.   The principal experiment was
made in the Firth of Forth on the 2d of June 1787.   The
vessel was double-hulled, and was worked by a capstan of
five bars.   The experiment was on the whole successful.
But the chief difficulty was in the propulsive power.   After
a spurt of an hour or so, the men became tired with their
laborious work.   Mr. Taylor, student of divinity, and tutor

of Mr. Miller's sons, was on board, and seeing the exhausted state of the men at the capstan, suggested the employment of steam-power. Mr. Miller was pleased with the idea, and resolved to make inquiry upon the subject.

At that time William Symington, a young engineer from Wanlockhead, was exhibiting a road locomotive in Edinburgh. He was a friend of Taylor's, and Mr. Miller went to see the Symington model. In the course of his conversation with the inventor, he informed the latter of his own project, and described the difficulty which he had experienced in getting his paddle-wheels turned round. On which Symington immediately asked, "Why don't you use the steam-engine?" The model that Symington exhibited, produced rotary motion by the employment of ratchet-wheels. The rectilinear motion of the piston-rod was thus converted into rotary motion. Mr. Miller was pleased with the action of the ratchet-wheel contrivance, and gave Symington an order to make a pair of engines of that construction. They were to be used on a small pleasure-boat on Dalswinton Lake.

The boat was constructed on the double-hull or twin plan, so that the paddle should be used in the space between the hulls.[1] After much vexatious delay, arising from the entire novelty of the experiment, the boat and engines were at length completed, and removed to Dalswinton Lake. This, the first steamer that ever "trod the waters like a thing of life," the herald of a new and mighty power, was tried on the 14th of October 1788. The vessel steamed delightfully, at the rate of from four to five miles an hour, though this was not her extreme rate of speed. I append a copy of a sketch made by my father of this, the first actual steamboat, with her remarkable crew.

---

[1] This steam twin boat was in fact the progenitor of the *Castalia*, constructed about a hundred years later for the conveyance of passengers between Calais and Dover.

The persons on board consisted of Patrick Miller, William Symington, Sir William Monteith, Robert Burns (the poet, then a tenant of Mr. Miller's), William Taylor, and Alexander Nasmyth. There were also three of Mr. Miller's servants, who acted as assistants. On the edge of the lake was a young gentleman, then on a visit to Dalswinton. He

THE FIRST STEAMBOAT. BY ALEXANDER NASMYTH.[1]

was no less a person than Henry Brougham, afterwards Lord Chancellor of England. The assemblage of so many remarkable men was well worthy of the occasion.

---

[1] The original drawing of the steamer was done by my father, and lent by me to Mr. Woodcroft, who inserted it in his *Origin and Progress of Steam Navigation*. He omitted my father's name, and inserted only that of the lithographer, although it is a document of almost national importance in the history of Steam Navigation.

Taking into account the extraordinary results which have issued from this first trial of an actual steamboat, it may well be considered that this was one of the most important circumstances which ever occurred in the history of navigation. It ought, at the same time, to be remembered that all that was afterwards done by Symington, Fulton, and Bell followed long after the performance of this ever-memorable achievement.

I may also mention, as worthy of special record, that the hull of this first steamboat was of iron. It was constructed of tinned iron plate. It was therefore the first iron steamboat, if not the first iron ship, that had ever been made. I may also add that the engines, constructed by Symington, which propelled this first iron steamboat are now carefully preserved at the Patent Museum at South Kensington, where they may be seen by everybody.[1]

To return to my father's profession as a portrait painter. He had given so much assistance to Mr. Miller, while acting as his chief draughtsman in connection with the triple and twin ships, and also while attending him at Leith and elsewhere, that it had considerably interfered with his practice ; though everything was done by him *con amore*, in the best sense of the term. In return for this, however, Mr. Miller made my father the generous offer of a loan to enable him to visit Italy, and pursue his studies there. It was the most graceful mode in which Mr. Miller could express his obligations. It was an offer pure and simple, without security, and as such was thankfully accepted by my father.

In those days an artist was scarcely considered to have

---

[1] The original engines of the boat, with the ratchet-wheel contrivance of Symington, are there : the very engine that propelled the first steamer on Dalswinton Lake. It may be added that Mr. Miller expended about £30,000 on naval improvements, and, as is often the case, he was wholly overlooked by the Government.

completed his education until he had studied the works of the great masters at Florence and Rome. My father left England for Italy on the 30th of December 1782. He reached Rome in safety, and earnestly devoted himself to the study of art. He remained in Italy for the greater part of two years. He visited Florence, Bologna, Padua, and other cities where the finest works of art were to be found. He made studies and drawings of the best of them, besides making sketches from nature of the most remarkable places he had visited. He returned to Edinburgh at the end of 1784, and immediately resumed his profession of a portrait painter. He was so successful that in a short time he was enabled to repay his excellent friend Miller the £500 which he had so generously lent him a few years before.

The satisfactory results of his zealous practice, and of his skill and industry in his profession, together with the prospect of increasing artistic work, enabled him to bring to a happy conclusion an engagement he had entered into before leaving Edinburgh for Italy. I mean his marriage to my mother—one of the greatest events of his life—which took place on the 3rd of January 1786. Barbara Foulis was a distant relation of his own. She was the daughter of William Foulis, Esq., of Woodhall and Colinton, near Edinburgh. Her brother, the late Sir James Foulis, my uncle, succeeded to the ancient baronetcy of the family.[1]

My mother did not bring with her any fortune, so to speak, in the way of gold or acres; but she brought something far better into my father's home—a sweetness of

---

[1] In Burke's *Peerage and Baronetage* an account is given of the Foulis family. They are of Norman origin. A branch settled in Scotland in the reign of Malcolm Canmore. By various intermarriages, the Foulises are connected with the Hopetoun, Bute, and Rosebery families. The present holder of the title represents the houses of Colinton, Woodhall, and Ravelstone.

disposition, and a large measure of common sense, which made her, in all respects, the devoted helpmate of her husband. Her happy cheerful temperament, and her constant industry and attention, shed an influence upon all around her. By her example she inbred in her children the love of truth, excellence, and goodness. That was indeed the best fortune she could bring into a good man's home.

During the first year of my father's married life, when he lived in St James' Square, he painted the well-known portrait of Robert Burns the poet. Burns had been introduced to him by Mr. Miller at Dalswinton. An intimate friendship sprang up between the artist and the poet. The love of nature and of natural objects was common to both. They also warmly sympathised in their political views. When Burns visited Edinburgh my father often met him. Burns had a strange aversion to sit for his portrait, though often urgently requested to do so. But when at my father's studio, Burns at last consented, and his portrait was rapidly painted. It was done in the course of a few hours, and my father made a present of it to Mrs. Burns.[1] A mezzotint engraving of it was afterwards published by William Walker, son-in-law of the famous Samuel Reynolds. When the first proof impression was submitted to my father, he said to Mr. Walker : " I cannot better express to you my opinion of your admirable engraving, than by telling you that it conveys to me a more true and lively remembrance of Burns than my own picture of him does ; it so perfectly renders the spirit of his expression, as well as the details of his every feature."

While Burns was in Edinburgh, my father had many interesting walks with him in the neighbourhood of the romantic city. The Calton Hill, Arthur's Seat, Salisbury Crags, Habbie's How, and the nooks in the Pentlands, were always full of interest ; and Burns, with his brilliant and

---

[1] The portrait is now in the Royal Scottish Academy at Edinburgh.

humorous conversation, made the miles very short as they strode along.   Lockhart says, in his *Life of Burns*, that "the magnificent scenery of the Scottish capital filled the poet with extraordinary delight.   In the spring mornings he walked very often to the top of Arthur's Seat, and, lying prostrate on the turf, surveyed the rising of the sun out of the sea in silent admiration; his chosen companion on such occasions being that learned artist and ardent lover of nature, Alexander Nasmyth."

A visit which the two paid to Roslin Castle is worthy of commemoration.   On one occasion my father and a few choice spirits had been spending a "nicht wi' Burns."   The place of resort was a tavern in the High Street, Edinburgh. As Burns was a brilliant talker, full of spirit and humour, time fled until the "wee sma' hours ayont the twal'" arrived.   The party broke up about three o'clock.   At that time of the year (the 13th of June) the night is very short, and morning comes early.   Burns, on reaching the street, looked up to the sky.   It was perfectly clear, and the rising sun was beginning to brighten the mural crown of St. Giles's Cathedral.

Burns was so much struck with the beauty of the morning that he put his hand on my father's arm and said, "It'll never do to go to bed in such a lovely morning as this!   Let's awa' to Roslin Castle."   No sooner said than done.   The poet and the painter set out.   Nature lay bright and lovely before them in that delicious summer morning.   After an eight-miles walk they reached the castle at Roslin.   Burns went down under the great Norman arch, where he stood rapt in speechless admiration of the scene. The thought of the eternal renewal of youth and freshness of nature, contrasted with the crumbling decay of man's efforts to perpetuate his work, even when founded upon a rock, as Roslin Castle is, seemed greatly to affect him.

My father was so much impressed with the scene that,

while Burns was standing under the arch, he took out his
pencil and a scrap of paper and made a hasty sketch of the
subject. This sketch was highly treasured by my father, in
remembrance of what must have been one of the most
memorable days of his life.

Talking of clubs reminds me that there was a good deal
of club life in Edinburgh in those days. The most notable
were those in which the members were drawn together by
occupations, habits, or tastes. They met in the evenings,
and conversed upon congenial subjects. The clubs were
generally held in one or other of the taverns situated in
or near the High Street. Every one will remember the
Lawyers' Club, held in an Edinburgh close, presided over by
Pleydell, so well described by Scott in his *Guy Mannering*.

In my father's early days he was a member of a very
jovial club, called the Poker Club. It was so-called because
the first chairman, immediately on his election, in a spirit
of drollery, laid hold of the poker at the fireplace, and
adopted it as his insignia of office. He made a humorous
address from the chair, or "the throne," as he called it,
with sceptre or poker in hand ; and the club was thereupon
styled by acclamation " The Poker Club." I have seen my
father's diploma of membership ; it was tastefully drawn on
parchment, with the poker duly emblazoned on it as the
regalia of the club.

In my own time, the club that he was most connected
with was the Dilletanti Club. Its meetings were held every
fortnight, on Thursday evenings, in a commodious tavern in
the High Street. The members were chiefly artists, or men
known for their love of art. Among them were Henry
Raeburn, Hugh Williams (the Grecian), Andrew Geddes,
William Thomson, John Shetkay, William Nicholson, William
Allan, Alexander Nasymth, the Rev. John Thomson of
Duddingston, George Thomson, Sir Walter Scott, John
Lockhart, Dr. Brewster, David Wilkie, Henry Cockburn,

Francis Jeffrey, John A. Murray, Professor Wilson, John Ballantyne, James Ballantyne, James Hogg (the Ettrick Shepherd), and David Bridges, the secretary.[1] The drinks were restricted to Edinburgh ale and whisky toddy.

An admirable picture of the club in full meeting was painted by William Allan, in which characteristic portraits of all the leading members were introduced in full social converse. Among the more prominent portraits is one of my father, who is represented as illustrating some subject he is describing, by drawing it on the part of the table before him, with his finger dipped in toddy. Other marked and well-known characteristics of the members are skilfully introduced in the picture. The artist afterwards sold it to Mr. Horrocks of Preston, in Lancashire.

Besides portrait painting, my father was much employed in assisting the noblemen and landed gentry of Scotland in improving the landscape appearance of their estates, especially when seen from their mansion windows. His fine taste, and his love of natural scenery, gave him great advantages in this respect. He selected the finest sites for the new mansions, when they were erected in lieu of the old

---

[1] Davie Bridges was a character. In my early days he was a cloth merchant in the High Street. His shop was very near that gigantic lounge, the old Parliament House, and was often resorted to by non-business visitors. Bridges had a good taste for pictures. He had a small but choice collection by the Old Masters, which he kept arranged in the warehouse under his shop. He took great pride in exhibiting them to his visitors, and expatiating upon their excellence. I remember being present in his warehouse with my father when a very beautiful small picture by Richard Wilson was under review. Davie burst out emphatically with, "Eh, man, did ye ever see *such glorious buttery touches* as on these clouds!" His joking friends dubbed him "Director-General of the Fine Arts for Scotland," a title which he complacently accepted. Besides showing off his pictures, Davie was an art critic, and wrote articles for the newspapers and magazines. Unfortunately, however, his attention to pictures prevented him from attending to his shop, and his customers (who were not artists) forsook him, and bought their clothes elsewhere. He accordingly shut up his shop, and devoted himself to art criticism, in which, for a time, he possessed a monopoly.

towers and crenellated castles. Or, he designed alterations of the old buildings so as to preserve their romantic features, and at the same time to fit them for the requirements of modern domestic life.

In those early days of art - knowledge, there scarcely existed any artistic feeling for the landscape beauty of nature. There was an utter want of appreciation of the dignified beauty of the old castles and mansions, the remnants of which were in too many instances carted away as material for new buildings. There was also at that time an utter ignorance of the beauty and majesty of old trees. A forest of venerable oaks or beeches was a thing to be done away with. They were merely cut down as useless timber; even when they so finely embellished the landscape. My father exerted himself successfully to preserve these grand old forest trees. His fine sketches served to open the eyes of their possessors to the priceless treasures they were about to destroy; and he thus preserved the existence of many a picturesque old tree. He even took the pains in many cases to model the part of the estate he was dealing with; and he also modelled the old trees he wished to preserve. Thus, by a judicious clearing out of the intercepting young timber, he opened out distant views of the landscape, and at the same time preserved many a monarch of the forest.[1]

---

[1] It is even now to be deeply deplored that those who inherit or come into possession of landed estates do not feel sufficiently impressed with the possession of such grand memorials of the past. Alas! how often have we to lament the want of taste that leads to the sacrifice of these venerable treasures. Would that the young men at our universities—especially those likely to inherit estates—were impressed with the importance of preserving them. They would thus confer an inestimable benefit to thousands. About forty years ago Lord Cockburn published a pamphlet on *How to Destroy the Beauty of Edinburgh!* He enforced the charm of green foliage in combination with street architecture. The burgesses were then cutting down trees. His lordship went so far as to say " that he would as soon *cut down a burgess as a tree!*" Since then the growth of trees in Edinburgh, especially in what was once the North Loch, has been greatly improved.

My father modelled old castles, old trees, and suchlike objects as he wished to introduce into his landscapes. I append an illustration, which may perhaps give a slight idea of his artistic skill as a modeller. The one I specially refer

THE FAMILY TREE.

to, he called "The Family Tree," as he required each of his family to assist in its production. We each made a twig or small branch, which he cleverly fixed into its place as a part of the whole. The model tree in question was constructed of

wire slightly twisted together, so as to form the main body
of a branch. It was then subdivided into branchlets, and
finally into individual twigs. All these, combined together
by his dexterous hand, resulted in the model of an old leaf-
less tree, so true and correct, that any one would have
thought that it had been modelled direct from nature.

The Duke of Athol consulted my father as to the improvements which he desired to make in his woodland
scenery near Dunkeld. The Duke was desirous that a rocky
crag, called Craigybarns, should be planted with trees, to relieve the grim barrenness of its appearance. But it was impossible for any man to climb the crag in order to set seeds
or plants in the clefts of the rocks. A happy idea struck
my father. Having observed in front of the castle a pair of
small cannon used for firing salutes on great days, it occurred
to him to turn them to account. His object was to deposit
the seeds of the various trees amongst the soil in the clefts of
the crag. A tinsmith in the village was ordered to make a
number of canisters with covers. The canisters were filled
with all sorts of suitable tree seeds. The cannon was loaded,
and the cannisters were fired up against the high face of the
rock. They burst and scattered the seed in all directions.
Some years after, when my father revisited the place, he was
delighted to find that his scheme of planting by artillery
had proved completely successful; for the trees were flourish-
ing luxuriantly in all the recesses of the cliff. This was
another instance of my father's happy faculty of resourcefulness.

Certain circumstances about this time compelled my
father almost entirely to give up portrait painting and betake
himself to another branch of the fine arts. The earnest and
lively interest which he took in the state of public affairs,
and the necessity which then existed for reforming the
glaring abuses of the State, led him to speak out his mind
freely on the subject. Edinburgh was then under the reign

of the Dundases ; and scarcely anybody dared to mutter
his objections to anything perpetrated by the "powers that
be." Edinburgh was then a much smaller place than it is
now. There was more gossip, and perhaps more espionage,
among the better classes, who were few in number.

At all events, my father's frank opinions on political subjects began to be known. He attended Fox dinners. He
was intimate with men of known reforming views. All this
was made the subject of general talk. Accordingly, my
father received many hints from aristocratic and wealthy personages, that "if this went on any longer they would with-
draw from him their employment." My father did not alter
his course; it was right and honest. But he suffered nevertheless. His income from portrait painting fell off rapidly.

At length he devoted himself to landscape painting. It
was a freer and more enjoyable life. Instead of painting
the faces of those who were perhaps without character or
attractiveness, he painted the fresh and ever-beautiful face of
nature. The field of his employment in this respect was
almost inexhaustible. His artistic talent in this delightful
branch of art was in the highest sense congenial to his mind
and feelings; and in course of time the results of his new
field of occupation proved thoroughly satisfactory. In fact,
men of the highest rank with justice entitled him the
"Father of landscape painting in Scotland."

At the same time, when changing his branch of art, he
opened a class in his own house for giving practical instruction in the art of landscape painting. He removed his house
and studio from St. James's Square to No. 47 York Place.
There was at the upper part of this house a noble and commodious room. There he held his class. The house was
his own, and was built after his own designs. A splendid
prospect was seen from the upper windows; and especially
from the Belvidere, which he had constructed on the summit
of the roof. It extended from Stirling in the west to the

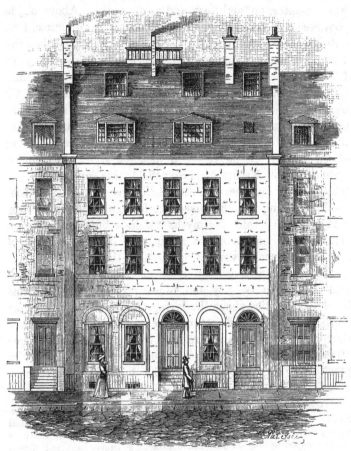

NO. 47 YORK PLACE, EDINBURGH.

Bass in the east. In fine summer evenings the sun was often seen setting behind Ben Lomond and the more conspicuous of the Perthshire mountains.

My father did not confine himself to landscape painting, or to the instruction of his classes. He was an all-round man. He had something of the Universal about him. He was a painter, an architect, and a mechanic. Above all, he possessed a powerful store of common sense. Of course, I am naturally a partial judge of my father's character; but this I may say, that during my experience of over seventy years I have never known a more incessantly industrious man. His hand and mind were always at work from morn till night. During the time that he was losing his business in portrait painting, he set to work and painted scenery for the theatres. The late David Roberts—himself a scene painter of the highest character—said that his style was founded upon that of Nasmyth.[1] Stansfield was another of his friends. On one occasion Stansfield showed him his sketch-book, observing that he wished to form a style of his own. " Young man," said Nasmyth, " there's but one style an artist should endeavour to attain, and that is *the style of nature;* the nearer you can get to that the better."

My father was greatly interested in the architectural beauty of his native city, and he was professionally consulted by the authorities about the laying out of the streets of the New Town. The subject occupied much of his time and thought, especially when resting from the mental fatigue arising from a long sitting at his easel. It was his regular practice to stroll about where the building work was in

---

[1] David Roberts, R.A., in his Autobiography, gives the following recollections of Alexander Nasmyth :—" In 1819 I commenced my career as principal scene painter in the Theatre Royal, Glasgow. This theatre was immense in its size and appointments—in magnitude exceeding Drury Lane and Covent Garden. The stock scenery had been painted by Alexander Nasmyth, and consisted of a series of pictures far surpassing anything of the kind I had ever seen. These included chambers, palaces, streets, landscapes, and forest scenery. One, I remember particularly, was the outside of a Norman castle, and another of a cottage charmingly painted, and of which I have a sketch. But the act scene, which was a view on the Clyde looking towards the Highland mountains, with Dumbarton Castle in the middle distance, was such a combination of magnificent scenery, so wonderfully painted, that it excited universal admiration. These productions I studied incessantly ; and on them my style, if I have any, was originally founded."

progress, or new roads were being laid out, and watch the proceedings with keen interest. This was probably due to the taste which he had inherited from his forebears—more especially from his father, who had begun the buildings of the New Town. My father took pleasure in modelling any improvement that occurred to him ; and to discuss the subject with the architects and builders who were professionally engaged in the works. His admirable knack of modelling the contour of the natural surface of the ground, and applying it to the proposed new roads or new buildings, was striking and characteristic. His efforts in this direction were so thoroughly disinterested that those in office were all the more anxious to carry out his views. He sought for no reward ; but his excellent advice was not unrecognised. In testimony of the regard which the Magistrates of Edinburgh had for his counsel and services, they presented him in 1815 with a sum of £200, together with a most complimentary letter acknowledging the value of his disinterested advice. It was addressed to him under cover, directed to " Alexander Nasmyth, Architect."

He was, indeed, not unworthy of the name. He was the architect of the Dean Bridge, which spans the deep valley of the Water of Leith, north-west of the New Town. Sir John Nesbit, the owner of the property north of the stream, employed my father to make a design for the extension of the city to his estate. The result was the construction of the Dean Bridge, and the roads approaching it from both sides. The Dean Estate was thus rendered as easy and convenient to reach as any of the level streets of Edinburgh. The construction of the bridge was superintended by the late James Jardine, C.E.[1]

From the Dean Bridge another of my father's architec-

---

[1] Mr. Telford was afterwards called upon to widen the bridge. He threw out parapets on each side, but it did not improve the original design.

tural buildings may be seen, at St. Bernard's Well.   It was
constructed at the instance of his friend Lord Gardenstone.
The design consists of a graceful circular temple, built over
a spring of mineral water, which issues from the rock below.
It was dedicated to Hygeia, the Goddess of Health.   The
whole of the details are beautifully finished, and the base-
ment of the design will be admired by every true artist.

ST. BERNARD'S WELL.

It is regarded as a great ornament, and is thoroughly in
keeping with the beauty of the surrounding scenery.

Shortly after the death of Lord Nelson it was proposed
to erect a monument to his memory on the Calton Hill.
My father supplied a design, which was laid before the
Monument Committee.   It was so much approved that the
required sum was rapidly subscribed.   But as the estimated
cost of this erection was found slightly to exceed the amount

subscribed, a nomin-
ally cheaper design
was privately adopted.
It was literally a job.
The vulgar, churn-like
monument was thus
thrust on the public
and actually erected;
and there it stands to
this day, a piteous
sight to beholders. It
was eventually found
greatly to exceed in
cost the amount of
the estimate for my
father's design. I give
a sketch of my father's
memorial; and I am
led to do this because
it is erroneously al-
leged that he was the
architect of the pre-
sent inverted spy
glass, called "Nelson's
Monument."

Then, with respect
to my father's powers
as a mechanic. This
was an inherited fa-
culty, and I leave my
readers to infer from
the following pages
whether I have not
had my fair share of
this inheritance. Be-

NELSON'S MONUMENT AS IT SHOULD HAVE BEEN.

sides his painting room, my father had a workroom fitted up
with all sorts of mechanical tools.   It was one of his greatest
pleasures to occupy himself there as a relief from sitting at
his easel, or while within doors from the inclemency of the
weather.   The walls and shelves of his workroom were
crowded with a multitude of artistic and ingenious mechanical objects, nearly all of which were the production of his
own hands.   Many of them were associated with the most
eventful incidents in his life.   He only admitted his most
intimate friends, or such as could understand and appreciate the variety of objects connected with art and mechanism, to his workroom.   His natural taste for order and
arrangement gave it a very orderly aspect, however crowded
its walls and shelves might be.   Everything was in its place,
and there was a place for everything.   It was in this workroom that I first began to handle mechanical tools.   It was
my primary technical school—the very foreground of my life.

I may mention one or two of my father's mechanical
efforts, or rather his inventions in applied science.   One of
the most important was the " bow-and-string bridge," as he
first called it, to which he early directed his attention.   He
invented this important method of construction about the
year 1794.   The first bow-and-string bridge was erected
in the island of St. Helena over a deep ravine.   Many
considered, from its apparent slightness, that it was not fitted
to sustain any considerable load.   A remarkable and con-
vincing proof was, however, given of its stability by the
passage over it of a herd of wild oxen, that rushed across
without the slightest damage to its structure.   After so
severe a test it was for many succeeding years employed as
a most valuable addition to the accessibility of an important
portion of the island.   The bow-and-string bridge has since
been largely employed in spanning wide spaces over which
suburban and other railways pass, and in roofing over such
stations as those at Birmingham, Charing Cross, and other

Great Metropolitan centres, as well as in bow-and-string
bridges over rivers. I append a fac-simile of his original

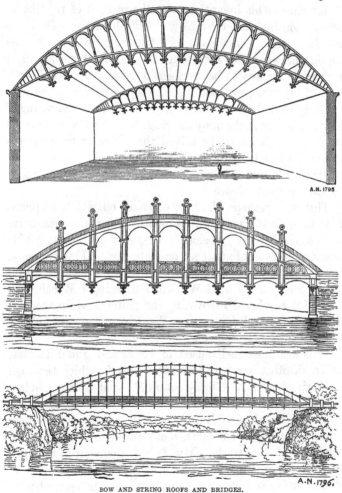

BOW AND STRING ROOFS AND BRIDGES.

drawings for the purpose of showing our great railway
engineers the originator of the graceful and economical

method of spanning wide spaces, now practised in every part of the civilised world.

Another of his inventions was the method of riveting by compression instead of by blows of the hammer. It originated in a slight circumstance. One wet, wintry Sunday morning he went into his workroom. There was some slight mechanical repairs to be performed upon a beautiful little stove of his own construction. To repair it, iron rivets were necessary to make it serviceable. But as the hammering of the hot rivets would annoy his neighbours by the unwelcome sound of the hammer, he solved the difficulty by using the jaws of his bench vice to *squeeze* the hot rivets in when put into their places. The stove was thus quickly repaired in the most perfect silence.

This was, perhaps, the first occasion on which a squeeze or compressive action was substituted for the percussive action of the hammer, in closing red-hot rivets, for combining together pieces of stout sheet or plate iron. This system of riveting was long afterwards patented by Smith of Deanston in combination with William Fairbairn of Manchester; and it was employed in riveting the plates used in the construction of the bridges over the River Conway and the Menai Straits.

It is also universally used in boiler and girder making, and in all other wrought-iron structures in which thorough sound riveting is absolutely essential; and by the employment of hydraulic power in a portable form a considerable portion of iron shipbuilding is effected by the silent *squeeze* system in place of hammers, much to the advantage of the soundness of the work. My father frequently, in aftertimes, practised this mode of riveting by compression in place of using the blow of a hammer; and in remembrance of the special circumstances under which he contrived this silent and most effective method of riveting, he named it " The Sunday Rivet."

CHAPTER III.

AN ARTIST'S FAMILY.

ALTHOUGH Alexander Nasmyth had to a considerable extent lost his aristocratic connection as a portrait painter, yet many kind and generous friends gathered around him. During his sojourn in Italy, in 1783, he had the good fortune to make the acquaintance of Sir James Hall of Dunglass, Haddingtonshire. The acquaintance afterwards ripened into a deeply-rooted friendship.

During the winter season Sir James resided with his family in his town house in George Street. He was passionately attached to the pursuit of art and science. He practised the art of painting in my father's room, and was greatly helped by him in the requisite manipulative skill. Sir James was at that time engaged in writing his well-known essay "On the Origin of Gothic Architecture," and in this my father was of important help to him. He executed the greater number of the illustrations to this beautiful work. The book when published had a considerable influence in restoring the taste of architects to a style which they had heretofore either neglected or degraded.

Besides his enthusiasm in art and architecture, Sir James devoted a great deal of time to the study of geology. The science was then in its infancy. Being an acute observer, Hall's attention was first attracted to the subject by

E

the singular geological features of the sea-coast near his
mansion at Dunglass.   The neighbourhood of Edinburgh also
excited his interest.   The upheaval of the rocks by volcanic
heat—as seen in the Castle Hill, the Calton Hill, and
Arthur's Seat—formed in a great measure the foundation of
the picturesque beauty of the city.   Those were the days of
the Wernerian and Huttonian controversy as to the origin
of the changes on the surface of the earth.   Sir James Hall
was President of the Edinburgh Royal Society, and necessarily took an anxious interest in the discussion.   He
observed and experimented, and established the true volcanic
nature of the composition and formation of the rocks and
mountains which surround Edinburgh.

I have been led to speak of this subject, because when
a boy I was often present at the discussions of these great
principles.   My father, Sir James Hall, Professors Playfair and
Leslie, took their accustomed walks round Edinburgh, and I
clung eagerly to their words.   Though unable to understand
everything that was said, these walks had a great influence
upon my education.   Indeed, what education can compare
with that of listening attentively to the conversation and
interchange of thought of men of the highest intelligence ?
It is on such occasions that *ideas*, not mere *words*, take hold
of the memory, and abide there until the close of life.

Besides mixing in the society of scientific men, my
father enjoyed a friendly intercourse with the artists of his
day.   He was often able to give substantial help and assistance to young students ; and he was most liberal in giving
them valuable practical instruction, and in assisting them
over the manipulative difficulties which lay in their way.
He was especially assiduous when he saw them inspired by
the true spirit of art, and full of application and industry,—
without which nothing can be accomplished.   Amongst
these young men were David Wilkie, Francis Grant, David
Roberts, Clarkson Stansfield, William Allan, Andrew Geddes,

"Grecian" Williams, Lizars the engraver, and the Rev. John Thomson of Duddingston.

Henry Raeburn was one of his most intimate friends and companions. He considered Raeburn's broad and masterly style of portrait painting as an era in Scottish art. Raeburn, with innate tact, discerned the character of his sitters, and he imported so much of their individuality into his portraits as to make them admirable *likenesses* in the highest sense. In connection with Raeburn, I may mention that when he was knighted by George IV. in 1822, my father, who was then at the head of his profession in Scotland, was appointed chairman at the dinner held to do honour to the great Scottish portrait painter.

Raeburn often joined my father in his afternoon walks round Edinburgh—a relaxation so very desirable after hours of close attention to artistic work. They took delight in the wonderful variety of picturesque scenery by which the city is surrounded. The walks about Arthur's Seat were the most enjoyable of all. When a boy I had often the pleasure of accompanying them, and of listening to their conversation. I thus picked up many an idea that served me well in after life. Indeed, I may say, after a long experience, that there is no class of men whose company I more delight in than that of artists. Their innate and highly-cultivated power of observation, not only as regards the ever-varying aspects of nature, but also as regards the quaint, droll, and humorous varieties of character, concur in rendering their conversation most delightful. I look back on these events as among the brightest points in my existence. I have been led to digress on this subject. Although more correctly belonging to my father's life, yet it is so amalgamated with my own that it almost forms part of it, and it is difficult for me to separate the one from the other.

And then there were the pleasant evenings at home.

When the day's work was over, friends looked in to have a fireside crack—sometimes scientific men, sometimes artists, often both. They were all made welcome. There was no formality about their visits. Had they been formal, there would have been comparatively little pleasure. The visitor came in with his "Good e'en," and seated himself. The family went on with their work as before. The girls were usually busy with the needles, and others with pen and pencil. My father would go on with the artistic work he had in hand, for his industry was incessant. He would model a castle or a tree, or proceed with some proposed improvement of the streets or approaches of the rapidly-expanding city. Among the most agreeable visitors were Professor Leslie, James Jardine, C.E., and Dr. Brewster. Their conversation was specially interesting. They brought up the last new thing in science, in discovery, in history, or in campaigning, for the war was then raging throughout Europe.

The artists were a most welcome addition to the family group. Many a time did they set the table in a roar with their quaint and droll delineations of character. These unostentatious gatherings of friends about our fireside were a delightful social institution. The remembrance of them lights up my recollection of the happiest period of a generally happy life. Could I have been able to set forth the brightness and cheerfulness of these happy evenings at my father's house, I am fain to think that my description might have been well worth reading. But all the record of them that remains is a most cherished recollection of their genial tone and harmony, which makes me think that, although in these days of rapid transit over earth and ocean, and surrounded as we are with the results of applied scientific knowledge, we are not a bit more happy than when all the vaunted triumphs of science and so-called education were in embryo.

The supper usually followed, for my father would not allow his visitors to go away empty-mouthed. The supper did not amount to much. Rizard or Finnan haddies, or a dish of oysters, with a glass of Edinburgh ale, and a rummer of toddy, concluded these friendly evenings. The cry of "Caller Aou" was constantly heard in the streets below of an evening. When the letter r was in the name of the month, the supply of oysters was abundant. The freshest oysters, of the most glorious quality, were to be had at 2s. 6d. the hundred! And what could be more refreshing food for my father's guests? These unostentatious and inexpensive gatherings of friends were a most delightful social institution among the best middle-class people of Edinburgh some sixty or seventy years ago. What they are now I cannot tell. But I fear they have disappeared in the more showy and costly tastes that have sprung up in the progress of what is called "modern society."

No part of my father's character was more admirable than his utter unselfishness. He denied himself many things, that he might give the more pleasure to his wife and children. He would scarcely take part in any enjoyment, unless they could have their fair share of it. In all this he was faithfully followed by my mother. The admirable example of well-sustained industry that was always before her, sustained her in her efforts for the good of her family. She was intelligently interested in all that related to her husband's business and interests, as well as in his recreative enjoyments. The household affairs were under her skilful guidance. She conducted them with economy, and yet with generous liberality, free from the least taint of ostentation or extravagance. The home fireside was the scene of cheerfulness. And most of our family have been blest with this sunny gift. Indeed, a merrier family circle I have never seen. There were twelve persons round the table to be provided for, besides two servants. This required, on my mother's part, a great deal of management, as every housekeeper will know. Yet everything was provided and paid for within the year's income.

The family result of my father and mother's happy marriage was four sons and seven daughters. Patrick, the eldest, was born in 1787. He was called after my father's dear and constant friend, Patrick Miller of Dalswinton. I will speak by-and-by of his artistic reputation. Then followed a long succession of daughters—Jane, the eldest, was born in 1778; Barbara in 1790; Margaret in 1791; Elizabeth in 1793; Anne in 1798; Charlotte in 1804. Then came a succession of three sons—Alexander, George, and James. There followed another daughter, Mary; but as she only lived for about eighteen months, I remained the youngest child of the family.

My sisters all possessed, in a greater or less degree, an innate love of art, and by their diligent application they acquired the practice of painting landscape in oils. My father's admirable system and method of teaching rendered them expert in making accurate sketches from nature, which, as will afterwards be seen, they turned to good account. My eldest sister, Jane, was in all respects a most estimable character, and a great help to my mother in the upbringing of the children. Jane was full of sound common sense; her judgment seemed to be beyond her years. Because of this the younger members of the family jokingly nicknamed her " Old Solid " ! Even my father consulted her in every case of importance in reference to domestic and financial affairs. I had the great good fortune, when a child, to be placed under her special protection, and I have reason to be thankful for the affectionate care which she took of me during the first six years of my life.

Besides their early education in art, my mother was equally earnest in her desire to give her daughters a thorough practical knowledge in every department and

detail of household management. When they had attained
a suitable age they were in succession put in charge of all the
household duties for two weeks at a time. The keys were
given over to them, together with the household books, and
at the end of their time their books were balanced to a
farthing. They were then passed on to the next in succession. One of the most important branches of female education—the management of the domestic affairs of a family,
the superintendence of the cooking so as to avoid waste of
food, the regularity of the meals, and the general cleaning
up of the rooms—was thus thoroughly attained in its best
and most practical forms. And under the admirable
superintendence of my mother everything in our family
went on like clockwork.

My father's object was to render each and all of his
children—whether boys or girls—independent on their
arrival at mature years. Accordingly, he sedulously kept
up the attention of his daughters to fine art. By this
means he enabled them to assist in the maintenance of the
family while at home, and afterwards to maintain themselves
by the exercise of their own abilities and industry after
they had left. To accomplish this object, as already described,
he set on foot drawing classes, which were managed by his
six daughters, superintended by himself.

Edinburgh was at that time the resort of many county
families. The war which raged abroad prevented them
going to the Continent. They therefore remained at home,
and the Scotch families for the most part took up their
residence in Edinburgh. There were many young ladies
desiring to complete their accomplishments, and hence the
establishment of my sisters' art class. It was held in the
large painting-room in the upper part of the house. It soon
became one of the most successful institutions in Edinburgh.
When not engaged in drawing and oil painting, the young
ladies were occupied in sketching from nature, under the

superintendence of my sisters, in the outskirts of Edinburgh. This was one of the most delightful exercises in which they could be engaged ; and it also formed the foundation for many friendships which only terminated with life.

My father increased the interest of the classes by giving little art lectures. They were familiar but practical. He never gave lectures *as such*, but rather demonstrations. It was only when a pupil encountered some technical difficulty, or was adopting some wrong method of proceeding, that he undertook to guide them by his words and practical illustrations. His object was to embue the minds of the pupils with high principles of art. He would take up their brushes and show by his dexterous and effective touches how to bring out, with marvellous ease, the right effects of the landscape. The other pupils would come and stand behind him, to see and hear his clear instructions carried into actual practice on the work before him. He often illustrated his little special lessons by his stores of instructive and interesting anecdotes, which no doubt helped to rivet his practice all the deeper into their minds. Thus the Nasmyth classes soon became the fashion. In many cases both mothers and daughters might be seen at work together in that delightful painting-room. I have occasionally met with some of them in after years, who referred to those pleasant hours as among the most delightful they had ever spent.

These classes were continued for many years. In the meantime my sisters' diligence and constant practice enabled them in course of time to exhibit their works in the fine art exhibitions of Edinburgh. Each had her own individuality of style and manner, by which their several works were easily distinguished from each other. Indeed, whoever works after Nature will have a style of their own. They all continued the practice of oil painting until an advanced age. The average duration of their lives was about seventy-eight.

There was one point which my father diligently impressed

upon his pupils, and that was the felicity and the happiness attendant upon pencil drawing. He was a master of the pencil, and in his off-hand sketches communicated his ideas to others in a way that mere words could never have done. It was his Graphic Language. A few strokes of the pencil can convey ideas which quires of writing would fail to impart. This is one of the most valuable gifts which a man who has to do with practical subjects can possess. "The language of the pencil" is truly a universal one, especially in communicating ideas which have reference to material forms. And yet it is in a great measure neglected in our modern system of so-called education.

The language of the tongue is often used to disguise our thoughts, whereas the language of the pencil is clear and explicit. Who that possesses this language can fail to look back with pleasure on the course of a journey illustrated by pencil drawings ? They bring back to you the landscapes you have seen, the old streets, the pointed gables, the entrances to the old churches, even the bits of tracery, with a vividness of association such as mere words could never convey. Thus, looking at an old sketch-book brings back to you the recollection of a tour, however varied, and you virtually make the journey over again with its picturesque and beautiful associations.

On many a fine summer's day did my sisters make a picnic excursion into the neighbourhood of Edinburgh. They were accompanied by their pupils, sketch-book and pencil in hand. As I have already said, there is no such scenery near any city that I know of.   Arthur's Seat. and Salisbury Crags, Duddingston Loch, the Braid Hills, Craigmillar Castle, Hawthornden, Roslin, Habbie's How, and the many valleys and rifts in the Pentlands, with Edinburgh and its Castle in the distance ; or the scenery by the seashore, all round the coast from Newhaven to Gullane and North Berwick Law.

The excursionists came home laden with sketches. I
have still by me a multitude of these graphic records made
by my sisters. Each sketch, however slight, strikes the
keynote, as it were, to many happy recollections of the
circumstances, and the persons who were present at the
time they were made. I know not of any such effective
stimulant to the recollection of past events as these graphic
memoranda. Written words may be forgotten, but these
slight pencil recollections imprint themselves on the mind
with a force that can never be effaced. Everything that
occurred at the time rises up as fresh in the memory as if
hours and not years had passed since then. They bring to
the mind's eye many dear ones who have passed away, and
remind us that we too must follow them.

It is much to be regretted that this valuable art of
graphic memoranda is not more generally practised. It is
not merely a most valuable help to the memory, but it
educates the eye and the hand, and enables us to cultivate
the faculty of definite observation. This is one of the most
valuable accomplishments that I know of, being the means
of storing up ideas, and not mere words, in the mental recollection of both men and women.

Before I proceed to record the recollections of my own
life, I wish to say something about my eldest brother Patrick,
the well-known landscape painter. He was twenty-one years
older than myself. My father was his best and almost
his only instructor. At a very early age he manifested a
decided taste for drawing and painting. His bent was landscape. This gave my father great pleasure, as it was his
own favourite branch of art. The boy acquired great skill
in sketching trees, clouds, plants, and foregrounds. He
studied with wonderful assiduity and success. I possess many
of his graphic memoranda, which show the care and industry
with which he educated his eye and hand in rendering with
truth and fidelity the intimate details of his art. The wild

plants which he introduced into the foregrounds of his
pictures were his favourite objects of study. But of all
portions of landscape nature, the Sky was the one that most
delighted him. He studied the form and character of clouds
—the resting cloud, the driving cloud, and the rain cloud—
and the sky portions of his paintings were thus rendered so
beautifully attractive.

He was so earnest in his devotion to the study of landscape that in some respects he neglected the ordinary routine
of school education. He successfully accomplished the three
R.'s, but after that his School was in the fields, in the face of
nature. He was by no means a Romantic painter. His
taste was essentially for Home subjects. In his landscapes he
introduced picturesque farm-houses and cottages, with their
rural surroundings ; and his advancement and success were
commensurate with his devotion to this fine branch of art. The
perfect truth with which he represented English scenery,
associated as it is with so many home-loving feelings, forms
the special attractiveness of his works. This has caused them
to be eagerly sought after, and purchased at high prices.

Patrick had a keen sense of humour, though in other
respects he was simple and unpretending. He was a great
reader of old-fashioned novels, which indeed in those days
were the only works of the kind to be met with. *The
Arabian Nights, Robinson Crusoe, The Mysteries of Udolpho*,
and suchlike, were his favourites, and gave a healthy filip
to his imagination. He had also a keen relish for music,
and used to whistle melodies and overtures as he went
along with his work. He acquired a fair skill in violin
playing. While tired with sitting or standing he would
take up his violin, play a few passages, and then go to work
again.

Patrick removed to London in 1808, and exhibited at
the Royal Academy in the following year. He made excursions to various parts of England, where he found subjects

congenial to his ideas of rural beauty. The immediate
neighbourhood of London, however, abounded with the most
charming and appropriate subjects for his pencil. These
consisted of rural "bits" of the most picturesque but homely
description — decayed pollard trees and old moss-grown
orchards, combined with cottages and farm-houses in the
most *paintable* state of decay, with tangled hedges and
neglected fences, overrun with vegetation clinging to them
with all " the careless grace of nature." However neglected
these might be by the farmer, they were always tit-bits for
Patrick. When sketching such subjects he was in his glory,
and he returned to his easel loaded with sketch-book treasures,
which when painted form the gems of many a collection.

In some of these charming subjects glimpses of the
distant capital may be observed, the dome of St. Paul's
towering over all ; but they are introduced with such skill
and correctness as in no way to interfere with the rural
character of his subject. When he went farther afield—to
Windsor Forest, Hampshire, the New Forest, or the Isle of
Wight—he was equally diligent with his pencil, and came
home laden with sketches of the old monarchs of the forest.
When in a state of partial decay his skilful touch brought
them to life again, laden with branches and lichen, with
leaves and twigs and bark, and with every feature that
gives such a charm to these important elements in true
English landscape scenery.

On my brother's first visit to London, accompanied by
my father, he visited many collections where the old Dutch
masters were to be seen, and he doubtless derived much
advantage from his careful studies, more particularly from
the works of Hobbema, Ruysdael, and Wynants. These
came home to him as representations of Nature as she is.
They were more free from the traditional modes of representing her. The works of Claude Lorraine and Richard Wilson
were also the objects of his admiration, though the influence

of the time for classicality of treatment to a certain extent
vitiated these noble works. When a glorious sunset was
observed, the usual expression among the lovers of art was,
" What a magnificent Claudish effect !" thus setting up the
result of man's feeble attempt at representation as the
standard of comparison, in place of the far grander original !

My brother carefully studied Nature herself. His works,
following those of my father, led back the public taste to a
more healthy and true condition, and by the aid of a noble
army of modern British landscape painters, this department
of art has been elevated to a very high standard of truth
and excellence.

I find some letters from Patrick to my father, after his
settlement as an artist in London. My father seems to
have supplied him with money during the early part of his
career, and afterwards until he had received the amount of
his commissions for pictures. In one of his letters he says :
" That was an unlucky business, the loss of that order
which you were so good as send me on my account." It
turned out that the order had dropt out of the letter enclosing it, and was not recovered. In fact, Patrick was
very careless about all money transactions.

In 1814 he made the acquaintance of Mr. Barnes, and
accompanied him to Bure Cottage, Ringwood, near Southampton, where he remained for some time. He went into
the New Forest, and brought home " lots of sketches." In
1815 he exhibited his works at the Royal Academy. He
writes to his father that " the prices of my pictures in the
Gallery are—two at fourteen guineas each (small views in
Hampshire), one at twelve guineas, and two at fourteen
guineas. They are all sold but one." These pictures would
now fetch in the open market from two to three hundred
guineas each. But in those days good work was little
known, and landscapes especially were very little sought
after.

Patrick Nasmyth's admirable rendering of the finer
portions of landscape nature attracted the attention of col-
lectors, and he received many commissions from them at very
low prices.   There was at that time a wretched system of
delaying the payment for pictures painted on commission, as
well as considerable loss of time by the constant applications
made for the settlement of the balance.   My brother was
accordingly under the necessity of painting his pictures for
the Dealers, who gave him at once the price which he required for his works.   The influence of this system was not
always satisfactory.   The Middlemen or Dealers, who stood
between the artist and the final possessor of the works,
were not generous.   They higgled about prices, and the sums
which they gave were almost infinitesimal compared with
the value of Patrick Nasmyth's pictures at the present time.

The Dealers were frequent visitors at his little painting-
room in his lodgings.   They took undue advantage of my
brother's simplicity and innate modesty in regard to the
commercial value of his works.   When he had sketched in
a beautiful subject, and when it was clear that in its highest
state of development it must prove a fine work, the Dealer
would pile up before him a row of guineas, or sovereigns,
and say, " Now, Peter, that picture's to be mine !"   The
real presence of cash proved too much for him.   He never
was a practical man.   He agreed to the proposal, and
thus he parted with his pictures for much less than they
were worth.   He was often remonstrated with by his brother
artists for letting them slip out of his hands in that way—
works that he would not surrender until he had completed
them, and brought them up to the highest point of his
fastidious taste and standard of excellence.

Among his dearest friends were David Roberts and
Clarkson Stansfield.   He usually replied to their friendly
remonstrances by laughingly pointing to his bursting port-
folios of sketches, and saying, " There's lots of money in

these banks to draw from." He thus warded off their earnest and often-repeated friendly remonstrances. Being a single man, and his habits and style of living of the most simple kind, he had very little regard for money except as it ministered to his immediate necessities. His evenings were generally spent at a club of brother artists "over the water;" and in their company he enjoyed many a pleasant hour. His days were spent at his easel. They were occasionally varied by long walks into the country near London, for the purpose of refilling his sketch-book.

It was on one of such occasions—when he was sketching the details of some picturesque pollard old willows up the Thames, and standing all the time in wet ground—that he caught a severe cold which confined him to the house. He rapidly became worse. Two of his sisters, who happened to be in London at the time, nursed him with devoted attention. But it was too late. The disease had taken fatal hold of him. On the evening of the 17th August 1831, there had been a violent thunderstorm. At length the peals of thunder ceased, the rain passed away, and the clouds dispersed. The setting sun burst forth in a golden glow. The patient turned round on his couch and asked that the curtains might be drawn. It was done. A blaze of sunset lit up his weary and worn-out face. "How glorious it is!" he said. Then, as the glow vanished he fell into a deep and tranquil sleep, from which he never awoke. Such was the peaceful end of my brother Patrick, at the comparatively early age of forty-four years.

CHAPTER IV.

I WAS born on the morning of the 19th of August 1808, at my father's house, No. 47 York Place, Edinburgh. I was named James Hall after my father's dear friend, Sir James Hall of Dunglass. My mother afterwards told me that I must have been " a very noticin' bairn," as she observed me, when I was only a few days old, following with my little eyes any one who happened to be in the room, as if I had been thinking to my little self, " Who are you ? "

After a suitable time I was put under the care of a nursemaid. I remember her well—Mary Peterkin—a truly Scandinavian name. She came from Haddingtonshire, where most of the people are of Scandinavian origin. Her hair was of a bright yellow tint. She was a cheerful young woman, and sang to me like a nightingale. She could not only sing old Scotch songs, but had a wonderful memory for fairy tales. When under the influence of a merry laugh, you could scarcely see her eyes ; their twinkle was hidden by her eyelids and lashes. She was a willing worker, and was always ready to lend a helping hand at everything about the house. She took great pride in me, calling me her " laddie."

When I was toddling about the house, another sister was born, the last of the family. Little Mary was very delicate ; and to improve her health she was sent to a small farm-

house at Braid Hills, about four miles south of Edinburgh.
It was one of the most rural and beautiful surroundings of
the city at that time.   One of my earliest recollections is
that of being taken to see poor little Mary at the farmer's
house.   While my nursemaid was occupied in inquiring
after my sister, I was attracted by the bright red poppies in
a neighbouring field.   When they made search for me I could
not be found.   I was lost for more than an hour.   At last,
seeing a slight local disturbance among the stalks of the corn,
they rushed to the spot, and brought me out with an armful
of brilliant red poppies.   To this day poppies continue to be
my greatest favourites.

When I was about four or five years old, I was observed
to give a decided preference to the use of my left hand.
Everything was done to prevent my using it in preference to
the right.   My mother thought that it arose from my being
carried on the wrong arm by my nurse while an infant.
The right hand was thus confined, and the left hand was used.
I was constantly corrected, but " on the sly " I always used
it, especially in drawing my first little sketches.   At last my
father, after viewing with pleasure one of my artistic efforts,
done with the forbidden hand, granted it liberty and independence for all time coming.   " Well," he said, " you may
go on in your own way in the use of your left hand, but I
fear you will be an awkward fellow in everything that requires handiness in life."   I used my right hand in all that
was necessary, and my left in all sorts of practical manipulative
affairs.   My left hand has accordingly been my most willing
and obedient servant in transmitting my will through my
fingers into material or visible forms.   In this way I became
ambidexter.

When I was about four years old, I often followed my
father into his workshop when he had occasion to show
to his visitors some of his mechanical contrivances or artistic models.   The persons present usually expressed their

F

admiration in warm terms of what was shown to them. On
one occasion I gently pulled the coat-tail of one of the
listeners, and *confidentially* said to him, as if I knew all
about it, " My papa's a kevie Fellae !"   My father was so
greatly amused by this remark that he often referred to it
as " the last good thing" from that old-fashioned creature
little Jamie.

One of my earliest recollections is the annual celebration
of my brother Patrick's birthday.   Being the eldest of the
family, his birthday was held in special honour.   My father
invited about twenty of his most intimate friends to dinner.
My mother brought her culinary powers into full operation.
The younger members of the family also took a lively interest
in all that was going on, with certain reversionary views as
to " the day after the feast."   We took a great interest in
the Trifle, which was no trifle in reality, in so far as regarded
the care and anxiety involved in its preparation.   In connection with this celebration, it was an established institution that a large hamper always arrived in good time from
the farm attached to my mother's old home at Woodhall,
near Edinburgh.   It contained many substantial elements
for the entertainment—a fine turkey, fowls, duck, and suchlike ; with two magnums of the richest cream.   There never
was such cream !   It established a standard of cream in my
memory ; and since then I have always been hypercritical
about the article.

On one of these occasions, when I was about four years
old, and being the youngest of the family, I was taken into
the company after the dinner was over, and held up by my
sister Jane to sing a verse from a little song which my nurse
Mary Peterkin had taught me, and which ran thus :

> " I'll no bide till Saturday,
> But I 'll awa' the morn,
> An' follow Donald Hielandman,
> An' carry his poother-horn."

This was my first and last vocal performance. It was received with great applause. In fact, it was encored. The word "poother," which I pronounced "pootle," excited the enthusiasm of the audience. I was then sent to bed with a bit of plum-cake, and was doubtless awakened early next morning by the irritation of the dried crumbs of the previous night's feast.

I am reminded, by reading over a letter of my brother Patrick's, of an awkward circumstance which happened to me when I was six years old. In his letter to my father, dated London, 22d September 1814, he says : " I did get a surprise when Margaret's letter informed me of my little brother Jamie's fall. It was a wonderful escape. For God's sake keep an eye upon him !" Like other strong and healthy boys, I had a turn for amusing myself in my own way. When sliding down the railing of the stairs I lost my grip and fell suddenly off. The steps were of stone. Fortunately, the servants were just coming in laden with carpets which they had been beating. I fell into their midst and knocked them out of their hands. I was thus saved from cracking my poor little skull. But for that there might have been no steam-hammer—at least of my contrivance !

Everything connected with war and warlike exploits is interesting to a boy. The war with France was then in full progress. Troops and bands paraded the streets. Recruits were sent away as fast as they could be drilled. The whole air was filled with war. Everybody was full of excitement about the progress of events in Spain. When the great guns boomed forth from the Castle, the people were first startled. Then they were surprised and anxious. There had been a battle and a victory ! " Who had fallen ? " was the first thought in many minds. Where had the battle been, and what was the victory? Business was suspended. People rushed about the streets to ascertain the facts. It might have been at Salamanca, Talavera, or Vittoria. But a long time elapsed

before the details could be received; and during that time
sad suspense and anxiety prevailed in almost every household. There was no telegraph then. It was only after the
Gazette had been published that people knew who had fallen
and who had survived.

The war proceeded. The volunteering which went on at
the time gave quite a military aspect to the city. I remember how odd it appeared to me to see some well-known faces
and figures metamorphosed into soldiers. It was considered
a test of loyalty as well as of patriotism, to give time, money,
and leisure to take up the arms of defence, and to practise
daily in military uniform in the Meadows or on Bruntsfield
Links. Windows were thrown up to hear the bands playing
at the head of the troops, and crowds of boys, full of military
ardour, went, as usual, hand to hand in front of the drums
and fifes. The most interesting part of the procession to my
mind was the pioneers in front, with their leather aprons,
their axes and saws, and their big hairy caps and beards.
They were to me so suggestive of clearing the way through
hedges and forests, and of what war was in its actual progress.

Every victory was followed by the importation of large
numbers of French prisoners. Many of these were sent to
Edinburgh Castle. They were permitted to relieve the
tedium of their confinement by manufacturing and selling
toys, workboxes, brooches, and carved work of different kinds.
In the construction of these they exhibited great skill, taste,
and judgment. They carved them out of bits of bone and
wood. The patterns were most beautiful, and they were
ingeniously and tastefully ornamented. The articles were to
be had for a mere trifle, although fit to be placed along
with the most choice objects of artistic skill.

These poor prisoners of war were allowed to work at their
tasteful handicrafts in small sheds or temporary workshops
at the Castle, behind the palisades which separated them

from their free customers outside. There was just room between the bars of the palisades for them to hand through their exquisite works, and to receive in return the modest prices which they charged. The front of these palisades became a favourite resort for the inhabitants of Edinburgh; and especially for the young folks. I well remember being impressed with the contrast between the almost savage aspect of these dark-haired foreigners, and the neat and delicate produce of their skilful fingers.

At the peace of Amiens, which was proclaimed in 1814, great rejoicings and illuminations took place, in the belief that the war was at an end. The French prisoners were sent back to their own country, alas! to appear again before us at Waterloo. The liberation of those confined in Edinburgh Castle was accompanied by an extraordinary scene. The French prisoners marched down to the transport ships at Leith by torchlight. All the town was out to see them. They passed in military procession through the principal streets, singing as they marched along their revolutionary airs, "Ca Ira" and "The Marseillaise." The wild enthusiasm of these haggard-looking men, lit up by torchlight and accompanied by the cheers of the dense crowd which lined the streets and filled the windows, made an impression on my mind that I can never forget.

A year passed. Napoleon returned from Elba, and was rejoined by nearly all his old fighting-men. I well remember, young as I was, an assembly of the inhabitants of Edinburgh in Charlotte Square, to bid farewell to the troops and officers then in garrison. It was a fine summer evening when this sad meeting took place. The bands were playing as their last performance, "Go where glory waits thee!" The air brought tears to many eyes; for many who were in the ranks might never return. After many a handshaking the troops marched to the Castle, previous to their early embarkation for the Low Countries on the following morning.

Then came Waterloo and the victory! The Castle guns boomed forth again; and the streets were filled with people anxious to hear the news. At last came the *Gazette* filled with the details of the killed and wounded. Many a heart was broken, many a fireside was made desolate. It was indeed a sad time. The terrible anxiety that pervaded so many families; the dreadful sacrifice of lives on so many battlefields; and the enormously increased taxation, which caused so many families to stint themselves to even the barest necessaries of life;—such was the inglorious side of war.

But there was also the glory, which almost compensated for the sorrow. I cannot resist narrating the entry of the Forty-Second Regiment into Edinburgh shortly after the battle of Waterloo. The old "Black Watch" is a regiment dear to every Scottish heart. It has fought and struggled when resistance was almost certain death. At Quatre Bras two flank companies were cut to pieces by Piré's cavalry. The rest of the regiment was assailed by Reillé's furious cannonade, and suffered severely. The French were beaten back, and the remnant of the Forty-Second retired to Waterloo, where they formed part of the brigade under Major-General Pack. At the first grand charge of the French, Picton fell and many were killed. Then the charge of the Greys took place, and the Highland regiments rushed forward, with cries of "Scotland for ever!" Only a remnant of the Forty-Second survived. They were however recruited, and marched into France with the rest of the army.

Towards the end of the year the Forty-Second returned to England, and in the beginning of 1816 they set out on their march towards Edinburgh. They were everywhere welcomed with enthusiasm. Crowds turned out to meet them and cheer them. When the first division of the regiment approached Edinburgh, almost the entire population turned out to welcome them. At Musselburgh, six miles off, the road was thronged with people. When the soldiers reached Piershill, two miles off, the road was so crowded that it took them two hours to reach the Castle.   I was on a balcony in the upper part of the High Street, and my father, mother, and sisters were with me.   We had waited very long; but at last we heard the distant sound of the cheers, which came on and on, louder and louder.

The High Street was wedged with people excited and anxious.   There seemed scarcely room for a regiment to march through them.   The house-tops and windows were crowded with spectators.   It was a grand sight.   The high-gabled houses reaching as far as the eye could see, St. Giles' with its mural crown, the Tron Kirk in the distance, and the picturesque details of the buildings, all added to the effectiveness of the scene.

At last the head of the gallant band appeared.   The red coats gradually wedged their way through the crowd, amidst the ringing of bells and the cheers of the spectators.   Every window was in a wave of gladness, and every house-top was in a fever of excitement.   As the red line passed our balcony, with Colonel Dick at its head, we saw a sight that can never be forgotten.   The red-and-white plumes, the tattered colours riddled with bullets, the glittering bayonets, were seen amidst the crowd that thronged round the gallant heroes, amidst tears and cheers and hand-shakings and shouts of excitement. The mass of men appeared like a solid body moving slowly along; the soldiers being almost hidden amongst the crowd. At last they passed, the pipers and drums playing a Highland march; and the Forty-Second slowly entered the Castle. It was perhaps the most extraordinary scene ever witnessed in Edinburgh.

One of my greatest enjoyments when a child was in going out with the servants to the Calton, and wait while the " claes " bleached in the sun on the grassy slopes of the hill. The air was bright and fresh and pure.   The lasses regarded these occasions as a sort of holiday.   One or two of the

children usually accompanied them. They sat together, and
the servants told us their auld-warld stories ; common enough
in those days, but which have now, in a measure, been forgotten. "Steam" and "progress" have made the world
much less youthful and joyous than it was then.

The women brought their work and their needles with
them, and when they had told their stories, the children ran
about the hill making bunches of the wild flowers. They
ran after the butterflies and the bumbees, and made acquaintance in a small way with the beauties of nature. Then the
servants opened their baskets of provisions, and we had a
delightful picnic. Though I am now writing about seventy
years after the date of these events, I can almost believe
that I am enjoying the delightful perfume of the wild thyme
and the fragrant plants and flowers, wafted around me by
the warm breezes of the Calton hillside.

In the days I refer to, there was always a most
cheerful and intimate intercourse kept up between the
children and the servants. They were members of the same
family, and were treated as such. The servants were for
the most part country-bred—daughters of farm servants or
small farmers. They were fairly educated at their parish
schools ; they could read and write, and had an abundant
store of old recollections. Many a pleasant crack we had
with them as to their native places, their families, and all
that was connected with them. They became lastingly
attached to their masters and mistresses, as well as to the
children. All this led to true attachment ; and when they
left us, for the most part to be married, we continued to
keep up a correspondence with them, which lasted for
many years.

While enjoying these delightful holidays, before my
school-days began, my practical education was in progress,
especially in the way of acquaintance with the habits of
nature in a vast variety of its phases, always so attractive

to the minds of healthy children.   It happened that close to
the Calton Hill, in the valley at its northern side, there were
many workshops where interesting trades were carried on,
such as those of coppersmiths, tinsmiths, brassfounders, gold-
beaters, and blacksmiths.   Their shops were all gathered
together in a busy group at the foot of the hill, in a place
called Greenside.   The workshops were open to the inspection of passers by.   Little boys looked in and saw the men
at work amidst the blaze of fires and the beatings of hammers.

Amongst others, I was an ardent admirer.   I may almost
say that this row of busy workshops was my first school of
practical education.   I observed the mechanical manipulation of the men, their dexterous use of the hammer, the
chisel, and the file; and I imbibed many lessons which
proved of use to me in my later years.   Then I had tools at
home in my father's workshop.   I tried to follow their
methods; I became greatly interested in the use of tools
and their appliances; I could make things for myself.   In
short, I became so skilled that the people about the house
called me "a little Jack-of-all-trades."

While sitting on the grassy slopes of the Calton Hill I
would often hear the chimes sounding from the grand old
tower of St. Giles.   The cathedral lay on the other side of
the valley which divides the Old Town from the New.   The
sounds came over the murmur of the traffic in the streets
below.

The chime-bells were played every day from twelve till
one—the old-fashioned dinner-hour of the citizens.   The
practice had been in existence for more than a hundred and
fifty years.   The pleasing effect of the merry airs, which
came wafted to me by the warm summer breezes, made me
long to see them as well as hear them.

My father was always anxious to give pleasure to his
children.   Accordingly, he took me one day, as a special
treat, to the top of the grand old tower, to *see* the chimes

played. As we passed up the tower, a strong vaulted room
was pointed out to me, where the witches used to be
imprisoned. I was told that the poor old women were
often taken down from this dark vault to be burnt alive!
Such terrible tales enveloped the tower with a horrible
fascination to my young mind. What a fearful contrast to
the merry sound of the
chimes issuing from its
roof on a bright summer
day.

On my way up to
the top flat, where the
chimes were played, I
had to pass through the
vault in which the great
pendulum was slowly
swinging in its ghostly-
like *tick-tack, tick-tack ;*
while the great ancient
clock was keeping time
by its sudden and start-
ling movement. The
whole scene was almost
as uncanny as the wit-
ches' cell underneath.

MURAL CROWN OF ST. GILES', EDINBURGH.

There was also a wild rumbling thumping sound overhead.
I soon discovered the cause of this, when I entered the flat
where the musician was at work. He was seen in violent
action, beating or hammering on the keys of a gigantic piano-
forte-like apparatus. The instruments he used were two
great leather-faced mallets, one of which he held in each
hand. Each key was connected by iron rods to the chime-
bells above. The frantic and mad-like movements of the
musician, as he energetically rushed from one key to another,
often widely apart, gave me the idea that the man was

daft—especially as the noise of the mallets was such that I heard no music emitted from the chimes so far overhead. It was only when I had climbed up the stair of the tower to where the bells were rung that I understood the performance, and comprehended the beating of the chimes which gave me so much pleasure when I heard them at a distance.

Another source of enjoyment in my early days was to accompany my mother to the market. As I have said before, my mother, though generous in her hospitality, was necessarily thrifty and economical in the management of her household. There were no less than fourteen persons in the house to be fed, and this required a good deal of marketing. At the time I refer to (about 1816) it was the practice of every lady who took pride in managing economically the home department of her husband's affairs, to go to market in person. The principal markets in Edinburgh were then situated in the valley between the Old and New Towns, in what used to be called the Nor' Loch.

Dealers in fish and vegetables had their stalls there. The market for butcher-meat was near at hand : and each were in their several locations. It was a very lively and bustling sight to see the marketing going on. When a lady was observed approaching, likely to be a customer, she was at once surrounded by the " caddies." They were a set of sturdy hard-working women, each with a creel on her back. Their competition for the employer sometimes took a rather energetic form. The rival candidates pointed to her with violent exclamations ; " She's *my* ledie ! she's *my* ledie !" ejaculated one and all. To dispel the disorder, a selection of *one* of the caddies would be made, and then all was quiet again until another customer appeared.

There was a regular order in which the purchases were deposited in the creel. First, there came the fish, which were carefully deposited in the lowest part, with a clean deal board over them. The fishwives were a most sturdy and independent class, both in manners and language. When at home, at Newhaven or Fisherrow, they made and mended their husbands' nets, put their fishing tackle to rights, and when the fishing boats came in they took the fish to market at Edinburgh. To see the groups of these hard-working women, trudging along with their heavy creels on their backs, clothed in their remarkable costume, with their striped petticoats, kilted up and showing their sturdy legs, was indeed a remarkable sight. They were cheerful and good-humoured, but very outspoken. Their skins were clear and ruddy, and many of the young fishwives were handsome and pretty. They were, in fact, the incarnation of health.

In dealing with them at the Fish Market there was a good deal of higgling. They often asked two or three times more than the fish were worth—at least, according to the then market price. After a stormy night, during which the husbands and sons had toiled to catch the fish, on the usual question being asked, " Weel, Janet, hoo's haddies the day ? " " Haddies, mem ? Ou, *haddies is men's lives the day !* " which was often true, as haddocks were often caught at the risk of their husbands' lives. After the usual amount of higgling, the haddies were brought down to their proper market price,—sometimes a penny for a good haddock, or, when herrings were rife, a dozen herrings for twopence, crabs for a penny, and lobsters for threepence. For there were no railways then to convey the fish to England, and thus equalise the price for all classes of the community.

Let me mention here a controversy between a fishwife and a buyer called Thomson. The buyer offered a price so ridiculously small for a parcel of fish that the seller became quite indignant, and she terminated at once all further higgling. Looking up to him, she said, " Lord help yer e'e-sight, Maister Tamson ! " " Lord help my e'e-sight, woman ! What has that to do with it ? " " Ou," said she, "because ye ha'e nae nose to put spectacles on!" As it happened, poor Mr. Thomson had, by some accident or disease, so little of a nose left, if any at all, that the bridge of the nose for holding up the spectacles was almost entirely wanting. And thus did the fishwife retaliate on her niggardly customer.

When my mother had got her fish laid at the bottom of the creel, she next went to the flesher for her butcher-meat. There was no higgling here, for the meat was sold at the ordinary market price. Then came the poultry stratum; then the vegetables, or fruits in their season; and, finally, there was "the floore"—a bunch of flowers; not a costly bouquet, but a large assortment of wallflowers, daffodils (with their early spring fragrance), polyanthuses, liliacs, gilly-flowers, and the glorious old-fashioned cabbage rose, as well as the even more gloriously fragrant moss rose. The caddy's creel was then topped up, and the marketing was completed. The lady was then followed home, the contents were placed in the larder, and the flowers distributed all over the house.

I have many curious traditional evidences of the great fondness for cats which distinguished the Nasmyth family for several generations. My father had always one or two of such domestic favourites, who were, in the best sense, his "familiars." Their quiet, companionable habits rendered them very acceptable company when engaged in his artistic work. I know of no sound so pleasantly tranquillising as the purring of a cat, or of anything more worthy of admiration in animal habit as the neat, compact, and elegant manner in which the cat adjusts itself at the fireside, or in a snug, cosy place, when it settles down for a long quiet sleep. Every spare moment that a cat has before lying down to rest is occupied in carefully cleaning itself, even under adverse circumstances. The cat is the true original inventor of a sanitary process, which has lately been patented and paraded before the public as a sanitary *novelty;* and yet it has been in practice ever since cats were created. Would that men and women were more alive to habitual cleanliness—even the cleanliness of cats. The kindly and gentle animal gives them all a lesson.

Then, nothing can be more beautiful in animal action than the exquisitely precise and graceful manner in which the cat exerts the exact amount of effort requisite to land it at the height and spot it wishes to reach at one bound. The neat and delicately precise manner in which cats use their paws when playing with those who habitually treat them with gentle kindness, is truly admirable. In these respects cats are entitled to the most kindly regard. There are, unfortunately, many who entertain a strong prejudice against this most perfect and beautiful member of the animal creation, and who abuse them because they resist ill treatment, which their innate feeling of independence causes them to resist. Cats have no doubt less personal attachments than dogs, but when kindly treated they become in many respects attached and affectionate animals.

My father, when a boy, made occasional visits to Hamilton, in the West of Scotland, where the descendants of his Covenanting ancestors still lived. One of them was an old bachelor—a recluse sort of man; and yet he had the Nasmyth love of cats. Being of pious pedigree and habits, he always ended the day by a long and audible prayer. My father and his companions used to go to the door of his house to listen to him, but especially to hear his culminating finale. He prayed that the Lord would help him to forgive his enemies and all those who had done him injury; and then, with a loud burst, he concluded, " Except John Anderson o' the Toonhead, for he killed my cat, and him I'll ne'er forgie! " In conclusion, I may again refer to Elspeth Nasmyth, who was burnt alive for witchcraft, because she had four black cats, and read her Bible through two pairs of spectacles !

CHAPTER V.

BEFORE I went to school it was my good fortune to be placed under the special care of my eldest sister, Jane. She was twenty years older than myself, and had acquired much practical experience in the management of the younger members of the family. I could not have had a more careful teacher. She initiated me into the depths of A B C, and by learning me to read she gave me the key to the greatest thoughts of the greatest thinkers who have ever lived.

But all this was accomplished at first in a humdrum and tentative way. About seventy years ago children's books were very uninteresting. In the little stories manufactured for children, the good boy ended in a coach-and-four, and the bad boy in a ride to Tyburn. The good boys must have been a set of little snobs and prigs, and I could scarcely imagine that they could ever have lived as they were represented in these goody books. If so, they must have been the most tiresome and uninteresting vermin that can possibly be imagined.

After my sister had done what she could for me, I was sent to school to learn English. I was placed under the tuition of a leading teacher called Knight, whose school-room was in the upper storey of a house in George Street.

Here I learned to read with ease.   But my primitive habit
of spelling by ear, in accordance with the simple sound of
the letters of the alphabet (phonetically, so to speak) brought
me into collision with my teacher.   I got many a cuff on
the side of the head, and many a "palmy" on my hands
with a thick strap of hard leather, which did not give me
very inviting views as to the pleasures of learning.   The
master was vicious and vindictive.   I think that it is a
very cowardly act to deal with a little boy in so cruel a
manner, and to send him home with his back and fingers
tingling and sometimes bleeding, because he cannot learn so
quickly as his fellows.

On one occasion Knight got out of temper with my
stupidity or dulness in not comprehending something about
a "preter-pluperfect tense," or some mystery of that sort.
He seized me by the ears, and beat my head against the
wall behind me, with such savage violence that when he let
me go, stunned and unable to stand, I fell forward on the
floor bleeding violently at the nose, and with a terrific head-
ache.   The wretch might have ruined my brain for life.   I
was carried home and put to bed, where I lay helpless for
more than a week.   My father threatened to summon the
teacher before the magistrates for what might have been a
fatal assault on poor little me ; but on making a humble
apology for his brutal usage he was let off.   Of course I
was not sent back to *his* school.   I have ever since entertained a hatred against grammatical rules.

There was at that time an excellent system of teaching
young folks the value of thrift.   This consisted in saving for
some purpose or another the Saturday's penny—one penny
being our weekly allowance of pocket-money.   The feats
we could perform in the way of procuring toys, picture-
books, or the materials for constructing flying kites, would
amaze the youngsters of the present day, who are generally
spoiled by extravagance.   And yet we obtained far more

pleasure from our purchases. We had in my time "penny pigs," or thrift boxes. They were made in a vase form, of brown glazed earthenware, the only entrance to which was a slit—enough to give entrance to a penny. When the Saturday's penny was not required for any immediate purposes, it was dropped through the slit, and remained there until the box was full. The maximum of pennies it could contain was about forty-eight. When that was accomplished, the penny pig was broken with a hammer, and its rich contents flowed forth. The breaking of the pig was quite an event. The fine fat old George the Third penny pieces looked thoroughly substantial in our eyes. And then there was the spending of the money—in some longlooked-for toy, or pencils, or book, or painting materials.

One of the ways in which I used my Saturday pennies was in going with some of my choice friends into the country to have a picnic. We used to light a fire behind a hedge or a dyke, or in the corner of some ruin, and there roast our potatoes, or broil a herring on some extempore gridiron we had contrived for the purpose. We lit the fire by means of a flint and steel and a tinder-box, which in those days every boy used to possess. The bramble-berries gave us our dessert. We thoroughly enjoyed these glorious Saturday afternoons. It gave us quite a Robinson Crusoe sort of feeling to be thus secluded from the world. Then the beauty of the scenery amidst which we took our repast was such as I cannot attempt to describe. A walk of an hour or so would bring us into the presence of an old castle, or amongst the rocky furze and heather-clad hills, amidst clear rapid streams, so that, but for the distant peeps of the city, one might think that he was far from the busy haunts of men and boys.

To return to my school-days. Shortly after I left the school in George Street, where the schoolmaster had almost split my skull in battering it upon the wall behind me, I was

G

entered as a pupil at the Edinburgh High School, in October 1817. The school was situated near the old Infirmary. Professor Pillans was the rector, and under him were four masters. I was set to study Latin under Mr. Irvine. He was a mere schoolmaster in the narrowest sense of the term. He was not endowed with the best of tempers, and it was often put to the breaking strain by the tricks and negligence of the lower form portion of his class. It consisted of nearly two hundred boys; the other three masters had about the same number of scholars. They each had a separate class-room.

I began to learn the elementary rudiments of Latin grammar. But not having any natural aptitude for acquiring classic learning so called, I fear I made but little progress during the three years that I remained at the High School. Had the master explained to us how nearly allied many of the Latin and Greek roots were to our familiar English words, I feel assured that so interesting and valuable a department of instruction would not have been neglected. But our memories were strained by being made to say off "by heart," as it was absurdly called, whole batches of grammatical rules, with all the botheration of irregular verbs and suchlike. So far as I was concerned, I derived little benefit from my High School teaching, except that I derived one lesson which is of great use in after life. I mean as regards the performance of duty. I did my tasks punctually and cheerfully, though they were far from agreeable. This is an exercise in early life that is very useful in later years.

In my walks to and from the High School, the usual way was along the North and South Bridges—the first over the Nor' Loch, now the railway station, and the second over the Cowgate. That was the main street between the Old Town and the New. But there were numerous wynds and closes (as the narrow streets are called) which led down from the High Street and the upper part of the Canongate to the High School, through which I often preferred to wander. So long as Old Edinburgh was confined within its walls the nobles lived in those narrow streets; and the old houses are full of historical incident. My father often pointed these houses out to me, and I loved to keep up my recollections. I must have had a little of the antiquarian spirit even then. I got to know all the most remarkable of these ancient houses —many of which were distinguished by the inscriptions on the lintel of the entrance, as well as the arms of the former possessors. Some had mottoes such as this : " BLESIT BE GOD AND HYS GIFTIS. 1584."

DOORHEAD, FROM AN OLD MANSION.

There was often a tower-shaped projection from the main front of the house, up which a spiral stair proceeded. This is usually a feature in old Scotch buildings. But in these closes the entrance to the houses was through a ponderous door, studded with great broad-headed nails, with loopholes at each side of the door, as if to present the strongest possible resistance to any attempt at forcible entrance. Indeed, in the old times before the Union the nobles were often as strong as the King, and many a time the High Street was reddened by the blood of the noblest and bravest of the land. In 1588 there was a cry of " A Naesmyth," " A Scott," in the High Street. It was followed by a clash of arms, and two of Sir Michael Naesmyth's sons were killed in that bloody feud. Edinburgh was often the scene of such disasters. Hence the strengthening of their houses, so as to resist the inroads of feudal enemies.

The mason-work of the doors was executed with great care and dexterity. It was chamfered at the edges in a bold manner, and ornamented with an O.G. bordering, which had

a fine effect, while it rendered the entrance more pleasant
by the absence of sharp angles. The same style of ornamentation was generally found round the edges of the stone-
work of the windows, most commonly by chamfering off the
square angle of the stone-work. This not only added a grim
grace to the appearance of the windows, but allowed a more
free entrance of light into the apartments, while it permitted
the inmates to have a better range of view up and down
the Close. These gloomy-looking mansions were grim in a
terrible sense, and they reminded one of the fearful transactions of "the good old times"! On many occasions, when
I was taking a daunder through these historic houses in the
wynds and closes of the Old Town, I have met Sir Walter
Scott showing them to his visitors, and listened to his deep,
earnest voice while narrating to them some terrible incident
in regard to their former inhabitants.

On other occasions I have frequently met Sir Walter
sturdily limping along over the North Bridge, while on his
way from the Court of Session (where he acted as Clerk of
the Records) to his house in Castle Street. In the same
way I saw most of the public characters connected with the
Law Courts or the University. Sir Walter was easily distinguished by his height, as well as his limp or halt in his
walk. My father was intimate with most, if not all, of the
remarkable Edinburgh characters, and when I had the pleasure of accompanying him in his afternoon walks I could
look at them and hear them in the conversations that took
place.

I remember, when I was with my father in one of his
walks, that a young English artist accompanied us. He
had come across the Border to be married at Gretna Green,
and he brought his bride onward to Edinburgh. My father
wished to show him some of the most remarkable old buildings of the town. It was about the end of 1817, when one
of the most interesting buildings in Edinburgh was about to

THE OLD TOLBOOTH, EDINBURGH. BY ALEXANDER NASMYTH. FROM THE DRAWING IN THE POSSESSION OF LORD INGLIS, LORD JUSTICE-GENERAL.

be demolished. This was no less a place than the Old Tol-booth in the High Street,—a grand but gloomy old build-ing. It had been originally used as the city palace of the Scottish kings. There they held their councils and dispensed justice. But in course of time the King and Court abandoned the place, and it had sunk into a gaol or prison for the most abandoned of malefactors. After their trial the prisoners were kept there waiting for execution, and they were hanged on a flat-roofed portion of the building at its west end.

At one of the strongest parts of the building a strong oak chest, iron-plated, had been built in, held fast by a thick wall of stone and mortar on each side. The iron chest measured about nine feet square, and was closed by a strong iron door with heavy bolts and locks. This was the *Heart of Mid-lothian*, the condemned cell of the Tolbooth.[1] The iron chest was so heavy that the large body of workmen could not, with all their might, pull it out. After stripping it of its masonry, they endeavoured by strong levers to tumble it down into the street. At last, with a " Yo ! heave ho !" it fell down with a mighty crash. The iron chest was so strong that it held together, and only the narrow iron door, with its locks, bolts, and bars, was burst open, and jerked off amongst the bystanders.

It was quite a scene. A large crowd had assembled, and amongst them was Sir Walter Scott. Recognising my father, he stood by him, while both awaited the ponderous crash. Sir Walter was still The Great Unknown, but it was pretty well known who had given such an interest to the build-ing by his fascinating novel, *The Heart of Midlothian*. Sir Walter afterwards got the door and the key for his house at Abbotsford.

---

[1] Long after the condemned cell had been pulled down, an English Chartist went down to Edinburgh to address a large meeting of his brother politicians. He began by addressing them as " *Men of the Heart of Midlothian !*" There was a loud guffaw throughout the audience. He addressed them as if they were a body of condemned malefactors.

There was a rush of people towards the iron chest, to look into the dark interior of that veritable chamber of horrors. My father's artist friend went forward with the rest, to endeavour to pick up some remnant of the demolished structure. As soon as the clouds of dust had been dispersed, he observed, under the place where the iron box had stood, a number of skeletons of rats, as dry as mummies. He selected one of these, wrapped it in a newspaper, and put it in his pocket as a recollection of his first day in Edinburgh, and of the total destruction of the " Heart of Midlothian." This artist was no other than John Linnell, the afterwards famous landscape painter. He was then a young and unknown man. He brought a letter of introduction to my father. He also brought a landscape as a specimen of his young efforts, and it was so splendidly done that my father augured a brilliant career for this admirable artist.[1]

I had the pleasure of seeing Sir Walter Scott on another and, to me, a very memorable occasion. From an early period of my schoolboy days I had a great regard for every object that had reference to bygone times. They influenced my imagination, and conjured up in my mind dreamy visions of the people of olden days. It did not matter whether it was an old coin or an old castle. I took pleasure in rambling about the old castles near Edinburgh, many of them con-

---

[1] I was so much impressed with the events of the day, and also with the fact of the young artist having taken with him so repulsive a memento as a rat's skeleton, that I never forgot it. More than half a century later, when I was at a private view of the Royal Academy, I saw sitting on one of the sofas a remarkable and venerable-looking old gentleman. On inquiring of my friend Thomas Webster who he was, he answered, " Why, that's old Linnell!" I then took the liberty of sitting down beside him, and, apologising for my intrusion on his notice, I said it was just fifty-seven years since I had last seen him ! I mentioned the circumstance of the rat-skeleton which he had put in his pocket at Edinburgh. He was pleased and astonished to have the facts so vividly recalled to his mind. At last he said, " Well, I have that mummy rat, the relic of the Heart of Midlothian, safe in a cabinet of curiosities in my house at Redhill to this day."

nected with the times of Mary Queen of Scots. Craigmillar
Castle was within a few miles of the city; there was also
Crighton Castle, and above all Bothwell Castle. This grand
massive old ruin left a deep impression on my mind. The
sight of its gloomy interior, with the great hall lighted up
only by stray glints of sunshine, as if struggling for access
through the small deep-seated windows in its massive walls,
together with its connection with the life and times of Queen
Mary, had a far greater influence upon my mind than I experienced while standing amidst the Coliseum of Rome.

Like many earnest-minded boys, I had a severe attack at
the right time of life, say from 12 to 15, of what I would
call " the collecting period." This consisted, in my case, of
accumulating old coins, perhaps one of the most salutary
forms of this youthful passion. I made exchanges with my
school companions. Sometimes my father's friends, seeing
my anxiety to improve my collection, gave me choice specimens of bronze and other coins of the Roman emperors,
usually duplicates from their own collection. These coins
had the effect of promoting my knowledge of Roman history.
I read up in order to find out the acts and deeds of the old
rulers of the civilised world. Besides collecting the coins,
I used to make careful drawings of the obverse and reverse
faces of each in an illustrated catalogue which I kept in my
little coin cabinet.

I remember one day, when sitting beside my father,
making a very careful drawing of a fine bronze coin of
Augustus, that Sir Walter Scott entered the room. He frequently called upon my father in order to consult him with
respect to his architectural arrangements. Sir Walter caught
sight of me, and came forward to look over the work I was
engaged in. At his request I had the pleasure of showing
him my little store of coin treasures, after which he took
out of his waistcoat pocket a beautiful silver coin of the
reign of Mary Queen of Scots, and gave it to me as being

his "young brother antiquarian." I shall never forget the
kind fatherly way in which he presented it.   I considered
it a great honour to be spoken to in so friendly a way by
such a man; besides, it vastly enriched my little collection
of coins and medals.

It was in the year 1817 that I had the pleasure, never
to be forgotten, of seeing the great engineer, James Watt.
He was then close upon his eighty-second year.   His visit
to Edinburgh was welcomed by the most distinguished scientific and literary men of the city.   My father had the
honour of meeting him at a dinner given by the Earl of
Buchan, at his residence in George Street.   There were present, Sir James Hall, President of the Royal Society; Francis
Jeffrey, Editor of the *Edinburgh Review;* Walter Scott, still
the Great Unknown; and many other distinguished nota-
bilities.   The cheerful old man delighted them with his
kindly talk, as well as astonished them with the extent and
profundity of his information.

On the following day Mr. Watt paid my father a visit.
He carefully examined his artistic and other works.   Having
inspected with great pleasure some landscape paintings of
various scenes in Scotland executed by my sisters, who were
then highly efficient artists, he purchased a specimen from
each of them, as well as three landscapes painted by my
father, as a record of his pleasant visit to the capital of his
native country.   I well remember the sight I then got of
the Great Engineer.   I had just returned from the High
School when he was leaving my father's house.   It was but
a glimpse I had of him.   But his benevolent countenance
and his tall but bent figure made an impression on my
mind that I can never forget.   It was even something
to have seen for a few seconds so truly great and noble a
man.

I did not long continue my passion for the collection of
coins.   I felt a greater interest in mechanical pursuits.   I

have a most cherished and grateful remembrance of the happy hours and days that I spent in my father's workroom. When the weather was ungenial he took refuge amongst his lathes and tools, and then I followed and watched him. He took the greatest pleasure in instructing me. Even in the most humble mechanical job he was sure to direct my attention to the action of the tools and to the construction of the work he had in hand, and pointed out the manipulative processes requisite for its being effectually carried out. My hearty zeal in assisting him was well rewarded by his implanting in my mind the great fundamental principles on which the practice of engineering in its grandest forms is based. But I did not learn all this at once. It only came gradually, and by dint of constant repetition and inculcation. In the meantime I made a beginning by doing some little mechanical work on my own account.

While attending the High School, from 1817 to 1820, there was the usual rage amongst boys for spinning-tops, "peeries," and "young cannon." By means of my father's excellent foot-lathe I turned out the spinning-tops in capital style, so much so that I became quite noted amongst my school companions. They all wanted to have specimens of my productions. They would give any price for them. The peeries were turned with perfect accuracy, and the steel shod, or spinning pivot, was centred so as to correspond with the heaviest diameter at the top. They could spin twice as long as the bought peeries. When at full speed they would "sleep," that is, turn round without a particle of waving. This was considered high art as regarded top-spinning.

Flying-kites and tissue paper balloons were articles that I was somewhat famed for producing. There was a good deal of special skill required for the production of a flying-kite. It must be perfectly still and steady when at its highest flight in the air. Paper messengers were sent up to it along the string which held it to the ground. The top of the Calton Hill was the most favourite place for enjoying this pleasant amusement.

Another article for which I became equally famous was the manufacture of small brass cannon. These I cast and bored, and mounted on their appropriate gun-carriages. They proved very effective, especially in the loudness of the report when fired. I also converted large cellar-keys into a sort of hand-cannon. A touch-hole was bored into the barrel of the key, with a sliding brass collar that allowed the key-guns to be loaded and primed and ready for firing.

The principal occasion on which the brass cannon and hand-guns were used was on the 4th of June—King George the Third's birthday. This was always celebrated with exuberant and noisy loyalty. The guns of the Castle were fired at noon, and the number of shots corresponded with the number of years that the king had reigned. The grand old Castle was enveloped in smoke, and the discharges reverberated along the streets and among the surrounding hills. Everything was in holiday order. The coaches were hung with garlands, the shops were ornamented, the troops were reviewed on Bruntsfield Links, and the citizens drank the king's health at the Cross, throwing the glasses over their backs. The boys fired off gunpowder, or threw squibs or crackers from morning till night. It was one of the greatest schoolboy events of the year.

My little brass cannon and hand-guns were very busy that day. They were fired until they became quite hot. These were the pre-lucifer days. The fire to light the powder at the touch-hole was obtained by the use of a flint, a steel, and a tinder-box. The flint was struck sharply on the steel; a drop of fire fell into the tinder-box, and the match of hemp string, soaked in saltpetre, was readily lit, and fired off the little guns.

I carried on quite a trade in forging beautiful little steels. I forged them out of old files, which proved excellent material for the purpose. I filed them up into neat and correct forms, and then hardened and tempered them, *secundum artem*, at the little furnace stove in my father's workroom, where of course there were also a suitable anvil, hammer, and tongs. I often made potent use of these steels in escaping from the ordeal of some severe task imposed upon me at school. The schoolmaster often deputed his authority to the monitors to hear us say our lessons. But when I slyly exhibited a beautiful steel the monitor could not maintain his grim sense of duty, and he often let me escape the ordeal of repeating some passage from a Latin school-book by obtaining possession of the article. I thus bought myself off. This system of bribery and corruption was no doubt shockingly improper, but as I was not naturally endowed with the taste for learning Latin and Greek, I continued my little diplomatic tricks until I left school.

As I have said, I did not learn much at the High School. My mind was never opened up by what was taught me there. It was a mere matter of rote and cram. I learnt by heart a number of Latin rules and phrases, but what I learnt soon slipped from my memory. My young mind was tormented by the tasks set before me. At the same time my hungry mind thirsted for knowledge of another kind.

There was one thing, however, that I *did* learn at the High School. That was the blessings and advantages of friendship. There were several of my schoolfellows of a like disposition with myself, with whom I formed attachments which ended only with life. I may mention two of them in particular—Jemmy Patterson and Tom Smith. The former was the son of one of the largest iron founders in Edinburgh. He was kind, good, and intelligent. He and I were great cronies. He took me to his father's workshops. Nothing could have been more agreeable to my tastes. For there I saw how iron castings were made. Mill-work and steam-engines were repaired there, and I could see the

way in which power was produced and communicated. To
me it was a most instructive school of practical mechanics.
Although I was only about thirteen at the time, I used to
lend a hand, in which hearty zeal made up for want of
strength. I look back to these days, especially to the
Saturday afternoons spent in the workshops of this admirably conducted iron foundry, as a most important part of my
education as a mechanical engineer. I did not *read* about
such things; for words were of little use. But I saw and
handled, and thus all the ideas in connection with them
became permanently rooted in my mind.

Each department of the iron foundry was superintended
by an able and intelligent man. He was distinguished not
only by his ability but for his steadiness and sobriety. The
men were for the most part promoted to their foremanship
from the ranks, and had been brought up in the concern
from their boyhood. They possessed a strong individuality
of character, and served their employer faithfully and loyally.
One of these excellent men, with whom I was frequently
brought into contact, was William Watson. He took special
charge of all that related to the construction and repairs of
steam-engines, water-wheels, and mill work generally. He
was a skilful designer and draughtsman and an excellent
pattern maker. His designs were drawn in a bold and
distinct style, on large deal boards, and were passed into
the hands of the mechanics to be translated by them into
actual work.

It was no small privilege to me to stand by, and now and
then hold the end of the long straight edge, or by some humble
but zealous genuine help of mine contribute to the progress
of these substantial and most effective mechanical drawings.
Watson explained to me, in the most common-sense manner,
his reasons for the various forms, arrangements, and proportions of the details of his designs. He was an enthusiast
on the subject of Euclid; and to see the beautiful problems

applied by him in working out his excellent drawings was to me a lesson beyond all price.

Watson was effectively assisted by his two sons, who carried out their father's designs in the form of the wood patterns by which the foundry-men or moulders reproduced their forms in cast iron, and the smiths by their craft realised the wrought - iron portions. These sons of Mr. Watson were of that special class of workmen called mill- wrights—a class now almost extinct, though many of the best known engineers originally belonged to them. They could work with equal effectiveness in wood or iron.

Another foreman in Mr. Patterson's foundry was called Lewis. He had special charge of the iron castings designed for architectural and ornamental purposes. He was a man of great taste and artistic feeling, and I was able even at that time to appreciate the beauty of his designs. One of the most original characters about the foundry, however, was Johnie Syme. He took charge of the old Boulton and Watt steam-engine, which gave motion to the machinery of the works. It also produced the blast for the cupolas, in which the pig and cast iron scrap was daily melted and cast into the various objects produced in the foundry. Johnie was a complete incarnation of technical knowledge. He was the Jack-of-all-trades of the establishment; and the standing counsel in every out-of-the-way case of managing and overcoming mechanical difficulties. He was the super-intendent of the boring machines. In those days the boring of a steam-engine cylinder was considered high art *in excelsis!* Patterson's firm was celebrated for the accuracy of its boring.

I owe Johnie Syme a special debt of gratitude, as it was he who first initiated me into that most important of all technical processes in practical mechanism—the art of hard-ening and tempering steel. It is, perhaps, not saying too much to assert that the successful practice of the mechanical

arts, by means of which the civilised man rises above the
savage condition, is due to that wonderful change. Man
began with wood, and stone, and bone; he proceeded to
bronze and iron; but it was only by means of hardened
steel that he could accomplish anything in arms, in agriculture, or in architecture. The instant hardening which
occurs on plunging a red-hot piece of steel into cold water
may well be described as mysterious. Even in these days,
when science has defined the causes of so many phenomena,
the reason of steel becoming hard on suddenly cooling it
down from a red-heat, is a fact that no one has yet explained!
The steel may be *tempered* by modifying the degrees of heat
to which it is subsequently subjected. It may thus be
toughened by slightly reheating the hardened steel; the re-
softening course is indicated by certain prismatic tints, which
appear in a peculiar mode of succession on its surface.
The skilful artisan knows by experience the exact point at
which it is necessary again to plunge it into cold water in
order to realise the requisite toughness or hardness of the
material required for his purposes.

In all these matters, my early instructor, Johnie Syme,
gave me such information as proved of the greatest use to
me in the after history of my mechanical career. ｜ Johnie
Syme was also the very incarnation of quaint sly humour;
and when communicating some of his most valued arcana of
practical mechanical knowledge he always reminded me of
some of Ostade's Dutchmen, by an almost indescribable sly
humorous twinkle of the eye, which in that droll way stamped
his information on the memory.

Tom Smith was another of my attached cronies. Our
friendship began at the High School in 1818. A similarity
of disposition bound us together. Smith was the son of an
enterprising general merchant at Leith. His father had a
special genius for practical chemistry. He had established
an extensive colour manufactory at Portobello, near Edin-

burgh, where he produced white lead, red lead, and a great variety of colours—in the preparation of which he required a thorough knowledge of chemistry. Tom Smith inherited his father's tastes, and admitted me to share in his experiments, which were carried on in a chemical laboratory situated behind his father's house at the bottom of Leith Walk.

We had a special means of communication. When anything particular was going on at the laboratory, Tom hoisted a white flag on the top of a high pole in his father's garden. Though I was more than a mile apart, I kept a look-out in the direction of the laboratory with a spy-glass. My father's house was at the top of Leith Walk, and Smith's house was at the bottom of it. When the flag was hoisted I could clearly see the invitation to me to come down. I was only too glad to run down the Walk and join my chum; to take part in some interesting chemical process. Mr. Smith, the father, made me heartily welcome. He was pleased to see his son so much attached to me, and he perhaps believed that I was worthy of his friendship. We took zealous part in all the chemical proceedings, and in that way Tom was fitting himself for the business of his life.

Mr. Smith was a most genial tempered man. He was shrewd and quick-witted, like a native of York, as he was. I received the greatest kindness from him as well as from his family. His house was like a museum. It was full of cabinets, in which were placed choice and interesting objects in natural history, geology, mineralogy, and metallurgy. All were represented. Many of these specimens had been brought to him from abroad by his ship captains who transported his colour manufactures and other commodities to foreign parts.

My friend Tom Smith and I made it a rule—and in this we were encouraged by his father—that, so far as was possible, we ourselves should actually *make* the acids and other substances used in our experiments. We were

not to buy them ready made, as this would have taken the
zest out of our enjoyment. We should have lost the
pleasure and instruction of producing them by means of
our own wits and energies. To encounter and overcome a
difficulty is the most interesting of all things. Hence,
though often baffled, we eventually produced perfect specimens of nitrous, nitric, and muriatic acids. We distilled
alcohol from duly fermented sugar and water, and rectified
the resultant spirit from fusel oil by passing the alcoholic
vapour through animal charcoal before it entered the worm
of the still. We converted part of the alcohol into sul-
phuric ether. We produced phosphorus from old bones,
and elaborated many of the mysteries of chemistry.

The amount of practical information which we obtained
by this system of making our own chemical agents was
such as to reward us, in many respects, for the labour we
underwent. To outsiders it might appear a very troublesome and roundabout way of getting at the finally desired
result. But I feel certain that there is no better method of
rooting chemical, or any other instruction, deeply in our
minds. Indeed, I regret that the same system is not
pursued by the youth of the present day. They are
seldom, if ever, called upon to exert their own wits and
industry to obtain the requisites for their instruction. A
great deal is now said about technical education; but how
little there is of technical handiness or head work! Everything is *bought ready made* to their hands; and hence there
is no call for individual ingenuity.

I often observe, in shop-windows, every detail of model
ships and model steam-engines, supplied ready made for
those who are "said to be" of an ingenious and mechanical
turn. Thus the vital uses of resourcefulness are done away
with, and a sham exhibition of mechanical genius is paraded
before you by the young impostors—the result, for the
most part, of too free a supply of pocket money. I have

known too many instances of parents, being led by such false evidence of constructive skill, to apprentice their sons to some engineering firm ; and, after paying vast sums, finding out that the pretender comes out of the engineering shop with no other practical accomplishment than that of glove-wearing and cigar-smoking !

The truth is that the eyes and the fingers—*the bare fingers*—are the two principal inlets to sound practical instruction.   They are the chief sources of trustworthy knowledge in all the materials and operations which the engineer has to deal with.   No *book* knowledge can avail for that purpose.   The nature and properties of the materials must come in through the finger ends.   Hence, I have no faith in young engineers who are addicted to wearing gloves.   Gloves, especially kid gloves, are perfect non-conductors of technical knowledge.   This has really more to do with the efficiency of young aspirants for engineering success than most people are aware of.   Yet kid gloves are now considered the genteel thing.

H

CHAPTER VI.

I LEFT the High School at the end of 1820. I carried with me a small amount of Latin, and no Greek. I do not think I was much the better for my small acquaintance with the dead languages. I wanted something more living and quickening. I continued my studies at private classes. Arithmetic and geometry were my favourite branches. The three first books of Euclid were to me a new intellectual life. They brought out my power of reasoning. They trained me mentally. They enabled me to arrive at correct conclusions, and to acquire a knowledge of absolute truths. It is because of this that I have ever since held the beautifully perfect method of reasoning, as exhibited in the exact method of arriving at Q.E.D., to be one of the most satisfactory efforts and exercises of the human intellect.

Besides visiting and taking part in the works at Patterson's foundry, and joining in the chemical experiments at Smith's laboratory, my father gave me every opportunity for practising the art of drawing. He taught me to sketch with exactness every object, whether natural or artificial, so as to enable the hand to accurately reproduce what the eye had seen. In order to acquire this almost invaluable art, which can serve so many valuable purposes in life, he was careful to educate my eye, so that I might

perceive the relative proportions of the objects placed before
me. He would throw down at random a number of bricks,
or pieces of wood representing them, and set me to copy
their forms, their proportions, their lights and shadows
respectively. I have often heard him say that any one who
could make a correct drawing in regard to outline, and also
indicate by a few effective touches the variation of lights
and shadows of such a group of model objects, might not
despair of making a good and correct sketch of the exterior
of York Minster.

My father was an enthusiast in praise of this *graphic
language*, and I have followed his example. In fact, it
formed a principal part of my own education. It gave me
the power of recording observations with a few graphic
strokes of the pencil ; and far surpassed in expression any
number of mere words. This graphic eloquence is one of
the highest gifts in conveying clear and correct ideas as to
the forms of objects—whether they be those of a simple and
familiar kind, or of some form of mechanical construction, or
of the details of a fine building, or the characteristic features
of a wide-stretching landscape. This accomplishment of
accurate drawing, which I achieved for the most part in
my father's workroom, served me many a good turn in
future years with reference to the engineering work which
became the business of my life.

I was constantly busy ; mind, hands, and body were kept
in a state of delightful and instructive activity. When not
drawing, I occupied myself in my father's workshop at the
lathe, the furnace, or the bench. I gradually became initiated into every variety of mechanical and chemical manipulation. I made my own tools and constructed my chemical
apparatus, as far as lay in my power. With respect to the
latter, I constructed a very handy and effective blowpipe
apparatus, consisting of a small air force-pump, connected
with a cylindrical vessel of tin plate. By means of an

occasional use of the handy pump, it yielded such a fine steady blowpipe blast, as enabled me to bend glass tubes and blow bulbs for thermometers, to analyse metals or mineral substances, or to do any other work for which intense heat was necessary. My natural aptitude for manipulation, whether in mechanical or chemical operations, proved very serviceable to myself as well as to others; and (as will be shown hereafter), it gained for me the friendship of many distinguished scientific men.

But I did not devote myself altogether to experiments. Exercise is as necessary for the body as the mind. Without full health a man cannot enjoy comfort, nor can he possess endurance. I therefore took plenty of exercise out of doors. I accompanied my father in his walks round Edinburgh. My intellect was kept alive during these delightful excursions. For sometimes my father was accompanied by brother-artists, whose conversation is always so attractive; and sometimes by scientific men, such as Sir James Hall, Professor Leslie, Dr. Brewster, and others. Whatever may have been my opportunities for education so-called, nothing could have better served the purpose of real education (the evolution of the mental faculties) than the opportunities I enjoyed while accompanying and listening to the conversation of men distinguished for their originality of thought and their high intellectual capacity. This was a mental culture of the best kind.

The volcanic origin of the beautiful scenery round Edinburgh was often the subject of their conversation. Probably few visitors are aware that all those remarkable eminences, which give to the city and its surroundings so peculiar and romantic an aspect, are the results of the operation, during inconceivably remote ages, of volcanic force penetrating the earth's crust by disruptive power, and pouring forth streams of molten lava, now shrunk and cooled into volcanic rock. The observant eye, opened by the light of Science, can see unmistakable evidences of a condition of things which were in action at periods so remote as, in comparison, to shrink up the oldest of human records into events of yesterday.

I had often the privilege of standing by and hearing the philosophic Leslie, Brewster, and Hall, discussing these volcanic remains in their actual presence; sometimes at Arthur's Seat or on the Calton Hill, or at the rock on which Edinburgh Castle stands. Their observations sank indelibly into my memory, and gave me the key to the origin of this grand class of terrestrial phenomena. When standing at the "Giant's Ribs," on the south side of Arthur's Seat, I felt as if one of the grandest pages of the earth's history lay open before me. The evidences of similar volcanic action abound in many other places near Edinburgh; and they may be traced right across Scotland from the Bass Rock to Fingal's Cave, the Giant's Causeway in Antrim, and Slievh League on the south-west coast of Donegal in Ireland.

Volcanic action, in some inconceivably remote period of the earth's crust history, has been the *Plough*, and after denudation by water, has been the *Harrow*, by which the originally deep-seated mineral treasures of the globe have been brought within the reach of man's industrial efforts. It has thus yielded him inexhaustible mineral harvests, and helped him to some of the most important material elements in his progress towards civilisation. It is from this consideration that, while enjoying the results of these grand fundamental actions of the Creator's mighty agencies in their picturesque aspect, the knowledge of their useful results to man adds vastly to the grandeur of the contemplation of their aspect and nature. This great subject caused me, even at this early period of my life, to behold with special interest the first peep at the structure of the moon's surface, as revealed to me by an excellent Ramsden "spy-glass," which my father possessed, and thus planted the seed of that earnest desire to scrutinise more minutely the moon's

wonderful surface, which in after years I pursued by means
of the powerful reflecting telescopes constructed by myself.
To turn to another subject. In 1822 the loyalty of
Scotland was greatly excited when George the Fourth paid
his well-known visit to Edinburgh. It was then the second
greatest city in the kingdom, and had not been visited by
royalty for about 170 years. The civic authorities, and
the inhabitants generally, exerted themselves to the utmost
to give the king a cordial welcome, in spite of a certain
feeling of dissatisfaction as to his personal character. The
recent trial and death of Queen Caroline had not been forgotten, yet all such recollections were suppressed in the
earnest desire to give every respect to the royal visitor.
Edinburgh was crowded with people from all parts of the
country; heather was arrayed on every bonnet and hat;
and the reception was on the whole magnificent. Perhaps
the most impressive spectacle was the orderliness of the
multitude, all arrayed in their Sunday clothes. The streets,
windows, and house-tops were crowded; and the Calton
Hill, Salisbury Crags, and even Arthur's Seat itself, were
covered with people. On the night before the arrival a
gigantic bonfire on Arthur's Seat lit up with a tremendous
blaze the whole city, as well as the surrounding country.
It formed a magnificent and picturesque sight, illuminating
the adjacent mountains as well as the prominent features of
the city. It made one imagine that the grand old volcanic
mountain had once more, after a rest of some hundreds of
thousands of years, burst out again in its former vehemence
of eruptive activity.

There were, of course, many very distinguished men who
took part in the pageant of the king's entry into Edinburgh,
but none of them had their presence more cordially acknowledged than Sir Walter Scott, who never felt more proud of
" his own romantic town " than he did upon this occasion.
It is unnecessary to mention the many interesting features

of the royal reception. The king's visit lasted for seven or eight days, and everything passed off loyally, orderly, happily, and successfully.

Shortly after this time there was a great deal of distress among the labouring classes. All the manufacturing towns were short of employment, and the weavers and factory-workers were thrown upon the public. Many of the work-men thought that politics was the cause of their suffering. Radical clubs were formed, and the Glasgow weavers began to drill at nights in the hopes of setting things to rights by means of physical force. A large number of the starving weavers came to Edinburgh. A committee was formed, and contributions were collected, for the purpose of giving them temporary employment. They were set to work to make roads and walks round the Calton Hill and Salisbury Crags. The fine walk immediately under the precipitous crags, which opens out such perfect panoramic views of Edinburgh, was made by these poor fellows. It was hard work for their delicate hands and fingers, which before had been accustomed only to deal with threads and soft fabrics. They were very badly suited for handling the mattock, shovel, and hand-barrow. The result of their labours, however, proved of great advantage to Edinburgh in opening up the beauties of its scenery. The road round the crags is still called " The Radical Road."

Let me here mention one of the most memorable inci-dents of the year 1824. I refer to the destructive fire which took place in the old town of Edinburgh. It broke out in an apartment situated in one of the highest piles of houses in the High Street. In spite of every effort of the firemen the entire pile was gutted and destroyed. The fire was thought to be effectually arrested; but towards the afternoon of the next day smoke was observed issuing from the upper part of the steeple of the Tron Church. The steeple was built of timber, covered with lead. There is never

smoke but there is fire; and at last the flames burst forth.
The height of the spire was so lofty that all attempts to
extinguish the fire were hopeless.   The lead was soon
melted, and rushed in streams into the street below.   At
length the whole steeple fell down with a frightful crash.

I happened to see the first outbreak of this extraordinary
fire, and I watched its progress to its close.   Burning embers
were carried by the wind and communicated the fire to
neighbouring houses, which broke out about 10 P.M.   All
the fire-engines of Edinburgh and from the towns of the
neighbouring country were collected round the fire, and
played water upon the flames, but without effect.   Whole
ranges of lofty old houses were roaring with fire.   In the
course of two or three hours, several acres, covered by the
loftiest and most densely crowded houses in the High Street,
were in a blaze.   Some of them were of thirteen stories.
Floor after floor came crashing down, throwing out a blaze
of embers.   The walls of each house acted as an enormous
chimney—the windows acting as draught-holes.   The walls,
under the intense heat, were fluxed and melted into a sort
of glass.   The only method of stopping the progress of the
fire was to pull down the neighbouring houses, so as to
isolate the remaining parts of the High Street.

As the parapet of the grand old tower of the High
Church, St. Giles, was near the site of the fire—so near
as to enable one to look down into it,—my father obtained
permission to ascend, and I with him.   When we emerged
from the long dark spiral stairs on to the platform on the
top of the tower, we found a select party of the most
distinguished inhabitants looking down into the vast area
of fire; and prominent among them was Sir Walter Scott.
At last, after three days of tremendous efforts, the fire was
subdued; but not till after a terrible destruction of property.
The great height of the ruined remains of the piles of houses
rendered it impossible to have them removed by the ordinary

means.   After several fruitless attempts with chains and
ropes, worked by capstans, to pull them down, gunpowder
was at last resorted to.   Mines were dug under each vast
pile ; one or two barrels of gunpowder were placed into them
and fired; and then the before solid masses came tumbling
down amidst clouds of dust.   The management of this
hazardous but eventually safe process was conducted by
Captain Basil Hall.   He ordered a crew of sailors to be
brought up from the man-of-war guardship in the Firth of
Forth ; and by their united efforts the destruction of the
ruined walls was at last successfully accomplished.

In the autumn of 1823, when I was fifteen years old, I
had a most delightful journey with my father.   It was the
first occasion on which I had been a considerable distance
from home.   And yet the journey was only to Stirling.
My father had received a commission to paint a view of the
castle as seen from the ruins of Cambuskenneth Abbey,
situated a few miles from the town.   We started from Newhaven by a small steamboat, passing, on our way up the
Firth, Queensferry, Culross, and Alloa.   We then entered
the windings of the river, from which I saw the Ochils,
a noble range of bright green mountains.   The passage of
the steamer through the turns and windings of the Forth
was most interesting.

We arrived at Stirling, and at once proceeded to Cambuskenneth Abbey, where there was a noble old Gothic
tower.   This formed the foreground of my father's careful
sketch, with Stirling Castle in the background, and Ben
Lomond with many other of the Highland mountains in the
distance.   As my father wished to make a model of the
Gothic tower, he desired me to draw it carefully, and
to take the dimensions of all the chief parts as well as to
make detailed sketches of its minor architectural features.
It was a delightful autumn afternoon, and, before the day

had closed, our work at the abbey was done. We returned
to Stirling and took a walk round the castle to see the effect
of the sun setting behind the Highland mountains.

Next morning we visited the castle. I was much interested with the interior, especially with a beautifully
decorated Gothic oratory or private chapel, used by the
Scottish kings when they resided at Stirling. The oratory
had been converted with great taste into an ante-drawing-
room of the governor's house. The exquisite decorations of
this chapel were the first specimens of Gothic carving in oak
that I had ever seen, and they seemed to put our modern
carvings to shame.[1] The Great Hall, where the Scottish
Parliament used to meet, was also very interesting as connected with the ancient history of the country.

From Stirling we walked to Alloa, passing the picturesque
cascades rushing down the clefts of the Ochils. We put up
for the night at Clackmannan, a very decayed and melan-
choly-looking village, though it possessed a fine specimen of
the Scottish castellated tower. It is said that Robert Bruce
slept here before the Battle of Bannockburn. But the
most interesting thing that I saw during the journey was
the Devon Ironworks. I had read and heard about the
processes carried on there in smelting iron ore and running
it into pig-iron. The origin of the familiar trade term "pig-
iron " is derived from the result of the arrangement most
suitable for distributing the molten iron as it rushes forth
from the opening made at the bottom part of the blast
furnace, when, after its reduction from the ore, it collects in

---

[1] This exquisite specimen of a carved oak Gothic apartment had a terrible
incident in Scottish history connected with it. It was in this place that The
Douglas intruded his presence on James the Third. He urged his demands
in a violent and threatening manner, and afterwards laid hands upon the
king. The latter, in defending himself with his dagger, wounded the
Douglas mortally ; and to get rid of the body the king cast it out of the
window of the chapel, where it fell down the precipitous rock underneath.
The chapel has since been destroyed by fire.

a fluid mass of several tons weight. Previous to "tapping" the furnace, a great central channel is made in the sand-covered floor of the forge; this central channel is then sub-divided into many lateral branches or canals, into which the molten iron flows, and eventually hardens.

The great steam-engine that worked the blast furnace was the largest that I had ever seen. A singular expedient was employed at these works, of using a vast vault hewn in the solid rock of the hillside, for the purpose of storing up the blast produced by the engines, and so equalising the pressure; thus turning a mountain side into a reservoir for the use of a blast-furnace. This seemed to me a daring and wonderful engineering feat.

We waited at the works until the usual time had arrived for letting out the molten iron which had been accumulating at the lower part of the blast-furnace. It was a fine sight to see the stream of white-hot iron flowing like water into the large gutter immediately before the opening. From this the molten iron flowed on until it filled the moulds of sand which branched off from the central gutter. The iron left in the centre, when cooled and broken up, was called *sow* metal, while that in the branches was called *pig* iron; the terms being derived from the appearance of a sow engaged in its maternal duties. The pig-iron is thus cast in handy-sized pieces for the purpose of being transported to other iron-foundries; while the clumsy sow metal is broken up and passes through another process of melting, or is reserved for foundry uses at the works where it is produced. After inspecting with great pleasure the machinery connected with the foundry, we took our leave and returned to Edinburgh by steamer from Alloa.

Shortly after, I had the good fortune to make the acquaintance of Robert Bald, the well-known mining engineer. He was one of the most kind-hearted men I have ever known. He was always ready to communicate his

knowledge to young and old.   His sound judgment and
long practical experience in regard to coal-mining and the
various machinery connected with it, rendered him a man of
great importance in the northern counties, where his advice
was eagerly sought for.   Besides his special knowledge, he
had a large acquaintance with literature and science.   He
was bright, lively, and energetic.   He was a living record of
good stories, and in every circle in which he moved he was
the focus of cheerfulness.   In fact, there was no greater
social favourite in Edinburgh than Robert Bald.

Bald was very fond of young people, and he became
much attached to me.   He used to come to my father's
house, and often came in to see what I was about in the
work-room.   He was rejoiced to see the earnest and industrious manner in which I was employed, in preparing
myself for my proposed business as an engineer.   He looked
over my tools, mostly of my own making, and gave me every
encouragement.   When he had any visitors he usually
brought them and introduced them to me.   In this way I
had the happiness to make the acquaintance of Robert
Napier, Nelson, and Cook, of Glasgow; and in after life I
continued to enjoy their friendship.   It would be difficult
for me to detail the acts of true disinterested kindness which
I continued to receive from this admirable man.

On several occasions he wished me to accompany him on
his business journeys, in order that I might see some works
that would supply me with valuable information.   He had
designed a powerful pumping engine to drain more effectually
a large colliery district situated near Bannockburn—close to
the site of the great battle in the time of Robert the Bruce.
He invited me to join him.   It was with the greatest
pleasure that I accepted his invitation; for there would be
not only the pleasure of seeing a noble piece of steam
machinery brought into action for the first time, but also the
enjoyment of visiting the celebrated Carron Ironworks.

The Carron Ironworks are classic ground to engineers. They are associated with the memory of Roebuck, Watt, and Miller of Dalswinton. For there Roebuck and Watt began the first working steam-engine; Miller applied the steam-engine to the purposes of navigation, and invented the Carronade gun. The works existed at an early period in the history of British iron manufacture. Much of the machinery continued to be of wood. Although effective in a general way it was monstrously cumbrous. It gave the idea of vast power and capability of resistance, while it was far from being so in reality. It was, however, truly imposing and impressive in its effect upon strangers. When seen partially lit up by the glowing masses of white-hot iron, with only the rays of bright sunshine gleaming through the holes in the roof, and the dark, black, smoky vaults in which the cumbrous machinery was heard rumbling away in the distance—while the moving parts were dimly seen through the murky atmosphere, mixed with the sounds of escaping steam and rushes of water; with the half-naked men darting about with masses of red-hot iron and ladles full of molten cast-iron—it made a powerful impression upon the mind.

I was afterwards greatly interested by a collection of old armour, dug up from the field of the Battle of Bannockburn close at hand. They were arranged on the walls of the house of the manager of the Carron Ironworks. There were swords, daggers, lances, battle-axes, shields, and coats of chain-armour. Some of the latter were whole, others in fragmentary portions. I was particularly interested with the admirable workmanship of the coats of mail. The iron links extended from the covering of the head to the end of the arms, and from the shoulders down to the hips, in one linked iron fabric. The beauty and exactness with which this chain-armour had been forged and built up were truly wonderful. There must have been "giants in those days." This grand style of armour was in use from the time of the Conquest, and was most effective in the way of protection, as it was fitted by its flexibility to give full play to the energetic action of the wearer. It was infinitely superior to the senseless plate-armour that was used, at a subsequent period, to encase soldiers like lobsters. The chain-armour I saw at Carron left a deep impression on my mind. I never see a bit of it, or of its representation in the figures on our grand tombs of the thirteenth century, but I think of my first sight of it at Carron and of the tremendous conflict at Bannockburn.

Remembering, also, the impressive sight of the picturesque fire-lit halls, and the terrible-looking, cumbrous machinery which I first beheld on a grand scale at Carron, I have often regretted that some of our artists do not follow up the example set them by that admirable painter, Wright of Derby, and treat us to the pictures of some of our great ironworks. They not only abound with the elements of the picturesque in its highest sense, but also set forth the glory of the useful arts in such a way as would worthily call forth the highest power of our artists.

To return to my life at Edinburgh. I was now seventeen years old. I had acquired a considerable amount of practical knowledge as to the use and handling of mechanical tools, and I desired to turn it to some account. I was able to construct working models of steam-engines and other apparatus required for the illustration of mechanical subjects. I began with making a small working steam-engine for the purpose of grinding the oil-colours used by my father in his artistic work. The result was quite satisfactory. Many persons came to see my active little steam-engine at work, and they were so pleased with it that I received several orders for small workshop engines, and also for some models of steam-engines to illustrate the subjects taught at Mechanics' Institutions.

I contrived a sectional model of a complete condensing

steam-engine of the beam and parallel motion construction. The model, as seen from one side, exhibited every external detail in full and due action when the flywheel was moved round by hand; while, on the other or sectional side, every detail of the interior was seen, with the steam-valves and air-pump, as well as the motion of the piston in the cylinder, with the construction of the piston and the stuffing

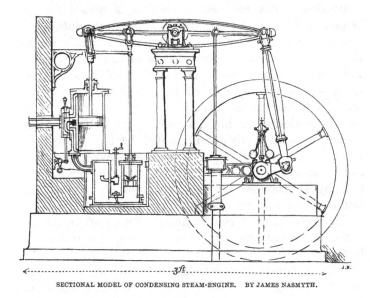

SECTIONAL MODEL OF CONDENSING STEAM-ENGINE.   BY JAMES NASMYTH.

box, together with the slide-valve and steam passages, all in due position and relative movement.

The first of these sectional models of the steam-engine was made for the Edinburgh School of Arts, where its uses in instructing mechanics and others in the application of steam were highly appreciated. The second was made for Professor Leslie, of the Edinburgh University, for use in his lectures on Natural Philosophy. The professor had, at his own private cost, provided a complete and excellent set of apparatus, which, for excellent workmanship and admirable utility, had never, I believe, been provided for the service of any university. He was so pleased with my addition to his class-room apparatus, that, besides expressing his great thanks for my services, he most handsomely presented me with a free ticket to his Natural Philosophy class as a regular student, so long as it suited me to make use of his instruction. But far beyond this, as a reward for my earnest endeavours to satisfy this truly great philosopher, was the kindly manner in which he on all occasions communicated to me conversationally his original and masterly views on the great fundamental principles of Natural Philosophy—especially as regarded the principles of Dynamics and the Philosophy of Mechanics. The clear views which he communicated in his conversation, as well as in his admirable lectures, vividly illustrated by the experiments which he had originated, proved of great advantage to me; and I had every reason to consider his friendship and his teaching as amongst the most important elements in my future success as a practical engineer.

Having referred to the Edinburgh School of Arts, I feel it necessary to say something about the origin of that excellent institution. A committee of the most distinguished citizens of Edinburgh was formed for the purpose of instituting a college, in which working men and mechanics might possess the advantages of instruction in the principles on which their various occupations were conducted. Among the committee were Leonard Horner, Francis Jeffrey, Henry Cockburn, John Murray of Henderland, Alexander Bryson, James Milne, John Miller, the Lord Provost, and various members of the Council. Their efforts succeeded, and the institution was founded. The classes were opened in 1821, in which year I became a student.

In order to supply the students, who were chiefly young men of the working-class, with sound instruction in the various branches of science, the lectures were delivered and the classes were superintended by men of established ability in their several departments. This course was regularly pursued from its fundamental and elementary principles to the highest point of scientific instruction. The consecutive lectures and examinations extended, as in the University, from October to May in each year's session. It was, in fact, our first technical college. In these later days, when so many of our so-called Mechanics' Institutes are merely cheap reading-clubs for the middle classes, and the lectures delivered are for the most part designed merely for a pleasant evening's amusement, it seems to me that we have departed greatly from the original design with which Mechanics' Institutions were founded.

As the Edinburgh School of Arts was intended for the benefit of mechanics, the lectures and classes were held in the evening after the day's work was over. The lectures on chemistry were given by Dr. Fyfe—a most able man. His clearness of exposition, his successful experiments, his careful analyses, and the easy and graphic method by which he carried his students from the first fundamental principles to the highest points of chemical science, attracted a crowded and attentive audience. Not less interesting were the lectures on Mechanical Philosophy, which in my time were delivered by Dr. Lees and Mr. Buchanan. The class of Geometry and Mathematics was equally well conducted, though the attendance was not so great.

The building which the directors had secured for the lecture-hall and class-rooms of the institution was situated at the lower end of Niddry Street, nearly under the great arch of the South Bridge. It had been built about a hundred years before, and was formerly used by an association of amateur musicians, who gave there periodical concerts of vocal and instrumental music. The orchestra was converted into a noble lecture table, with accommoda-

I

tion for any amount of apparatus that might be required for the purposes of illustration. The seats were arranged in the body of the hall in concentric segments, with the lecture table as their centre. In an alcove right opposite the lecturer might often be seen the directors of the institution — Jeffrey, Horner, Murray, and others — who took every opportunity of dignifying by their presence this noble gathering of earnest and intelligent working men.

A library of scientific books was soon added to the institution, by purchases or by gifts. Such was the eagerness to have a chance of getting the book you wanted that I remember standing on many occasions for some time amidst a crowd of applicants awaiting the opening of the door on an evening library night. It was as thick as if I had been standing at the gallery door of the theatre on a night when some distinguished star from London was about to make his appearance. There was the same eagerness to get a good place in the lecture-room, as near to the lecture table as possible, especially on the chemistry nights.

I. continued my regular attendance at this admirable institution from 1821 to 1826. I am glad to find that it still continues in active operation. In November 1880 the number of students attending the Edinburgh School of Arts amounted to two thousand five hundred ! I have been led to this prolix account of the beginning of the institution by the feeling that I owe a deep debt of gratitude to it, and because of the instructive and intellectually enjoyable evenings which I spent there, in fitting myself for entering upon the practical work of my life.

The successful establishment of the Edinburgh School of Arts had a considerable effect throughout the country. Similar institutions were established. Lectures were delivered, and the necessary illustrations were acquired — above all, the working models of the steam-engine. There

was quite a run upon me for supplying them. My third
working model was made to the order of Robert Bald, for
the purpose of presenting it to the Alloa Mechanics' Insti-
tute; the fourth was manufactured for Mr. G. Buchanan,
who lectured on mechanical subjects throughout the country ;
and the fifth was supplied to a Mr. Offley, an English
gentleman who took a fancy for the model when he came
to purchase some of my father's works.

The price I charged for my models was £10 ; and with
the pecuniary results I made over one-third to my father,
as a sort of help to remunerate him for my " keep," and
with the rest I purchased tickets of admission to certain
classes in the University.   I attended the Chemistry course
under Dr. Hope ; the Geometry and Mathematical course
under Professor Wallace ; and the Natural Philosophy
course under my valued friend and patron Professor Leslie.
What with my attendance upon the classes, and my workshop and drawing occupations, my time did not hang heavy
on my hands.

I got up early in the mornings to work at my father's
lathe, and I sat up late at night to do the brass castings in
my bedroom. Some of this, however, I did during the day-
time, when not attending the University classes.   The way
in which I converted my bedroom into a brass foundry was
as follows : I took up the carpet so that there might be
nothing but the bare boards to be injured by the heat. My
furnace in the grate was made of four plates of stout sheet
iron, lined with fire-brick, corner to corner. To get the
requisite sharp draught I bricked up with single bricks the
front of the fireplace, leaving a hole at the back of the
furnace for the short pipe just to fit into. The fuel was
generally gas coke and cinders saved from the kitchen. The
heat I raised was superb—a white heat, sufficient to melt
in a crucible six or eight pounds of brass.

Then I had a box of moulding sand, where the moulds

were gently rammed in around the pattern previous to the casting.   But how did I get my brass ?   All the old brass-works in my father's workshop drawers and boxes were laid under contribution.   This brass being for the most part soft and yellow, I made it extra hard by the addition of a due proportion of tin.   It was then capable of taking a pure finished edge.   When I had exhausted the stock of old brass, I had to buy old copper or new in the form of ingot or tile copper, and when melted I added to it one-seventh of its weight of pure tin, which yielded the strongest alloy of the two metals.   When cast into any required form this was a treat to work, so sound and close was the grain, and so durable in resisting wear and tear.   This is the true bronze or gun metal.

When melted, the liquid brass was let into the openings, until the whole of the moulds were filled.   After the metal cooled it was taken out; and when the room was sorted up no one could have known that my foundry operations had been carried on in my bedroom.   My brass foundry was right over my father's bedroom.   He had forbidden me to work late at night, as I did occasionally on the sly.   Sometimes when I ought to have been asleep I was detected by the sound of the ramming in of the sand of the moulding boxes. On such occasions my father let me know that I was dis-obeying his orders by rapping on the ceiling of his bedroom with a slight wooden rod of ten feet that he kept for mea-suring purposes.   But I got over that difficulty by placing a bit of old carpet under my moulding boxes as a non-con-ductor of sound, so that no ramming could afterwards be heard.   My dear mother also was afraid that I should damage my health by working so continuously.   She would come into the workroom late in the evening, when I was working at the lathe or the vice, and say, " Ye'll kill yer-self, laddie, by working so hard and so late."   Yet she took a great pride in seeing me so busy and so happy.

Nearly the whole of my steam-engine models were made in my father's workroom. His foot-lathe and stove, together with my brass casting arrangements in my bedroom, answered all my purposes in the way of model making. But I had at times to avail myself of the smithy and foundry that my kind and worthy friend, George Douglass, had established in the neighbourhood. He had begun business as " a jobbing smith," but being a most intelligent and energetic workman, he shot ahead and laid the foundations of a large trade in steam-engines. When I had any part of a job in hand that was beyond the capabilities of my father's lathe, or my bedroom casting apparatus, I immediately went to Douglass's smithy, where every opportunity was afforded me for carrying on my larger class of work.

His place was only about five minutes' walk from my father's house. I had the use of his large turning-lathe, which was much more suitable for big or heavy work than the lathe at home. When any considerable bit of steel or iron forging had to be done, a forge fire and anvil were always placed at my service. In making my flywheels for the sectional models of steam-engines I had a rather neat and handy way of constructing them. The boss of the wheel of brass was nicely bored; the arm-holes were carefully drilled and taped, so as to allow the arms which I had turned to be screwed in and appear like neat columns of round wrought iron or steel screwed into the boss of the flywheel.

In return for the great kindness of George Douglass in allowing me to have the use of his foundry, I resolved to present him with a specimen of my handiwork. I desired to try my powers in making a more powerful steam-engine than I had as yet attempted to construct, in order to drive the large turning-lathe and the other tools and machinery of his small foundry. I accordingly set to work and constructed a Direct-acting, high-pressure steam-engine, with a cylinder four inches in diameter. I use the term Direct-acting, because I dispensed with the beam and parallel motion, which was generally considered the correct mode of transferring the action of the piston to the crank.

The result of my labours was a very efficient steam-engine, which set all the lathes and mechanical tools in brisk activity of movement. It had such an enlivening effect upon the workmen that George Douglass afterwards told me that the busy hum of the wheels, and the active, smooth, rhythmic sound of the merry little engine had, through some sympathetic agency, so quickened the strokes of every hammer, chisel, and file in his workmen's hands, that it nearly doubled the output of work for the same wages !

The sympathy of activity acting upon the workmen's hands cannot be better illustrated than by a story told me by my father. A master tailor in a country town employed a number of workmen. They had been to see some tragic melodrama performed by some players in a booth at the fair. While there, a very slow, doleful, but catching air was played, which so laid hold of the tailors' fancy that for some time after they were found slowly whistling or humming the doleful ditty, the movement of their needles keeping time to it ; the result was that the clothing that should have been sent home on Saturday was not finished until the Wednesday following. The music had done it ! The master tailor, being something of a philosopher, sent his men to the play again ; but he arranged that they should be treated with lively merry airs. The result was that the lively airs displaced the doleful ditty ; and the tailors' needles again reverted to their accustomed quickness.

However true the story may be, it touches an important principle in regard to the stimulation of activity by the rapid movements or sounds of machinery, which influence every workman within their sight or hearing. We all know the influence of a quick merry air, played by fife and drum, upon the step and marching of a regiment of soldiers. It is the same with the quick movements of a steam-engine upon the activity of workmen.

I may add that my worthy friend, George Douglass, derived other advantages from the construction of my steam engine. Being of an enterprising disposition he added another iron foundry to his smaller shops ; he obtained many good engineering tools, and in course of time he began to make steam-engines for agicultural purposes. These were used in lieu of horse power for thrashing corn, and performing several operations that used to be done by hand labour in the farm-yards. Orders came in rapidly, and before long the chimneys of Douglass's steam-engines were as familiar in the country round Edinburgh as corn stalks. All the large farms, especially in Midlothian and East Lothian, were supplied with his steam-engines. The business of George Douglass became very large ; and in course of time he was enabled to retire with a considerable fortune.

In addition to the steam-engine which I presented to Douglass, I received an order to make another from a manufacturer of braiding. His machines had before been driven by hand labour ; but as his business extended, the manufacturer employed me to furnish him with an engine of two-horse power, which was duly constructed and set to work, and gave him the highest satisfaction.

I may here mention that one of my earliest attempts at original contrivance was an Expansometer—an instrument for measuring in bulk all metals and solid substances. The object to be experimented on was introduced into a tube of brass, with as much water round it as to fill the tube. The apparatus was then plunged into a vessel of boiling water, or heated to boiling point ; when the lengthening of the bar was measured by a multiplying index, as seen in the

JAMES NASMYTH'S EXPANSOMETER, 1826.

THE SOLID UNDER EXPERIMENT

annexed engraving.   By this simple means the expansion of any material might be ascertained under various increments of heat, say from 60° to 212°.   It was simply a thermometer, the mass marking its own temperature.   Dr. Brewster was so much pleased with the apparatus that he described it and figured it in the *Edinburgh Philosophical Journal*, of which he was then editor.

About the year 1827, when I was nineteen years old, the subject of steam carriages to run upon common roads occupied considerable attention.   Several engineers and mechanical schemers had tried their hands, but as yet no substantial results had come of their attempts to solve the problem.   Like others, I tried my hand. Having made a small working model of a steam carriage, I exhibited it before the members of the Scottish Society of Arts. The performance of this active little machine was so gratifying to the Society that they requested me to construct one of such power as to enable four or six persons to be conveyed along the ordinary roads.   The members of the Society, in their individual capacity, subscribed £60, which they placed in my hands as the means of carrying out their project.

I accordingly set to work at once.   I had the heavy parts of the engine and carriage done at Anderson's foundry at Leith. There was in Anderson's employment a most able general mechanic named Robert

Maclaughlan, who had served his time at Carmichael's, of Dundee. Anderson possessed some excellent tools, which enabled me to proceed rapidly with the work. Besides, he was most friendly, and took much delight in being concerned in my enterprise. This "big job" was executed in about four months. The steam carriage was completed

THE ROAD STEAM-CARRIAGE.   BY JAMES NASMYTH.

and exhibited before the members of the Society of Arts. Many successful trials were made with it on the Queensferry Road, near Edinburgh. The runs were generally of four or five miles, with a load of eight passengers sitting on benches about three feet from the ground.

The experiments were continued for nearly three months, to the great satisfaction of the members. I may mention that in my steam carriage I employed the waste steam to create a blast or draught by discharging it into the short chimney of the boiler at its lowest part, and found it most effective. I was not at that time aware that George Stephenson and others had adopted the same method; but it was afterwards gratifying to me to find that I had been correct as regards the important uses of the steam blast in the chimney. In fact, it is to this use of the waste steam that we owe the practical success of the locomotive-engine as a tractive power on railways, especially at high speeds.

The Society of Arts did not attach any commercial value to my steam road-carriage. It was merely as a matter of experiment that they had invited me to construct it. When it had proved successful they made me a present of the entire apparatus. As I was anxious to get on with my studies, and to prepare for the work of practical engineering, I proceeded no further. I broke up the steam-carriage and sold the two small high-pressure engines, provided with a compact and strong boiler, for £67, a sum which more than defrayed all the expenses of the construction and working of the machine.

I still continued to make investigations as to the powers and capabilities of the steam-engine. There were numerous breweries, distilleries, and other establishments, near Edinburgh, where such engines were at work. As they were made by different engineers, I was desirous of seeing them and making sketches of them, especially when there was any special peculiarity in their construction. I found this a most favourite and instructive occupation. The engine tenters became very friendly with me, and they were always glad to see me interested in them and their engines. They were especially delighted to see me make "drafts," as they called my sketches, of the engines under their charge.

My father sometimes feared that my too close and zealous application to engineering work might have a bad effect upon my health. My bedroom work at brass casting, my foundry work at the making of steam engines, and my studies at the University classes, were perhaps too much for a lad of

my age, just when I was in the hobbledehoy state—between
a boy and a man.   Whether his apprehensions were war-
ranted or not, it did so happen that I was attacked with
typhus fever in 1828, a disease that was then prevalent in
Edinburgh.   I had a narrow escape from its fatal influence.
But thanks to my good constitution, and to careful nursing,
I succeeded in throwing off the fever, and after due time
recovered my usual health and strength.

In the course of my inspection of the engines made by
different makers, I was impressed with the superiority of
those made by the Carmichaels of Dundee.   They were excellent both in design and in execution.   I afterwards found
that the Carmichaels were among the first of the Scottish
engine makers who gave due attention to the employment
of improved mechanical tools, with the object of producing
accurate work with greater ease, rapidity, and economy, than
could possibly be effected by the hand labour of even the most
skilful workmen.   I was told that the cause of the excellence
of the Carmichaels' work was not only in the ability of the
heads of the firm, but in their employment of the best engineers' tools.   Some of their leading men had worked at
Maudsley's machine shop in London, the fame of which had
already reached Dundee, and Maudsley's system of employing machine tools had been imported into the northern
steam factory.

I had on many occasions, when visiting the works where
steam-engines were employed, heard of the name and fame
of Maudsley.   I was told that his works were the very
centre and climax of all that was excellent in mechanical
workmanship.   These reports built up in my mind, at this
early period of my aspirations, an earnest and hopeful desire
that I might some day get a sight of Maudsley's celebrated
works in London.   In course of time it developed into a
passion.   I will now proceed to show how my inmost desires
were satisfied.

CHAPTER VII.

THE chief object of my ambition was now to be taken on at Henry Maudsley's works in London. I had heard so much of his engineering work, of his assortment of machine-making tools, and of the admirable organisation of his manufactory, that I longed to obtain employment there. I was willing to labour, in however humble a capacity, in that far-famed workshop.

I was aware that my father had not the means of paying the large premium required for placing me as an apprentice at Maudsley's firm. I was also informed that Maudsley had ceased to take pupils. After experience, he found that the premium apprentices caused him much annoyance and irritation. They came in "gloves;" their attendance was irregular; they spread a bad example amongst the regular apprentices and workmen; and on the whole they were found to be very disturbing elements in the work of the factory.

It therefore occurred to me that, by showing some specimens of my work and drawings, I might be able to satisfy Mr. Maudsley that I was not an amateur, but a regular working engineer. With this object I set to work, and made with special care, a most complete working model of a high-pressure engine. The cylinder was 2 inches diameter, and the stroke 6 inches. Every part of the engine, including

the patterns, the castings, the forgings, were the result of my
own individual handiwork.   I turned out this sample of my
ability as an engineer workman in such a style as even now
I should be proud to own.

In like manner I executed several specimens of my ability
as a mechanical draughtsman; for I knew that Maudsley
would thoroughly understand my ability to work after a
plan.   Mechanical drawing is the alphabet of the engineer.
Without this the workman is merely " a hand."   With it he
indicates the possession of " a head."   I also made some
samples of my skill in hand-sketching of machines, and parts
of machines, in perspective—that is, as such objects really
appear when set before us in their natural aspect.   I was
the more desirous of exhibiting the ability which I possessed
in mechanical draughtsmanship, as I knew it to be a somewhat rare and much-valued acquirement.   It was a branch
of delineative art that my father had carefully taught me.
Throughout my professional life I have found this art to be
of the utmost practical value.

Having thus provided myself with such visible and tan-
gible evidences of my capabilities as a young engineer, I
carefully packed up my working model and drawings, and
prepared to start for London.   On the 19th of May 1829,
accompanied by my father, I set sail by the Leith smack
Edinburgh Castle, Captain Orr, master.   After a pleasant
voyage of four days we reached the mouth of the Thames.
We sailed up from the Nore on Saturday afternoon, lifted up,
as it were, by the tide, for it was almost a dead calm all the
way.

The sight of the banks of the famous river, with the Kent
orchards in full blossom, and the frequent passages of
steamers with bands of music and their decks crowded with
pleasure-seekers, together with the sight of numbers of noble
merchant ships in the river, formed a most glorious and
exciting scene.   It was also enhanced by the thought that

I was nearing the great metropolis, around which so many bright but anxious hopes were centred, as the scene of my first important step into the anxious business of life.

The tide, which had lifted us up the river as far as Woolwich, suddenly turned; and we remained there during the night. Early next morning the tide rose, and we sailed away again. It was a bright mild morning. The sun came "dancing up the east" as we floated past wharfs and woodyards and old houses on the banks, past wherries and coal boats and merchant ships on the river, until we reached our destination at the Irongate Wharf, nearly opposite the Tower of London. I heard St. Paul's clock strike six just as we reached our mooring ground.

Captain Orr was kind enough to allow us to make the ship our hotel during the Sunday, as it was by no means convenient for us to remove our luggage on that day. My father took me on shore, and we went to Regent's Park. One of my sisters, who was visiting a friend in London, was living in that neighbourhood. My father so planned his route as to include many of the most remarkable streets and buildings and sights of London. He pointed out the principal objects, and gave me much information about their origin and history.

I was much struck with the beautiful freshness and luxuriant growth of the trees and shrubs in the squares; for spring was then in its first beauty. The loveliness of Regent's Park surprised me. The extent of the space, the brilliancy of the fresh-leaved trees, and the handsome buildings by which the park was surrounded, made it seem to me more splendid than a picture from the *Arabian Nights*. Under the happy aspect of a brilliant May forenoon, this first long walk through London, with all its happy attendant circumstances, rendered it one of the most vividly remembered incidents in my life.

After visiting my sister and giving her all the details of

the last news from home, she joined us in our walk down to Westminster Abbey. The first view of the interior stands out in my memory as one of the most impressive sights I ever beheld. I had before read, over and over again, the beautiful description of the Abbey given by Washington Irving in the *Sketch Book*, one of the most masterly pieces of writing that I know of. I now found my day-dreams realised.

We next proceeded over Westminster Bridge to call upon my brother Patrick. We found him surrounded by paintings from his beautiful sketches of Nature. Some of them were more or less advanced in the form of exquisite pictures, which now hang on many walls, and will long commemorate his artistic life. We closed this ever-memorable day by dining at a tavern at the Surrey end of Waterloo Bridge. We sat at an upper window which commanded a long stretch of the river, and from which we could see the many remarkable buildings, from St. Paul's to Westminster Abbey and the Houses of Parliament, which lay on the other side of the Thames.

On the following day my father and I set out in search of lodgings, hotels being at that time beyond our economical method of living. We succeeded in securing a tidy lodging at No. 14 Agnes Place, Waterloo Road. The locality had a special attraction for me, as it was not far from that focus of interest—Maudsley's factory. Our luggage was removed from the ship to the lodgings, and my ponderous cases, containing the examples of my skill as an engineer workman, were deposited in a carpenter's workshop close at hand.

I was now anxious for the interview with Maudsley. My father had been introduced to him by a mutual friend some two or three years before, and that was enough. On the morning of May the 26th we set out together, and reached his house in Westminster Road, Lambeth. It adjoined his factory. My father knocked at the door. My own heart beat fast. Would he be at home? Would he receive us? Yes! he was at home; and we were invited to enter.

Mr. Maudsley received us in the most kind and frank manner. After a little conversation my father explained the object of his visit. "My son," he said, pointing to me, "is very anxious to have the opportunity of acquiring a thorough practical knowledge of mechanical engineering, by serving as an apprentice in some such establishment as yours." "Well," replied Maudsley, "I must frankly confess to you that my experience of pupil apprentices has been so unsatisfactory that my partner and myself have determined to discontinue to receive them—no matter at what premium." This was a very painful blow to myself; for it seemed to put an end to my sanguine expectations.

Mr. Maudsley knew that my father was interested in all matters relating to mechanical engineering, and he courteously invited him to go round the works. Of course I accompanied them. The sight of the workshops astonished me. They excelled all that I had anticipated. The beautiful machine tools, the silent smooth whirl of the machinery, the active movements of the men, the excellent quality of the work in progress, and the admirable order and management that pervaded the whole establishment, rendered me more tremblingly anxious than ever to obtain some employment *there*, in however humble a capacity.

Mr. Maudsley observed the intense interest which I and my father took in everything going on, and explained the movements of the machinery and the rationale of the proceedings in the most lively and kindly manner. It was while we were passing from one part of the factory to another that I observed the beautiful steam-engine which gave motion to the tools and machinery of the workshops. The man who attended it was engaged in cleaning out the ashes from under the boiler furnace, in order to wheel them away to their place outside. On the spur of the moment I said to Mr. Maudsley, "If you would only permit me to do such a job as that in your service, I should consider myself most fortunate!" I shall never forget the keen but kindly look that he gave me. "So," said he, "you are one of that sort, are you?" I was inwardly delighted at his words.

When our round of the works was concluded, I ventured to say to Mr. Maudsley that "I had brought up with me from Edinburgh some working models of steam-engines and mechanical drawings, and I should feel truly obliged if he would allow me to show them to him." "By all means," said he; "bring them to me to-morrow at twelve o'clock." I need not say how much pleased I was at this permission to exhibit my handiwork, and how anxious I felt as to the result of Mr. Maudsley's inspection of it.

I carefully unpacked my working model of the steam-engine at the carpenter's shop, and had it conveyed, together with my drawings, on a hand-cart to Mr. Maudsley's next morning at the appointed hour. I was allowed to place my work for his inspection in a room next his office and counting-house. I then called at his residence close by, where he kindly received me in his library. He asked me to wait until he and his partner, Joshua Field, had inspected my handiwork.

I waited anxiously. Twenty long minutes passed. At last he entered the room, and from a lively expression in his countenance I observed in a moment that the great object of my long cherished ambition had been attained! He expressed, in good round terms, his satisfaction at my practical ability as a workman engineer and mechanical draughtsman. Then, opening the door which led from his library into his beautiful private workshop, he said, "This is where I wish you to work, beside me, as my assistant workman. From what I have seen there is no need of an apprenticeship in your case."

K

He then proceeded to show me the collection of exquisite tools of all sorts with which his private workshop was stored. They mostly bore the impress of his own clear-headedness and common-sense. They were very simple, and quite free from mere traditional forms and arrangements. At the same time they were perfect for the special purposes for which they had been designed. The workshop was surrounded with cabinets and drawers, filled with evidences of the master's skill and industry. Every tool had a purpose. It had been invented for some special reason. Sometimes it struck the keynote, as it were, to many of the important contrivances which enable man to obtain a complete mastery over materials.

There were also hung upon the walls, or placed upon shelves, many treasured relics of the first embodiments of his constructive genius. There were many models explaining, step by step, the gradual progress of his teeming inventions and contrivances. The workshop was thus quite a historical museum of mechanism. It exhibited his characteristic qualities in construction. I afterwards found out that many of the contrivances preserved in his private workshop were treasured as suggestive of some interesting early passage in his useful and active life. They were kept as relics of his progress towards mechanical perfection. When he brought them out from time to time, to serve for the execution of some job in hand, he was sure to dilate upon the occasion that led to their production, as well as upon the happy results that had followed their general employment in mechanical engineering.

It was one of his favourite maxims, "First, *get a clear notion* of what you desire to accomplish, and then in all probability you will succeed in doing it." Another was, "Keep a sharp look-out upon your materials; get rid of every pound of material you can *do without;* put to yourself the question, 'What business has it to be there?' avoid com-

plexities, and make everything as simple as possible." Mr.
Maudsley was full of quaint maxims and remarks, the result
of much shrewdness, keen observation, and great experience.
They were well worthy of being stored up in the mind, like
a set of proverbs, full of the life and experience of men.
His thoughts became compressed into pithy expressions exhibiting his force of character and intellect.   His quaint
remarks on my first visit to his workshop, and on subsequent occasions, proved to me invaluable guides to "right
thinking" in regard to all matters connected with mechanical
structure.

Mr. Maudsley seemed at once to take me into his confidence.   He treated me in the most kindly manner—not
as a workman or an apprentice, but as a friend.   I was an
anxious listener to everything that he said ; and it gave him
pleasure to observe that I understood and valued his conversation.   The greatest treat of all was in store for me.
He showed me his exquisite collection of taps and dies and
screw-tackle, which he had made with the utmost care for
his own service.   They rested in a succession of drawers
near to the bench where he worked.   There was a place for
every one, and every one was in its place.   There was a
look of tidiness about the collection which was very characteristic of the man.   Order was one of the rules which he
rigidly observed, and he endeavoured to enforce it upon all
who were in his employment.

He proceeded to dilate upon the importance of the
uniformity of screws.   Some may call it an improvement,
but it might almost be called a revolution in mechanical
engineering which Mr. Maudsley introduced.   Before his
time no system had been followed in proportioning the
number of threads of screws to their diameter.   Every bolt
and nut was thus a speciality in itself, and neither possessed
nor admitted of any community with its neighbours.   To
such an extent had this practice been carried that all bolts

and their corresponding nuts had to be specially marked as
belonging to each other.   Any intermixture that occurred
between them led to endless trouble and expense, as well as
inefficiency and confusion,—especially when parts of complex machines had to be taken to pieces for repairs.

None but those who lived in the comparatively early
days of machine manufacture can form an adequate idea of
the annoyance, delay, and cost, of this utter want of system,
or can appreciate the vast services rendered to mechanical
engineering by Mr. Maudsley, who was the first to introduce the practical measures necessary for its remedy.   In
his system of screw-cutting machinery, and in his taps
and dies, and screw-tackle generally, he set the example,
and in fact laid the foundation, of all that has since
been done in this most essential branch of machine construction.   Those who have had the good fortune to work
under him, and have experienced the benefits of his practice, have eagerly and ably followed him; and thus his
admirable system has become established throughout the
mechanical world.

Mr. Maudsley kept me with him for about three hours,
initiating me into his system.   It was with the greatest
delight that I listened to his wise instruction.   The sight of
his excellent tools, which he showed me one by one, filled
me with an almost painful feeling of earnest hope that I
might be able in any degree to practically express how
thankful I was to be admitted to so invaluable a privilege
as to be in close communication with this great master in
all that was most perfect in practical mechanics.

When he concluded his exposition, he told me in the
most kindly manner that it would be well for me to take
advantage of my father's presence in London to obtain some
general knowledge of the metropolis, to see the most remarkable buildings, and to obtain an introduction to some of my
father's friends.   He gave me a week for this purpose, and

said he should be glad to see me at his workshop on the following Monday week.

It singularly happened that on the first day my father went out with me, he encountered an old friend. He had first known him at Mr. Miller's of Dalswinton, when the first steamboat was tried, and afterwards at Edinburgh while he was walking the courts as an advocate, or writing articles for the *Edinburgh Review.* This was no other than Henry Brougham. He was descending the steps leading into St. James's Park, from the place where the Duke of York's monument now stands. Brougham immediately recognised my father. There was a hearty shaking of hands, and many inquiries on either side. "And what brings you to London now?" asked Brougham. My father told him that it was about his son here, who had obtained an important position at Maudsley's the engineer. "If I can do anything for you," said Brougham, addressing me, "let me know. It will afford me much pleasure to give you introductions to men of science in London." I ventured to say that "Of all the men of science in London that I most wished to see, was Mr. Faraday of the Royal Institution." "Well," said Brougham, "I will send you a letter of introduction." We then parted.

My father availed himself of the opportunity of introducing me to several of his brother artists. We first went to the house of David Wilkie, in Church Street, Kensington. We found him at home, and he received us most kindly. We next visited Clarkson Stanfield, David Roberts, and some other artists. They were much attached to my father, and had, in the early part of their career, received much kindness from him while living in Edinburgh. They all expressed the desire that I should visit them frequently. I had thus the privilege of *entrée* to a number of pleasant and happy homes, and my visits to them while in London was one of my principal sources of enjoyment.

On returning home to our lodgings that evening we found a note from Brougham, enclosing letters of introduction to Faraday and other scientific men; and stating that if at any time he could be of service to me he hoped that I would at once make use of him. My father was truly gratified with the substantial evidence of Brougham's kindly remembrance of him; and I? how could I be grateful enough? not only for my father's never-failing attention to my growth in knowledge and wisdom, but to his ever-willing readiness to help me onward in the path of scientific work- ing and mechanical engineering. And now I was fortunate in another respect, in being admitted to the school, and I may say the friendship, of the admirable Henry Maudsley. Everything now depended upon myself, and whether I was worthy of all these advantages or not.

One of the days of this most interesting and memorable week was devoted to accompanying Mr. Maudsley in a visit to Somerset House. In the Admiralty Museum, then occupying a portion of the building, was a complete set of the working models of the celebrated block-making machinery. Most of these were the result of Maudsley's own skilful handiwork. He also designed, for the most part, this wonderful and complete series of machines. Sir Samuel Bentham and Mr. Brunel had given the idea, and Maudsley realised it in all its mechanical details. These working models contained the prototypes of nearly all the modern engineer tools which have given us so complete a mastery over materials, and done so much for the age we live in.

It added no little to the enjoyment of this visit to hear Mr. Maudsley narrate, in his quaint and graphic language, the difficulties he had to encounter in solving so many mechanical problems. It occupied him nearly six years to design and complete these working models. They were forty-four in number—all masterly pieces of workmanship. To describe them was to him like living over again the most interesting and eventful part of his life.   And no doubt the experience which he had thus obtained formed the foundation of his engineering fortunes.

Mr. Maudsley next conducted us to the Royal Mint on Tower Hill.   Here we saw many of his admirable machines at work.   He had a happy knack, in his contrivances and inventions, of making "short cuts" to the object in view. He avoided complexities, did away with roundabout processes, however ingenious, and went direct to his point. "Simplicity" was his maxim in every mechanical contrivance.   His master mind enabled him to see through and attain the end he sought by the simplest possible means. The reputation which he had acquired by his minting machinery enabled him to supply it in its improved form to the principal Governments of the world.

Some of the other days of the week were occupied by my father in attending to his own professional affairs, more particularly in connection with the Earl of Cassilis—whose noble mansion in London, and whose castle at Colzean, on the coast of Ayrshire, contain some of my father's finest works.   The last day was most enjoyable.   Mr. Maudsley invited my father, my brother Patrick, and myself, to accompany him in his beautiful small steam yacht, *The Endeavour*, from Westminster to Richmond Bridge, and afterwards to dine with him at the Star and Garter.   I must first, however, say something of the origin of the *Endeavour*.

Mr. Maudsley's son, Joseph, inherited much of his father's constructive genius.   He had made a beautiful arrangement of William Murdoch's original invention of the vibrating cylinder steam-engine, and adapted it for the working of paddle-wheel steamers.   He first tried the action of the arrangement in a large working model, and its use was found to be in every respect satisfactory.   Mr. Maudsley resolved to give his son's design a full-sized trial.   He had a combined pair of vibrating engines constructed, of upwards of 20-horse power, which were placed in a beautiful small steam vessel, appropriately named the *Endeavour*. The result was perfectly successful. The steamer became a universal favourite. It was used to convey passengers and pleasure parties from Blackfriars Bridge to Richmond. Eventually it became the pioneer of a vast progeny of vessels propelled by similar engines, which still crowd the Thames. All these are the legitimate descendants of the bright and active little *Endeavour*.

To return to my trip to Richmond. We got on board the boat on the forenoon of May the 29th. It was one of the most beautiful days of the year. The spring was at its loveliest. The bright fresh green of the trees delighted me. I shall never forget the pleasure with which I beheld, for the first time, the beautiful banks of the Thames. There was at that time a noble avenue of elm trees extending along the southern bank of the river, from Westminster Bridge to Lambeth Palace ; while, on the northern side, many equally fine trees added picturesque grace to the then Houses of Parliament, while behind them were seen the great roof of Westminster Hall and the noble towers of Westminster Abbey. As we sped along we admired the ancient cedars, which gave dignity to the Bishop's grounds, on the one side, and the elms, laburnums, and lilacs, then in full bloom, which partially shaded the quaint old mansions of Cheyne Row, on the other. Alas ! the march of improvement and the inevitable extension of the metropolis is rapidly destroying these vestiges of the olden time.

The beautiful views that came into sight, as we glided up the river, kept my father and my brother in a state of constant excitement. There were so many truly picturesque and *paintable* objects. Patrick's deft pencil was constantly at work, taking graphic notes of "glorious bits." Dilapidated farm-buildings, old windmills, pollarded willows, were rapidly noted, to be afterwards revisited and made immortal

by his brush.   There were also the fine mansions and cozy
villas, partially shrouded by glorious trees, with their bright
velvety lawns sloping down towards the river; not forgetting the delicate streams of thin blue smoke rising lazily
through the trees in the tranquil summer air, and remind-
ing one of the hospitable preparations then in progress.

We landed at Richmond Bridge, and walked up past
the quaint old-fashioned mansions which gave so distinct a
character to Richmond at that time.   We then passed on
to the celebrated Richmond Terrace, at the top of the hill,
from which so glorious a view of the windings of the
Thames is seen, with the luxuriant happy-looking landscape around.   The enjoyment of this glorious day then
reached its climax.   We dined in the great dining-room,
from the large windows of which we observed a view almost
unmatched in the world, with the great tower of Windsor
seen in the distance.   I need not speak of the entertainment, which was everything that the kindest and most
genial hospitality could offer.   After a pleasant stroll in
the Park, amidst the noble and venerable oak trees, which
give such a dignity to the place, and after another visit to
the Terrace, where we saw the sun set in a blaze of glory
beyond the distant scenery, we strolled down the hill to
the boat, and descended the Thames in the cool of the
summer evening.

I must not, however, omit to mention the lodgings
taken for me by my father before he left London.   It was
necessary that they should be near Maudsley's works for
the convenience of going and coming.   We therefore looked
about in the neighbourhood of Waterloo Road.   One of the
houses we visited was situated immediately behind the
Surrey theatre.   It seemed a very nice tidy house, and my
father seemed to have a liking for it.   But when we were
introduced into the room where I was to sleep, he observed
an ultra-gay bonnet lying on the bed, with flashy bright

ribbons hanging from it. This sight seemed to alter his ideas, and he did not take the lodgings; but took another where there was no such bonnet.

I have no doubt about what passed through his mind at the time. We were in the neighbourhood of the theatre. There was evidently some gay young woman about the house. He thought the position might be dangerous for his son. I afterwards asked him why we had not taken that nice lodging. "Well," he said, "did not you see that ultra-gay bonnet lying on the bed? I think that looks rather suspicious!" Afterwards he added, "At all events, James, you will find that though there are many dirty roads in life, if you use your judgment you may always be able *to find a clean crossing!*" And so the good man left me. After an affectionate parting he returned to Edinburgh, and I remained in London to work out the plan of my life.

CHAPTER VIII.

ON the morning of Monday, the 30th of May 1829, I commenced my regular attendance at Mr. Maudsley's workshop. My first job was to assist him in making some modifications in the details of a machine which he had contrived some years before for generating original screws. I use the word " generating " as being most appropriate to express the objects and results of one of Mr. Maudsley's most original inventions.

It consisted in the employment of a knife-edged hardened steel instrument, so arranged as to be set at any required angle, and its edge caused to penetrate the surface of a cylindrical bar of soft steel or brass. This bar being revolved under the incisive action of the angularly placed knife-edged instrument, it thus received a continuous spiral groove cut into its surface. It was thus in the condition of a rudimentary screw ; the pitch, or interval between the threads, being determined by the greater or less angle of obliquity at which the knife-edged instrument was set with respect to the axis of the cylindrical bars revolving under its incisive action.

The spiral groove, thus generated, was deepened to the required extent by a suitable and pointed hard steel tool firmly held in the jaws of an adjustable slide made for the purpose, as part and parcel of the bed of the machine. In the case of square-threaded screws being required, a square-pointed tool was employed in place of the V or angle-threaded tool. And in order to generate or produce right hand or left hand screws, all that was necessary was to set the knife-edged instrument to a right or left hand inclina-tion in respect to the axis of the cylindrical bar at the outset of the operation.

This beautiful and truly original contrivance became, in the hands of its inventor, the parent of a vast progeny of perfect screws, whose descendants, whether legitimate or not, are to be found in every workshop throughout the world, where first class machinery is constructed. The production of perfect screws was one of Maudsley's highest ambitions and his principal technical achievement. It was a type of his invaluable faculty of solving the most difficult problems by the most direct and simple methods.

It was by the same method that he produced the Guide screw. His screw-cutting lathe was moved by combination wheels, and by its means he could, by the one Guide screw, obtain screws of every variety of pitch and diameter. As an illustration of its complete accuracy I may mention that by its means a screw of five feet in length and two inches in diameter was cut with fifty threads to the inch; the Nut to fit on to it being twelve inches long, and containing six hundred threads! This screw was principally used for dividing scales for astronomical and other metrical purposes of the highest class. By its means divisions were produced with such minuteness that they could only be made visual by a microscope.

This screw was sent for exhibition to the Society of Arts. It is still preserved with the utmost care at the Lambeth Works amongst the many admirable specimens of Henry Maudsley's inventive genius and delicate handiwork. Every skilled mechanic must thoroughly enjoy the sight of

it, especially when he knows that it was not produced by
an exceptional tool, but by the machine that was daily
employed in the ordinary work of the factory.

I must not, however, omit to say that I took an early
opportunity of presenting Brougham's letter of introduction
to Faraday at the Royal Institution. I was received most
cordially by that noble-minded man, whose face beamed
with goodness and kindness. After some pleasant conversation he said he would call upon me at Maudsley's, whom
he knew very well. Not long after Faraday called, and
found me working beside Maudsley in his beautiful little
workshop. A vice had been fitted up for me at the bench
where he himself daily worked. Faraday expressed himself
as delighted to find me in so enviable a position. He congratulated me on my special good fortune in having the
inestimable advantage of being associated as assistant workman with one of the greatest mechanical engineers of the
day.

Mr. Maudsley offered to conduct Faraday through his
workshops, and I was permitted to accompany them. I
was much impressed with the intelligent conversation of
Faraday, as well as with the quickness he exhibited in
appreciating not only the general excellence of the design
and execution of the works in progress, but his capacity for
entering into the technical details of the composite tools and
machinery which he saw during his progress through the
place. This most pleasant and memorable meeting with the
great philosopher initiated a friendship which I had the good
fortune to continue until the close of his life.

It was, of course, an immense advantage for me to be so
intimately associated with Mr. Maudsley in carrying on his
experimental work. I was not, however, his apprentice, but
his assistant workman. It was necessary, therefore, in his
opinion that I should receive some remuneration for my
services. Accordingly, at the conclusion of my first week

in his service, he desired me to go to his chief cashier and
arrange with him for receiving whatever amount of weekly
wages I might consider satisfactory. I went to the counting-house and had an interview with Mr. Young the cashier,
a most worthy man.[1] Knowing as I did the great advantages of my situation, and having a very modest notion of
my own worthiness to occupy it, I said, in answer to Mr.
Young's question as to the amount of wages I desired, that
" if he did not think ten shillings a week too much I could
do well enough with that." " Very well," said he, " let it
be so." And he handed me over half-a-sovereign !

I had determined, after I had obtained a situation, not
to cost my father another shilling. I knew how many calls
he had upon him, at the same time that he had his own
numerous household to maintain. I therefore resolved,
now that I had begun life on my own resources, to maintain myself, and to help him rather than be helped any
longer. Thus the first half-sovereign I received from Mr.
Young was a great event in my life. It was the first wages,
as such, that I had ever received. I well remember the
high satisfaction I felt as I carried it home to my lodging ;
and all the more so as I was quite certain that I could, by
strict economy and good management, contrive to make this
weekly sum of ten shillings meet all my current expenses.

I had already saved the sum of £20, which I placed in
the bank as a deposit account. It was the residue of the
sale of some of my model steam-engines at Edinburgh. My
readers will remember that I brought with me a model
steam-engine to show to Mr. Maudsley as a specimen of
my handiwork. It had gained for me the situation that I
desired, and I was now willing to dispose of it. I found a
purchaser in Mr. Watkins, optician at Charing Cross, who
supplied such apparatus to lecturers at Mechanics' Institu-

---

[1] I may mention that he was brother to Dr. Thomas Young, the celebrated
natural philosopher.

tions.   He gave me £35 for the model, and I added the
sum to my deposit account.   This little fund was quite
sufficient to meet any expenses beyond those of a current
weekly nature.

But I was resolved that my wages alone should maintain
me in food and lodging.   I therefore directed my attention
to economical living.   I found that a moderate dinner at
an eating-house would cost more than I could afford to spend.
In order to keep within my weekly income I bought the
raw materials and cooked them in my own way and to my
own taste.   I set to and made a drawing of a very simple,
compact, and handy
cooking apparatus.   I
took the drawing to a
tinsmith near at hand,
and in two days I had
it in full operation.   The
apparatus cost ten shil-
lings, including the lamp.
As it contributed in no
small degree to enable
me to carry out my re-
solution, and as it may
serve as a lesson to others
who have an earnest
desire to live economic-
ally, I think it may be useful to give a drawing and a
description of my cooking stove.

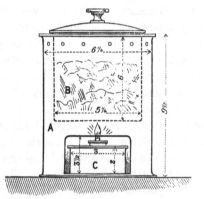

COOKING APPARATUS.
A. CYLINDRICAL OUTER CASE.
B. THE MEAT PAN, MOVABLE.
C. OIL LAMP.

The cooking or meat pan rested on the upper rim of the
external cylindrical case, and was easily removable in order
to be placed handy for service.   The requisite heat was
supplied by an oil lamp with three small single wicks,
though I found that one wick was enough.   I put the meat
in the pot, with the other comestibles, at nine o'clock in the
morning.   It simmered away all day, until half-past six in

the evening, when I came home with a healthy appetite to enjoy my dinner.   I well remember the first day that I set the apparatus to work.   I ran to my lodging, at about four P.M., to see how it was going on.   When I lifted the cover it was simmering beautifully, and such a savoury gusto came forth that I was almost tempted to fall to and discuss the contents.   But the time had not yet come, and I ran back to my work.

The meat I generally cooked in it was leg of beef, with sliced potato, bits of onion chopped down, and a modicum of white pepper and salt, with just enough of water to cover "the elements."   When stewed slowly the meat became very tender, and the whole yielded a capital dish, such as a very Soyer might envy.[1]   It was partaken of with a zest that, no doubt, was a very important element in its savouriness.   The whole cost of this capital dinner was about $4\frac{1}{2}$d. I sometimes varied the meat with rice boiled with a few raisins and a pennyworth of milk.   My breakfast and tea, with bread, cost me about fourpence each.   My lodgings cost 3s. 6d. a week.   A little multiplication will satisfy any one how it was that I contrived to live economically and comfortably on my ten shillings a week.   In the following year my wages were raised to fifteen shillings a week, and then I began to take butter to my bread.

To return to my employment under Mr. Maudsley.   One of the first jobs that I undertook was in assisting him to make a beautiful small model of a pair of 200 horse-power marine steam-engines.   The engines were then in course of construction in the factory.   They were considered a bold advance on the marine engines then in use, not only in regard to their great power, but in carrying out many

---

[1] I have this handy apparatus by me still ; and to prove its possession of its full original efficiency I recently set it in action after its rest of fifty years, and found that it yielded results quite equal to my grateful remembrance of its past services.

specialities in their details and general structure. Mr. Maudsley had embodied so much of his thought in the design that he desired to have an exact model of them placed in his library, so as to keep a visible record of his ideas constantly before him. In fact, the engines might be regarded as a culmination of his constructive abilities.

In preparing the model it was necessary that everything should be made in exact conformity with the original. There were about three hundred minute bolts and nuts to be reduced to the proportional size. I esteemed it a great compliment to be entrusted with their execution. They were all to be made of cast-steel, and the nuts had to be cut to exact hexagonal form. Many of them had collars. To produce them by the use of the file in the ordinary mode would not only have been diffi-cult and tedious, but in some cases practically impossible.

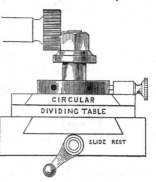

CONTRIVANCE To get rid of the difficulty I suggested to Mr. Maudsley a contrivance of my own by means of which the most rigid exact-ness in size as well as form could be given to these hexagonal nuts. He readily granted his permis-sion. I constructed a special apparatus, consisting of a hard

COLLAR-NUT CUTTING MACHINE.

steel circular cutter to act as a circular file. When brought into operation in the production of those minute six-sided collared nuts, held firm in the spindle of a small dividing plate and attached to the slide-rest, each side was brought in succession under the action of the circular file or cutter with the most exact precision in regard to the division of the six sides. The result was absolutely perfect as respects the exactness of the six equal sides of the hexagonal nut, as well

L

as their precise position in regard to the collar that was of one solid piece with it.

There was no great amount of ingenuity required in contriving this special tool, or in adapting it to the slide-rest of the lathe, to whose spindle end the file or cutter 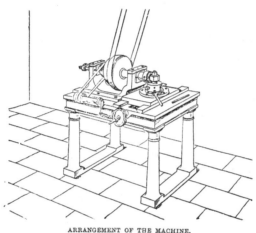 was fixed. But the result was so satisfactory, both as regards the accuracy and rapidity of execution in c o m pa r i so n with the usual process of hand filing, that Mr. Maudsley was greatly pleased with the ar-rangement as

ARRANGEMENT OF THE MACHINE.

well as with my zeal in contriving and executing this clever little tool. An enlarged edition of this collar-nut cutting machine was soon after introduced into the factory. It was one of the specialities that I adopted in my own workshop when I commenced business for myself, and it was eagerly adopted by mechanical engineers, whom we abundantly supplied with this special machine.

It was an inestimable advantage to me to be so intimately associated with this Great Mechanic. He was so invariably kind, pleasant, and congenial. He communicated an infinite number of what he humorously called " wrinkles," which afterwards proved of great use to me. My working hours usually terminated at six in the evening. But as many of the departments of the factory were often in full operation during busy times until eight o'clock, I went through them to observe the work while in progress. On these occasions I often met "the guv'nor," as the workmen called Mr. Maudsley. He was going his round of inspection, and when there was any special work in hand he would call me up to him and explain any point in connection with it that was worthy of particular notice. I found this valuable privilege most instructive, as I obtained from the chief mechanic himself a full insight into the methods, means, and processes by which the skilful workman advanced the various classes of work. I was also permitted to take notes and make rapid sketches of any object that specially interested me. The entire establishment thus became to me a school of practical engineering of the most instructive kind.

Mr. Maudsley took pleasure in showing me the right system and method of treating all manner of materials employed in mechanical structures. He showed how they might be made to obey your will, by changing them into the desired forms with the least expenditure of time and labour. This in fact is the true philosophy of construction. When clear ideas have been acquired upon the subject, after careful observation and practice, the comparative ease and certainty with which complete mastery over the most ob- durate materials is obtained, opens up the most direct road to the attainment of commercial as well as of professional success.

To be permitted to stand by and observe the systematic way in which Mr. Maudsley would first mark or line out his work, and the masterly manner in which he would deal with his materials, and cause them to assume the desired forms, was a treat beyond all expression. Every stroke of the hammer, chisel, or file, told as an effective step towards the intended result. It was a never-to-be-forgotten practical lesson in workmanship, in the most exalted sense of the term. Illustrating his often repeated maxim, " that there is a right way and a wrong way of doing everything," he would take the shortest and most direct cuts to accomplish his objects. The grand result of thoughtful practice is what we call experience : it is the power or faculty of seeing clearly, before you begin, what to avoid and what to select.

High-class workmanship, or technical knowledge, was in his hands quite a' science. Every piece of work was made subject to the soundest philosophical principles, as applied to the use and treatment of materials. It was this that gave such a charm of enjoyment to his dealing with tools and materials. He loved this sort of work for its own sake, far more than for its pecuniary results. At the same time he was not without regard for the substantial evidence of his supremacy in all that regarded first-class tools, admirable management, and thorough organisation of his factory.

The innate love of truth and accuracy which distinguished Mr. Maudsley, led him to value highly that class of technical dexterity in engineering workmen which enabled them to produce those details of mechanical structures in which per- fect flat or true plane surfaces were required. This was an essential condition for the effective and durable performance of their functions. Sometimes this was effected by the aid of the turning-lathe and slide-rest. But in most cases the object was attained by the dexterous use of the file, so that flat filing then was, as it still is, one of the highest qualities of the skilled workman. No one that I ever met with could go beyond Henry Maudsley himself in his dexterous use of the file. By a few masterly strokes he could produce plane surfaces so true that when their accuracy was tested by a standard plane surface of absolute truth, they were never found defective ; neither convex, nor concave, nor " cross winding,"—that is, twisted.

The importance of having such Standard Planes caused him to have many of them placed on the benches beside his workmen, by means of which they might at once conveniently test their work. Three of each were made at a time, so that

by the mutual rubbing of each on each the projecting surfaces were effaced. When the surfaces approached very near
to the true plane, the still projecting minute points were
carefully reduced by hard steel scrapers, until at last the
standard plane surface was secured. When placed over each
other they would float upon the thin stratum of air between
them until dislodged by time and pressure. When they
adhered closely to each other, they could only be separated
by sliding each off each. This art of producing absolutely
plane surfaces is, I believe, a very old mechanical "dodge."
But, as employed by Maudsley's men, it greatly contributed
to the improvement of the work turned out. It was used
for the surfaces of slide valves, or wherever absolute true
plane surfaces were essential to the attainment of the best
results, not only in the machinery turned out, but in educating the taste of his men towards first class workmanship.

Maudsley's love of accuracy also led him to distrust the
verdicts given by the employment of the ordinary callipers
and compasses in determining the absolute or relative dimensions of the refined mechanism which he delighted to construct with his own hands. So much depended upon the
manner in which the ordinary measuring instruments were
handled and applied that they sometimes failed to give
the required verdict as to accuracy. In order, therefore, to
get rid of all difficulties in this respect, he designed and
constructed a very compact and handy instrument which he
always had on his bench beside his vice. He could thus,
in a most accurate and rapid manner, obtain the most reliable
evidence as to the relative dimensions, in length, width, or
diameter, of any work which he had in hand. In consequence
of the absolute truth of the verdicts of the instrument, he
considered it as a Court of Final Appeal, and humorously
called it "The Lord Chancellor."

This trustworthy "Companion of the Bench" consisted
of a very substantial and inflexible bed or base of hard brass.

At one end of it was a perfectly hardened steel surface
plate, having an absolutely true flat or plane face, against
which one end or side of the object to be measured was
placed; whilst a similar absolutely true plane surface of
hardened steel was advanced by means of a suitable fine
thread screw, until the object to be measured was just
delicately in contact with it. The object was, as it were,
between the jaws of a vice, but without any squeeze—being

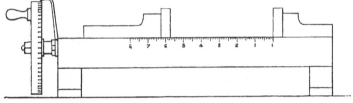

MAUDSLEY'S " LORD CHANCELLOR."

just free, which could be easily ascertained by feeling.
These two absolutely plane surfaces, between which the
object lay, had their distances apart easily read off from the
scale engraved on the bed of the instrument, in inches and
tenth parts of an inch, while the disk-head or handle of the
screw was divided on its edge rim into hundredth or thousandth parts, as these bore an exact metrical relation to
the pitch of the screw that moved the parallel steel faces of
the measuring vice (as I may term it) nearer or farther
apart.

Not only absolute measure could be obtained by this
means, but also the amount of minute differences could be
ascertained with a degree of exactness that went quite
beyond all the requirements of engineering mechanism;
such, for instance, as the thousandth part of an inch! It
might also have been divided so far as a millionth part of
an inch, but these infinitesimal fractions have really nothing
to do with the effective machinery that comes forth from
our workshops, and merely show the mastery we possess

over materials and mechanical forms. The original of this measuring machine of Maudsley's was exhibited at the Loan Collection at South Kensington in 1878. It is now treasured up, with other relics of his handiwork, in a cabinet at the Lambeth Works.

While writing upon this subject it may be worthy of remark, that the employment of a screw as the means of adjusting the points or reference marks of a measuring instrument, for the ascertainment of minute distances between objects, was first effected by William Gascoigne, about the year 1648. There can be no doubt that he was the inventor of the Micrometer—an instrument that, when applied (as he first did it) to the eye-piece of the Telescope, has been the means of advancing the science of astronomy to its present high position.[1]

I had abundant occupation for my leisure time after my regular attendance at the factory was over. I had not only the opportunity of studying mechanics, but of studying men. It is a great thing to know the character of those who are over you as well as those who are under you. It is also well to know the character of those who are associated with you in your daily work. I became intimate with the foremen and with many of the skilled workmen. From them I learnt a great deal. Let me first speak of the men of science who occasionally frequented Maudsley's private workshop. They often came to him to consult him on subjects with which he was specially acquainted.

Among Mr. Maudsley's most frequent visitors were General Sir Samuel Bentham, Mr. Barton, director of the Royal Mint, Mr. Bryan Donkin, Mr. Faraday, and Mr. Chantrey, the sculptor. As Mr. Maudsley wished me to be at hand to give him any necessary assistance, I had the opportunity of listening to the conversation between him and these distinguished visitors. Sir Samuel Bentham called

---

[1] See Grant's *History of Astronomy*, p. 453.

very often.   He had been associated with Maudsley during
the contrivance and construction of the block machinery. He
was brother of the celebrated Jeremy Bentham, and he applied
the same clear common - sense to mechanical subjects which
the other had done to legal, social, and political questions.

It was in the highest degree interesting and instructive
to hear these two great pioneers in the history and application
of mechanics discussing the events connected with the block-
making machinery.   In fact, Maudsley's connection with
the subject had led to the development of most of our
modern engineering tools.   They may since have been
somewhat altered in arrangement, but not in principle.
Scarcely a week passed without a visit from the General.
He sat in the beautiful workshop, where he always seemed
so happy.   It was a great treat to hear him and Maudsley
fight their battles over again, in recounting the difficulties,
both official and mechanical, over which they had so gloriously triumphed.

At the time when I listened to their conversation, the
great work in hand was the organisation of a systematic
series of experiments on the hulls of steam-ships, with the
view of determining the laws of resistance on their being
propelled through the sea by a power other than those of
winds and sails.   The subject was as complex as it was
interesting and important.   But it had to be put to the
test by actual experiment.   This was done in the first place
by large models of hulls, so as to ascertain at what point
the curves of least resistance could be applied.   Their
practical correctness was tested by careful experiment in
passing them through water at various velocities, to record
which conditions special instruments were contrived and
executed.   These, as well as the preparation of large models
of hulls, embodying the various improved "lines," occupied
a considerable portion of the time that I had the good
fortune to spend in Mr. Maudsley's private workshop.

Mr. Barton of the Royal Mint was quite a "crony" of Maudsley's. He called upon him often with respect to the improvements in stamping the current coin of the realm. Mr. Bryan Donkin was also associated with Maudsley and Barton on the subject of the national standard of the yard measure. But perhaps Mr. Chantrey was the most attractive visitor at the private workshop. He had many a long interview with Maudsley with respect to the planning and arranging of a small foundry at his studio, by means of which he might cast his bronze statues under his own superintendence. Mr. Maudsley entered *con amore* into the subject, and placed his skill and experience entirely at Chantrey's service. He constructed the requisite furnaces and cranes, and other apparatus, at Chantrey's studio; and it may be enough to state that, when brought into operation, they yielded the most satisfactory results.

Among my most intelligent private friends in London were George Cundell and his two brothers. They resided near my lodgings, and I often visited them on Saturday evenings. They were most kind, gentle, and genial. The eldest brother was in Sir William Forbes's bank. George was agent for Mr. Patrick Maxwell Stuart in connection with his West India estates, and the third brother was his assistant. The elder brother was an admirable performer on the violoncello, and he treated us during these Saturday evenings with noble music from Beethoven and Mozart. My special friend George was known amongst us as "the worthy master." He was thoroughly versed in general science, and was moreover a keen politician. He had the most happy faculty of treating complex subjects, both in science and politics, in a thoroughly common-sense manner. His two brothers had a fine feeling for art, and, indeed, possessed no small skill as practical artists. With companions such as these, gifted with a variety of tastes, I spent many of my Saturday evenings most pleasantly and profitably.

They were generally concluded with a glass of beer of "the worthy master's" own brewing.

When the season of the year and the state of the weather were suitable I often joined this happy fraternity in long and delightful Sunday walks to various interesting places round London.   Our walks included Waltham Abbey, Waltham Cross, Eltham Palace, Hampton Court, Epping Forest, and such like interesting places of resort.   When the weather was unfavourable my principal resort was West-minster Abbey, where, besides the beautifully - conducted service and the noble anthems, I could admire the glory of the architecture, and the venerable tombs, under which lay the best and bravest.   I used generally to sit at a point from which I could see the grand tomb of Aylmer de Vallance in its magnificent surroundings of quaint and glorious architecture.   It was solemn, and serious also, to think of the many generations who had filled the abbey, and of the numbers of the dead who lay beneath our feet.

I was so great an admirer of Norman and Gothic archi-tecture that there was scarcely a specimen of it in London which I did not frequently visit.   One of the most interesting examples I found in the Norman portion of St. Saviour's Church, near London Bridge, though some of it has since been destroyed by the so-called "restoration" in 1831. The new work has been executed in the worst taste and feeling.   I also greatly admired the Norman chapel of the Tower, and some Norman portions of the Church of St. Bartholomew the Less, near Smithfield.

No style of architecture that I have ever seen has so impressed me with its intrinsic gravity, and I may say solemnity, as that of the Norman.   There is a serious ear-nestness in its grave simplicity that has a peculiar influence upon the mind ; and I have little doubt but that this was felt and understood by those true architects who designed and built the noble cathedrals at Durham and elsewhere.

CHAPTER IX.

HOLIDAY IN THE MANUFACTURING DISTRICTS.

In the autumn of 1830 Mr. Maudsley went to Berlin for
the purpose of superintending the erection of machinery at
the Royal Mint there. He intended to be absent from London for about a month; and he kindly permitted me to take
a holiday during that period.

I had been greatly interested by the descriptions in the
newspapers of the locomotive competition at Rainhill, near
Liverpool. I was, therefore, exceedingly anxious to see
Stephenson's "Rocket," the engine that had won the prize.
Taking with me letters of introduction from Mr. Maudsley
to persons of influence at Liverpool, I left London for the
north on the afternoon of Saturday the 9th of September
1830. I took my place on the outside of the Liverpool
coach, which set out from "The Swan with Two Necks," in
Lad Lane, city, one of the most celebrated coach-offices in
those days.

The first part of the journey to Liverpool was very dis-
mal. The night was wet. The rain came pouring down,
and no sort of wrappings could keep it out. The outside
passengers became thoroughly soaked. On we went, however, as fast as four horses could carry us. Next morning
we reached Coventry, when the clouds cleared away, and the
sun at last burst forth. I could now enjoy this charming

part of old England.   Although I had only a hasty glimpse
in passing of the quaint streets and ancient buildings of the
town, I was perfectly delighted with the specimens of ancient
domestic architecture which I saw.   At that time Coventry
was quite a museum of that interesting class of buildings.
The greater part of them have since been swept away in the
so-called improvement of modern builders, none of whose
works can ever so attract the artistic eye.

During the rest of the day the journey was delightful.
Though the inside passengers had had the best of it during
the night, the outside passengers had the best of it now.   To
go scampering across the country on the top of the coach,
passing old villages, gentlemen's parks, under old trees, along
hedges tinged with autumn brown, up hill and down hill,
sometimes getting off the coach to lighten the load, and
walking along through the fields by a short cut to meet it
farther on ; all this was most enjoyable.   It gave me a new
interest in the happier aspects of English scenery, and of
rural and domestic life in the pretty old-fashioned farm
buildings that we passed on our way.   Indeed, there was
everything to delight the eye of the lover of the picturesque
during the course of that bright autumnal day.

The coach reached Liverpool on Sunday night.   I took
up my quarters at a commercial inn in Dale Street, where
I found every comfort which I desired at moderate charges.
Next morning, without loss of time, I made my way to the
then terminus of the Liverpool and Manchester Railway;
and there, for the first time, I saw the famous " Rocket."
The interest with which I beheld this distinguished and
celebrated engine was much enhanced by seeing it make
several short trial trips under the personal management of
George Stephenson, who acted as engineman, while his son
Robert acted as stoker.   During their trips of four or five
miles along the line the " Rocket " attained the speed of
thirty miles an hour—a speed then thought almost in-

credible! It was to me a most memorable and interesting
sight, especially to see the father and son so appropriately
engaged in working the engine that was to effect so great a
change in the future communications of the civilised world.

I spent the entire day in watching the trial trips, in
examining the railway works, and such portions of their
details as I could obtain access to. About mid-day the
" Rocket " was at rest for about an hour near where I stood ;
and I eagerly availed myself of the opportunity of making a
careful sketch of the engine, which I still preserve.   The
line was opened on the 15th of September, when the famous
" Rocket " led the way in conducting the first train of passengers from Liverpool to Manchester.   There were present
on that occasion thousands of spectators, many of whom had
come from distant parts of the kingdom to witness this
greatest of all events in the history of railway locomotion.

During my stay in Liverpool I visited the vast range of
magnificent docks which extend along the north bank of the
Mersey, all of which were crowded with noble merchant
ships, some taking in cargoes of British manufactures, and
others discharging immense stores of cotton, sugar, tobacco,
and foreign produce.   The sight was most interesting, and
gave me an impressive idea of the mighty functions of a
manufacturing nation — energy and intelligence, working
through machinery, increasing the value of raw materials and
enabling them to be transported for use to all parts of the
civilised world.

Mr. Maudsley having given me a letter of introduction to
his old friend William Fawcett, head of the firm of Fawcett,
Preston, and Company, engineers, I went over their factory.
They were engaged in producing sugar mills for the West
Indies, and also in manufacturing the steam-engines for
working them.   The firm had acquired great reputation for
their workmanship ; and their shops were crowded with
excellent specimens of their skill.   Everything was in good

order ; their assortment of machine tools was admirable. Mr. Fawcett, who accompanied me, was full in his praises of my master, whom he regarded as the great pioneer in the substitution of the unerring accuracy of machine tools, for the often untrustworthy results of mere manual labour.

I cannot resist referring to the personal appearance and manner of this excellent gentleman, William Fawcett. His peculiar courteous manner, both in speech and action, reminded me of " the grand old style " which I had observed in some of my father's oldest noble employers, and the representations given of them by some of our best actors. There was also a dignified kindliness about his manner that was quite peculiar to himself ; and when he conducted me through his busy workshops, the courtly yet kindly manner in which he addressed his various foremen and others, was especially cheering. When I first presented my letter of introduction from Henry Maudsley, he was sitting at a beautiful inlaid escritoire table with his letters arrayed before him in the most neat and perfect order. The writing-table stood on a small Turkey carpet apart from the clerks' desks in the room, but so near to them that he could readily communicate with them. His neat old-fashioned style of dress quite harmonised with his advanced age, and the kindly yet dignified grace of his manner left a lasting impression on me as a most interesting specimen of " the fine old English gentleman, quite of the olden time."

I spent another day in crossing the Mersey to Birkenhead—then a very small collection of buildings—and wandered about the neighbourhood. I had my sketch-book with me, and made a drawing of Liverpool from the other side of the river. Close to Birkenhead were some excellent bits of scenery, old and picturesque farmhouses, overshadowed with venerable oaks, with juttings-out of the New Red Sandstone rocks, covered with heather, furze, and broom, with pools of water edged with all manner of effective water plants. They formed capital subjects for the artistic pencil, especially when distant peeps of the Welsh hills came into the prospect. I made several sketches, and they kept company with my graphic memoranda of architectural and mechanical objects. I may here mention that on my return to London I showed them to my brother Patrick, and some of them so much met his fancy that he borrowed my sketch-book and painted some pictures from them, which at this day are hanging on the walls of some of his admirers.

With the desire of seeing as much as possible of all that was interesting in the mechanical, architectural, and picturesque line, on my return journey to London, I determined to walk, halting here or there by the way. The season of the year and the state of the weather were favourable for the purpose. I accordingly commenced my pedestrian tour on Saturday morning, the 17th September. I set out for Manchester. It was a long but pleasant walk. I well remember, when nearing Manchester, that I sat down to rest for a time on Patricroft Bridge. I was attracted by the rural aspect of the country, and the antique cottages that lay thereabouts. The Bridgewater Canal lay before me, and as I was told that it was the first mile of the waterway that the great Duke had made, it became quite classic ground in my eyes. I little thought at the time that I was so close to a piece of ground that should afterwards become my own, and where I should for twenty years carry on the most active and interesting business of my life.

I reached Manchester at seven in the evening, and took up my quarters at the King's Arms Inn, in Deansgate. Next day was Sunday. I attended service in the Cathedral, then called the Old Church. I was much interested by the service, as well as by the architecture of the building. Some of the details were well worthy of attention, being very original, and yet the whole was not of the best period of Gothic architecture. Some of the old buildings about the

Cathedral were very interesting. They were of a most quaint character, yet bold and effective. Much finely carved oak timber work was introduced into them ; and on the whole they gave a very striking illustration of the style of domestic architecture which prevailed in England some three or four centuries ago.

On the following day I called upon Mr. Edward Tootal, of York Street. He was a well-known man in Manchester. I had the happiness of meeting him in London a few months before. He then kindly invited me to call upon him should I ever visit Manchester, when he would endeavour to obtain for me a sight of some of the most remarkable manufacturing establishments. Mr. Tootal was as good as his word. He received me most cordially, and at once proceeded to take me to the extensive machine factory of Messrs. Sharp, Roberts, and Co. I found to my delight that a considerable portion of the establishment was devoted to the production of machine tools, a department of mechanical business that was then rising into the highest importance. Mr. Roberts, an admirable mechanic as well as inventor, had derived many of his ideas on the subject while working with Mr. Maudsley in London, and he had carried them out with many additions and improvements of his own contrivance. Indeed, Roberts was one of the most capable men of his time, and is entitled to be regarded as one of the true pioneers of modern mechanical mechanism.

Through the kindness of Mr. Tootal I had also the opportunity of visiting and inspecting some of the most extensive cotton mills in Manchester. I was greatly pleased with the beautiful contrivances displayed in the machinery. They were perfect examples of the highest order of ingenuity, combined with that kind of common-sense which casts aside all mere traditional forms and arrangements of parts, such as do not essentially contribute to the efficiency of the machine in the performance of its special and required purpose. I

found much to admire in the design as well as in the execution of the details of the machines. The arrangement
and management of the manufactories were admirable. The
whole of the buildings, howsoever extensive and apparently
complicated, worked like one grand and perfectly constructed
machine.

I was also much impressed by the keen interest which
the proprietors of these vast establishments took in the
minute details of their machinery, as well as by their intelligent and practical acquaintance with the technical minutiæ
of their business. Although many of them were men of
fortune, they continued to take as deep an interest in such
matters as if they were beginning life and had their fortunes
to make. Their chief pride and ambition was to be at the
head of a thoroughly well managed and prosperous establishment. And with this object, no detail, be it ever so small,
was beneath their care and attention. To a young man
like myself, then about to enter upon a similar career of
industry, these lessons were very important. They were
encouraging examples of carefully thought out designs, carried into admirable results by close attention to details, ever
watchful carefulness, and indomitable perseverance.

I brooded over these circumstances. They filled my
mind with hope. They encouraged me to go on in the path
which I had selected ; and I believed that at some time or
other I might be enabled to imitate the examples of zeal
and industry which I had witnessed during my stay in
Manchester. It was then that I bethought me of settling
down in this busy neighbourhood ; and as I plodded my
way back to London this thought continually occupied me.
It took root in my mind and grew, and at length the idea
became a reality.

I did not take the shortest route on my return journey
to London. I desired to pass through the most interesting
and picturesque places without unduly diverging from the

M

right direction. I wished to see the venerable buildings
and cathedrals of the olden time, as well as the engineering
establishments of the new. Notwithstanding my love for
mechanics I had always a spice of the antiquarian feeling in
me. It enabled me to look back to the remote past, into
the material records of man's efforts hundreds of years ago,
and contrast them with the modern progress of arts and
sciences. I was especially interested in the architecture of
bygone ages ; but here, alas ! arts and sciences have done
nothing. Modern Gothic architecture is merely an imitation
of the old, and often a very bad imitation. Even ancient
domestic architecture is much superior to the modern. We
can now only imitate it ; and often spoil when imitating.

I left Manchester and turned my steps in the direction
of Coalbrookdale. I passed through a highly picturesque
country, in which I enjoyed the sight of many old timber
houses, most attractive subjects for my pencil. My route
lay through Whitchurch, Wem, and Wellington ; then past
the Wrekin to Coalbrookdale. Before arriving there I saw
the first iron bridge constructed in England, an object of
historical interest in that class of structures. It was because
of the superb quality of the castings produced at Coalbrookdale, that the ironmasters there were able to accomplish the
building of a bridge of that material, which before had
baffled all projectors both at home and abroad.

I possessed a letter of introduction to the manager, and
was received by him most cordially. He permitted me to
examine the works. I was greatly interested at the sight
of the processes of casting. Many beautiful objects were
turned out for architectural, domestic, and other purposes.
I saw nothing particularly novel, however, in the methods
and processes of moulding and casting. The excellence
of the work depended for the most part upon the great
care and skill exercised by the workmen of the foundry.
They seemed to vie with each other in turning out the best

castings, and their models or patterns were made with the
utmost care.   I was particularly impressed with the cheerful
zeal and activity of the workmen and foremen of this justly
celebrated establishment.

On leaving Coalbrookdale I trudged my way towards
Wolverhampton. I rested at Shiffnal for the night. Next
day I was in the middle of the Black Country. I had no
letters of introduction to employers in Wolverhampton ; so
that, without stopping there, I proceeded at once to Dudley.
The Black Country is anything but picturesque. The earth
seems to have been turned inside out.   Its entrails are
strewn about ; nearly the entire surface of the ground is
covered with cinder-heaps and mounds of scoriæ.   The coal,
which has been drawn from below ground, is blazing on the
surface.   The district is crowded with iron furnaces, puddling furnaces, and coal-pit engine furnaces.   By day and by
night the country is glowing with fire, and the smoke of the
ironworks hovers over it.   There is a rumbling and clanking
of iron forges and rolling mills.   Workmen covered with
smut, and with fierce white eyes, are seen moving about
amongst the glowing iron and the dull thud of forge-
hammers.

Amidst these flaming, smoky, clanging works, I beheld
the remains of what had once been happy farmhouses, now
ruined and deserted.   The ground underneath them had
sunk by the working out of the coal, and they were falling
to pieces.   They had in former times been surrounded by
clumps of trees ; but only the skeletons of them remained,
dilapidated, black, and lifeless.   The grass had been parched
and killed by the vapours of sulphureous acid thrown out by
the chimneys ; and every herbaceous object was of a ghastly
gray—the emblem of vegetable death in its saddest aspect.
Vulcan had driven out Ceres.   In some places I heard a
sort of chirruping sound, as of some forlorn bird haunting the
ruins of the old farmsteads.   But no ! the chirrup was a

vile delusion. It proceeded from the shrill creaking of the coal-winding chains, which were placed in small tunnels beneath the hedgeless road.

I went into some of the forges to see the workmen at their labours. There was no need of introduction; the works were open to all, for they were unsurrounded by walls. I saw the white-hot iron run out from the furnace; I saw it spun, as it were, into bars and iron ribbands, with an ease and rapidity which seemed marvellous. There were also the ponderous hammers and clanking rolling-mills. I wandered from one to another without restraint. I lingered among the blast furnaces, seeing the flood of molten iron run out from time to time, and remained there until it was late. When it became dark the scene was still more impressive. The workmen within seemed to be running about amidst the flames as in a pandemonium; while around and outside the horizon was a glowing belt of fire, making even the stars look pale and feeble. At last I came away with reluctance, and made my way towards Dudley. I reached the town at a late hour. I was exhausted in mind and body, yet the day had been most interesting and exciting. A sound sleep refreshed me, and I was up in the morning early, to recommence my journey of inquiry.

I made my way to the impressive ruins of Dudley Castle, the remnant of a very ancient stronghold, originally built by Dud, the Saxon. The castle is situated on a finely wooded hill; it is so extensive that it more resembles the ruins of a town than of a single building. You enter through a treble gateway, and see the remnants of the moat, the court, and the keep. Here are the central hall, the guard-rooms, and the chapel. It must have been a magnificent structure. In the Midlands it was known as the "Castle of the Woods." Now it is abandoned by its owners, and surrounded by the Black Country. It is undermined by collieries, and even penetrated by a canal. The castle walls sometimes tremble when a blast occurs in the bowels of the mountain beneath. The town of Dudley lies quite close to the castle, and was doubtless protected by it in ancient times.

The architectural remains are of various degrees of antiquity, and are well worthy of study, as embodying the successive periods which they represent. Their melancholy grandeur is rendered all the more impressive by the coal and iron works with which they are surrounded—the olden type of buildings confronting the modern. The venerable trees struggle for existence under the destroying influence of sulphureous acid; while the grass is withered and the vegetation everywhere blighted. I sat down on an elevated part of the ruins, and looked down upon the extensive district, with its roaring and blazing furnaces, the smoke of which blackened the country as far as the eye could reach; and as I watched the decaying trees I thought of the price we had to pay for our vaunted supremacy in the manufacture of iron. We may fill our purses, but we pay a heavy price for it in the loss of picturesqueness and beauty.

I left the castle with reluctance, and proceeded to inspect the limestone quarries in the neighbourhood. The limestone has long been worked out from underneath the castle; but not far from it is Wren's Nest Hill, a mountain of limestone. The wrens have left, but the quarries are there. The walk to the hill is along green lanes and over quiet fields. I entered one of the quarries opened out in the sloping precipice, and penetrated as far as the glimmer of sunlight enabled me to see my way. But the sound of the dripping of water from the roof of the cave warned me that I was approaching some deep pool, into which a false step might plunge me. I therefore kept within the light of day. An occasional ray of the sun lit up the enormous rock pillars which the quarrymen had left to support the roof. It was a most impressive sight.

Having emerged from the subterranean cave, I proceeded

on my way to Birmingham. I reached the town in the evening, and found most comfortable quarters. On the following day I visited some of the factories where processes were carried on in connection with the Birmingham trade. I saw the mills where sheet brass and copper were rolled for the purpose of being plated with silver. There was nothing in these processes of novel interest, though I picked up many practical hints. I could not fail to be attracted by the dexterous and rapid manipulation of the work in hand, even by boys and girls whose quick sight and nimble fingers were educated to a high degree of perfection. I could have spent a month profitably among the vast variety of small traders in metal, of which Birmingham is the headquarters. Even in what is called " the toy trade," I found a vast amount of skill displayed in the production of goldsmith work, in earrings, brooches, gold chains, rings, beads, and glass eyes for stuffed birds, dolls, and men.

I was especially attracted by Soho, once the famous manufacturing establishment of Boulton and Watt. Although this was not the birthplace of the condensing steam-engine,[1] it was the place where it attained its full manhood of efficiency, and became the source and origin of English manufacturing power. Watt's engine has had a greater influence on the productive arts of mankind than any other that can be named. Boulton also was a thorough man of business, without whom, perhaps, Watt could never have made his way against the world, or perfected his

---

[1] The birthplace of the condensing engine of Watt was the workshop in the Glasgow University, where he first contrived and used a *separate* condenser— the true and vital element in Watt's invention. The condenser afterwards attained its true effective manhood at Soho. The Newcomen engine was in fact a condensing engine, but as the condensation was effected inside the steam cylinder it was a very costly source of power in respect of steam. Watt's happy idea of condensing in a separate vessel removed the defect. This was first done in his experimental engine in the Glasgow University workshop, and before he made the one at Kinniel for Dr. Roebuck.

magnificent invention. Not less interesting to my mind was
the memory of that incomparable mechanic, William Murdoch, a man of indomitable energy, and Watt's right-hand
man in the highest practical sense. Murdoch was the
inventor of the first model locomotive, and the inventor of
gas for lighting purposes ; and yet he always kept himself
in the background, for he was excessively modest. He was
happiest when he could best promote the welfare of the great
house of Boulton and Watt. Indeed he was a man whose
memory ought to be held in the highest regard by all true
engineers and mechanics.

The sight which I obtained of the vast series of workshops of this celebrated establishment—filled with evidences
of the mechanical genius of these master minds—made me
feel that I was indeed on classic ground, in regard to everything connected with steam-engine machinery. Some of the
engines designed by Watt—the prototypes of the powerful
condensing engines of the present day—were still performing
their daily quota of work. There was " Old Bess," a sort of
experimental engine, upon which Watt had tried many
adaptations and alterations, for the purpose of suiting it for
pumping water from coal mines. There was also the engine
with the sun-and-planet motion, an invention of William
Murdoch's. Both of these engines were still at work.

I went through the workshops, where I was specially
interested by seeing the action of the machine tools.
There I observed Murdoch's admirable system of transmitting power from one central engine to other small *vacuum*
engines attached to the individual machines they were set
to work. The power was communicated by pipes led from
the central air or exhaust pump to small *vacuum* or atmospheric engines devoted to the driving of each separate
machine, thus doing away with all shafting and leather
belts, the required speed being kept up or modified at
pleasure without in any way interfering with the other

machines. This *vacuum* method of transmitting power dates from the time of Papin; but until it received the masterly touch of Murdoch it remained a dead contrivance for more than a century.

I concluded my visits to the workshops of Birmingham by calling upon a little known but very ingenious man, whose work I had seen before I left Edinburgh, in a beautifully constructed foot turning-lathe made by John Drain. I was so much impressed with the exquisite design, execution, and completeness of the lathe, that I made it one of my chief objects to find out John Drain's workshop. It was with some difficulty that I found him. He was little known in Birmingham. His workshops were very small; they consisted of only one or two rooms. His exquisite lathes were not much in demand. They found their way chiefly to distant parts of the country, where they were highly esteemed.

I found that he had some exquisitely-finished lathes completed and in hand for engraving the steel plates for printing bank notes. They were provided with the means of producing such intricate ornamental patterns as to defy the utmost skill of the forger. Perkins had done a good deal in the same way; but Drain's exquisite mechanism enabled his engraving lathes to surpass anything that had before been attempted in the same line. I believe that Drain's earnest attention to his work, in which he had little or no assistance, undermined his health, and arrested the career of one who, had he lived, would have attained the highest position in his profession. I shall never forget the rare treat which his fine mechanism afforded me. Its prominent quality was absolute truth and accuracy in every part..

Having now had enough of the Black Country and of Birmingham workshops, I proceeded towards London. There were no more manufacturing districts to be visited. Everything now was to be green lanes, majestic trees, old man-

sions, venerable castles, and picturesque scenery.   There is
no way of seeing a country properly except on foot.   By
railway you whiz past and see nothing.   Even by coach
the best parts of the scenery are unseen.   "Shank's naig"
is the best of all methods, provided you have time.   I had
still some days to spare before the conclusion of my holiday.   I therefore desired to see some of the beautiful scenery
and objects of antiquarian interest before returning to work.

I made my way across country to Kenilworth.   The
weather was fine, and the walk was perfect.   The wayside
was bordered by grassy sward.   Wide and irregular margins
extended on each side of the road, and noble trees and
untrimmed hedges, in their glowing autumnal tint, extended
far and wide.   Everything was in the most gloriously
neglected and therefore highly picturesque condition.   Here
and there, old farmhouses and labourers' cottages peeped up
from amidst the trees and hedges—worthy of the landscape
painter's highest skill.

I reached Kenilworth about half an hour before sunset.
I made my way direct to the castle, glorious in its decay.
The fine mellow glow of the setting sun lit up the grand
and extensive ruins.   The massive Norman keep stood up
with melancholy dignity, and attracted my attention more
than any other part of the ruined building.   To me there
is an impressiveness in the simple massive dignity of
the Norman castles and cathedrals, which no other buildings possess.   There is an expression of terrible earnestness
about them.   The last look I had of the Norman keep was
grand.   The elevated part was richly tinted with the last
glow of the setting sun, while the outline of the buildings
beneath was shaded by a dark purply gray.   It was indeed
a sight never to be forgotten.   I waited until the sun had
descended beneath the horizon, still leaving its glimmer of
pink and crimson and gray, and then I betook me to the
little inn in the village, where I obtained comfortable

quarters for the night. I visited the ruins again in the morning. Although the glory of the previous evening had departed, I was much interested in observing the various styles of architecture adopted in different parts of the buildings—some old, some comparatively new. I found the older more grand and massive, and the newer, of the sixteenth century, wanting in dignity of design, and the workmanship very inferior. The reign of Shoddy had already begun before Cromwell laid the castle in ruins.

In the course of the day I proceeded to Warwick. I passed along the same delightful grass-bordered roads, shaded by noble trees. I reached the grand old town, with its antique buildings, and its noble castle — so famous in English history. Leaving the place with reluctance, I left it late in the afternoon to trudge on to Oxford. But soon after I started the rain began to fall. It was the first interruption to my walking journey which I had encountered during my three weeks' absence from London. As it appeared from the dark clouds overhead that a wet night had set in, I took shelter in a wayside inn at a place called Steeple Aston. My clothes were dripping wet; and after a glass of very hot rum and water I went to bed, and had a sound sleep. Next morning it was fair and bright. After a substantial homely breakfast I set out again. Nature was refreshed by the steady rain of the previous night, and the day was beautiful. I reached Deddington and stayed there for the night, and early next morning I set out for Oxford.

I was greatly excited by the first sight I had of the crowd of towers and spires of that learned and illustrious city. Nor were my expectations at all disappointed by a nearer approach to the colleges of Oxford. After a most interesting visit to the best of the buildings, I took in a fair idea of the admirable details of this noble city, and left in the afternoon of next day. I visited, on my way to Thame, the old church of Iffley. I was attracted to it by the fine old Norman work it contains, which I found most quaint and picturesque.

I slept at Thame for the night, and next day walked to Windsor. I arrived there at sunset, and had a fine view of the exterior of the castle and the surrounding buildings. I was, however, much disappointed on examining the architectural details. In sight of the noble trees about the castle, and the magnificent prospect from the terrace, I saw much that tended to make up for the disgust I felt at the way in which all that was so appropriate and characteristic in so historic a place as Windsor Castle should have been tampered with and rubbed out by the wretched conceit of the worst architects of our worst architectural period.

I left Windsor next morning, and walked direct for London. My time was up, but not my money. I had taken eight sovereigns on setting out from London to Liverpool by coach, and I brought one sovereign back with me. Rather than break into it I walked all the way from Windsor to London without halting for refreshment. My entire expenditure during my three weeks' journey was thus seven pounds.

When I look back upon that tour, I feel that I was amply rewarded. It was throughout delightful and instructive. The remembrance of it is as clear in my mind now as if I had performed the journey last year instead of fifty years ago. There are thousands of details that pass before my mind's eye that would take a volume to enumerate. I brought back a book full of sketches ; for graphic memoranda are much better fitted than written words to bring up a host of pleasant recollections and associations. I came back refreshed for work, and possessed by an anxious desire to press forward in the career of industry which I had set before me to accomplish.

CHAPTER X.

MR. MAUDSLEY arrived from Berlin two days after my return to London. He, too, had enjoyed his holiday. During his stay in Berlin he had made the friendship of the distinguished Humboldt. Shenkel, the architect, had been very kind to him, and presented him with a set of drawings and engravings of his great architectural works, which Mr. Maudsley exhibited to me with much delight. What he most admired in Shenkel was the great range of his talent in all matters of design, his minute attention to detail, and his fine artistic feeling.

Soon after Mr. Maudsley's return, a very interesting job was brought to him, in which he took even more than his usual interest. It was a machine that his friend Mr. Barton, of the Royal Mint, had obtained from France. It was intended to cut or engrave the steel dies used for stamping coin. It was a remarkable and interesting specimen of inventive ingenuity. It copied any object in relief which had been cast in plaster of Paris or brass from the artist's original wax model. The minutest detail was transferred to soft steel dies with absolute accuracy. This remarkable machine could copy and cut steel dies either in intaglio or in cameo of any size, and, in short, enabled the mechanic who managed it to transfer the most minute and characteristic touches of the original model to the steel dies for any variety of size of coin. Nevertheless, the execution of some of the details of the machine were so defective, that after giving the most tempting proof of its capabilities at the Royal Mint, Mr. Barton found it absolutely necessary to place it in Maudsley's hands, in order to have its details thoroughly overhauled, and made as mechanically perfect as its design and intention merited.

This interesting machine was accordingly brought to the private workshop, and placed in the hands of the leading mechanic, whom I had the pleasure of being associated with, namely James Sherriff, one of our most skilful workmen. We were both put to our mettle. It was a job quite to my taste, and being associated with so skilled a workman as Sherriff, and in constant communication with Mr. Maudsley, I had every opportunity of bringing my best manipulative ability into action and use while perfecting this beautiful machine. It is sufficient to say that by our united efforts, by the technical details suggested by Mr. Maudsley and carried out by us, and by the practical trials made under the superintendence of Mr. Wyon of the Mint, the apparatus was at length made perfect, and performed its duty to the satisfaction of every one concerned.

Mr. Maudsley had next a pair of 200 horse-power marine engines put in hand. His sons and partners were rather opposed to so expensive a piece of work being undertaken without an order. At that time such a power as 200 horse nominal was scarcely thought of; and the Admiralty Board were very cautious in ordering marine engines of any sort. Nevertheless, the engines were proceeded with and perfected. They formed a noble object in the great erecting shop. They embodied in every detail all Mr. Maudsley's latest improvements. In fact the work was the sum total of the great master's inventions and adaptations in marine engines. The Admiralty at last secured them for the purpose of being placed in a very fine vessel, the *Dee,* then in course of

construction.   Mr. Maudsley was so much pleased with the
result that he had a very beautiful model made of the
engines; and finding that I had some artistic skill as a
draughtsman, he set me to work to make a complete perspective drawing of them as they stood all perfect in the
erecting-shop.   This was a piece of work entirely to my
taste.   In due time I completed a graphic portrait of these
noble engines, treated, I hope, in an artistic spirit.   Indeed,
such a class of drawing was rarely to be had from an engineering draughtsman.   Mere geometrical drawing could
not give a proper idea, as a whole, of so grand a piece of
mechanism.   It required something of the artistic spirit to
fairly represent it.   At all events my performance won the
entire approval of my master.

Mr. Maudsley was a man of a wide range of mechanical
abilities.   He was always ready to enter upon any new work
requiring the exercise of special skill.   It did not matter
whether it was machine tools, engraving dies, block machinery, or astronomical instruments.   While at Berlin he
went to see the Royal Observatory.   He was naturally much
interested by the fine instruments there—the works of Rip-
sold and Moritz, the pioneers of improved astronomical
workmanship.   The continental instrument makers were
then far in advance of those of England.   Mr. Maudsley
was greatly impressed with the sight of the fine instruments
in the Berlin Observatory.   He was permitted to observe
some of the most striking and remarkable of the heavenly
bodies—Jupiter, Saturn, and the Moon.   It was almost a
new revelation to him; for the subject was entirely novel.
To be able to make such instruments seemed to him to be a
glorious achievement of refined mechanism and manipulative
skill.   He returned home full of the wonderful sights he had
seen.   It was a constant source of pleasure to him to dwell
upon the splendour and magnificence of the heavenly bodies.

He was anxious to possess a powerful telescope of his

own. His principal difficulty was in procuring a lens of considerable diameter, possessed of high perfection of defining power. I suggested to him the employment of a reflecting telescope, by means of which the difficulties connected with the employment of glass could be avoided. This suggestion was based upon some knowledge I had acquired respecting this department of refined mechanical art. I knew that the elder Herschel had by this means vastly advanced our knowledge of the heavenly bodies, indeed to an extent far beyond what had been achieved by the most perfect of glass lens instruments. Mr. Maudsley was interested in the idea I suggested; and he requested me to show him what I knew of the art of compounding the alloy called speculum metal. He wished to know how so brittle a material could be cast and ground and polished, and kept free from flaws or defects of every kind.

I accordingly cast for him a speculum of 8 inches diameter. I ground and polished it, and had it fitted up in a temporary manner to exhibit its optical capabilities, which were really of no mean order. But, as his ambition was to have a grand and powerful instrument of not less than 24 inches diameter, the preparation for such a speculum became a subject to him of the highest interest. He began to look out for a proper position for his projected observatory. He made inquiry about a residence at Norwood, where he thought his instrument might have fair play. It would there be free from the smoke and disturbing elements of such a place as Lambeth. His mind was full of this idea when he was called away by the claims of affection to visit a dear old friend at Boulogne. He remained there for more than a week, until assured of his friend's convalescence. But on his return voyage across the Channel he caught a severe cold. On reaching London he took to his bed, and never left it alive. After three or four weeks' suffering he died on the 14th of February 1831.

It was a very sad thing for me to lose my dear old
master. He was so good and so kind to me in all ways.
He treated me like a friend and companion. He was
always generous, manly, and upright in his dealings with
everybody. How his workmen loved him ; how his friends
lamented him ! He directed, before his death, that he
should be buried in Woolwich Churchyard, where a cast-
iron tomb, made to his own design, was erected over his
remains. He had ever a warm heart for Woolwich, where
he had been born and brought up. He began his life as a
mechanic there, and worked his way steadily upwards until
he reached the highest point of his profession. He often
returned to Woolwich after he had left it ; sometimes to
pay a share of his week's wages to his mother, while she
lived; sometimes to revisit the scenery of his youth. He
liked the green common, with the soldiers about it;
Shooter's Hill, with its wide look-out over Kent and down
the valley of the Thames; the river busy with shipping; the
Dockyard wharf, with the royal craft loading and unloading
their armaments. He liked the clangour of the arsenal
smithy, where he had first learned his art; and all the busy
industry of the place. It was natural, therefore, that being
so proud of his early connection with Woolwich he should
wish his remains to be laid there ; and Woolwich, on its
part, has equal reason to be proud of Henry Maudsley.

After the death of my master I passed over to the
service of his worthy partner, Joshua Field. I had an
equal pleasure in working under him. His kindness in
some degree mitigated the sad loss I had sustained by the
death of my lamented friend and employer. The first work
I had to perform for Mr. Field was to assist him in making
the working drawings of a 200 horse-power condensing
steam-engine, ordered by the Lambeth Waterworks Company.
The practical acquaintance which I by this time possessed
of the mechanism of steam power enabled me to serve Mr.

Field in a satisfactory manner. I drew out in full practical detail the rough but excellent hand sketches with which he supplied me. They were handed out for execution in the various parts of the factory ; and I communicated with the foremen as to the details and workmanship.

While I was occupied beside Mr. Field in making these working drawings, he gave me many most valuable hints as to the designing of machinery in general. In after years I had many opportunities of making good use of them. One point he often impressed upon me. It was, he said, most important to bear in mind the *get-at-ability* of parts—that is, when any part of a machine was out of repair, it was requisite to get at it easily without taking the machine to pieces. This may appear a very simple remark, but the neglect of such an arrangement occasions a vast amount of trouble, delay, and expense. None but those who have had to do with the repair of worn-out or damaged parts of machinery can adequately value the importance of this subject.

I found Mr. Field to be a most systematic man in all business affairs. I may specially name one of his arrangements which I was quick to take up and appreciate. I carried it out with great advantage in my after life. It was, to record subjects of conversation by means of graphic drawings. Almost daily, persons of note came to consult with him about machinery. On these occasions the consultations took place either with reference to proposed new work, or as to the progress of orders then in hand. Occasionally, some novel scheme of applying power was under discussion, or some new method of employing mechanism. On ordinary occasions rough and rapid sketches are made on any stray pieces of waste paper that are about, and after the conversation is over the papers are swept away into the waste basket and destroyed. And yet some of these rapid drawings involve matters of great interest and importance for after-consultations.

N

To avoid such losses, Mr. Field had always placed upon
his table a " talking book " or " graphic diary."   When his
visitors called and entered into conversation with him about
mechanical matters, he made rapid sketches on the successive pages of the book, and entered the brief particulars
and date of the conversation, together with the name and
address of the visitor.   So that a conversation, once begun,
might again be referred to, and, when the visitor called,
the graphic memoranda might be recalled without loss of
time, and the consultation again proceeded.   The pages of Mr.
Field's "talking books" were in many ways most interesting.
They contained data that, in future years, supplied valuable
evidence in respect to first suggestions of mechanical contrivances, and which sometimes were developed into very important results.   I may add that Mr. Field kept these "talking books " on a shelf in front of his drawing table.   The
back of each volume was marked with the year to which the
entries referred, and an index was appended to each.   A
general index book was also placed at the end of the goodly
range of these graphic records of his professional life.

The completion of the working drawings of the Lambeth
pumping engines occupied me until August 1831.   I had
then arrived at my twenty-third year.   I had no intention
of proceeding further with assistants' or journeymen's work.
I intended to begin business for myself.   Of course I could
only begin in a very small way.   I informed Mr. Field of my
intention, and he was gratified with my decision.   Not only so;
but he kindly permitted me to obtain castings of one of the
best turning-lathes in the workshops.   I knew that when I
had fitted it up it would become the parent of a vast progeny
of descendants—not only in the direct line, but in planing
machines, screw-cutting lathes, and many other minor tools.

At the end of the month, after taking a grateful farewell
of Mr. Field and his partners, I set sail for Leith with my
stock of castings, and reached Edinburgh in due time.   In

order to proceed with the construction of my machine tools,
I rented a small piece of land at Old Broughton.   It was
at the rear of my worthy friend George Douglass's small
foundry, and was only about five minutes' walk from my
father's house.   I erected a temporary workshop 24 feet
long by 16 feet wide.

I removed thither my father's foot-lathe, to which I had
previously added an excellent slide-rest of my own making.

MY TEMPORARY WORKSHOP AT EDINBURGH

I also added a " slow motion," which enabled me to turn
cast-iron and cast-steel portions of my great Maudsley lathe.
I soon had the latter complete and in action.   Its first child
was a planing machine capable of executing surfaces in the
most perfect style; it was 3 feet long by 1 foot 8 inches
wide.   Armed with these two most important and generally
useful tools, and by some special additions, such as boring
machines and drilling machines, I soon had a progeny of
legitimate descendants crowded about my little workshop, so
that I often did not know which way to turn.

I had one labourer to drive the wheel which gave motion to my big lathe; but I was very much in want of some one else to help me. One day a young hearty fellow called upon me. He had come from the Shotts Iron Company's Works in Edinburgh. Having heard of what I was about, he offered his services. When he told me that he had been bred as a millwright, and that he could handle the plane and the saw as well as the chisel and the file, I closed with him at once. He was to have fifteen shillings a week. I liked the young man very much, he was so hearty and cheerful. His name was Archibald Torry, or "Archie," as he was generally called during the twenty years that he remained in my service.

I obtained another assistant in the person of a young man whose father wished him to get an insight into practical engineering. I was offered a premium of £50 for twelve months' experience in my workshop. I arranged to take the young man, and to initiate him in the general principles and practice of engineering. The £50 premium was a very useful help to me, especially as I had engaged the millwright. It enabled me to pay Torry's wages during the time that he remained with me in Edinburgh. I found it necessary, however, to take in some work in the regular way of business, in order to supply me with the means of completing my proper supply of tools.

The chief of these extraneous and, I may say, disturbing jobs, was that of constructing a rotary steam-engine. Mr. Robert Steen had contrived and patented an engine of this sort. He was a dangerously enthusiastic man, and entertained the most visionary ideas as to steam power. He was of opinion that his own contrivance was more compact and simple, and possessed of more capability of producing power from the consumption of a given quantity of fuel, than the best steam-engines then in use. I warned him of his error; but nothing but an actual proof would satisfy him. He

urgently requested me to execute his order. He made me
a liberal and tempting offer of weekly payments for my
work during the progress of his engine. He only required
that I should give his invention the benefit of my careful
workmanship. He considered that that would be sufficient
to substantiate all his enthusiastic expectations. I was thus
seduced to accept his order.

I made the requisite drawings, and proceeded with
the work. At the same time my own machine tools were
in progress, though at a retarded pace. The weekly payments were regularly made, and I was kept in a sort of
financial ease. After three months the rotary engine was
finished to the inventor's complete satisfaction. But when
the power it gave out was compared with that of a good
ordinary steam-engine, the verdict as to consumption of fuel
was against the new rotary engine. Nevertheless, the enthusiastic projector, "tho' vanquished he would argue still,"
insisted that the merits of his contrivance would sooner or
later cause it to be a most formidable rival to the crank
steam-engines. As he was pleased with its performances, I
had no reason to be dissatisfied. I had done my part in
the matter, and Mr. Steen had done his. His punctual
weekly payments had assisted me in the completion of my
tools; and after a few months more labour I had everything ready for starting business on my own account.

My choice lay between Liverpool and Manchester. I had
seen both of these cities while on my visit to Lancashire
to witness the opening of the Liverpool and Manchester
Railway. I now proceeded to visit them again. I was
fortified with valuable introductions to leading men in both
places. I was received by them with great kindness and
hospitality. I have heard a great deal about the ingratitude
and selfishness of the world. It may have been my good
fortune, but I have never experienced either of those unfeel-
ing conditions. On the whole, I have found a great deal

of unselfish kindness among my fellow-beings. They have
often turned out of their way to do me a service; and I can
never be too grateful for the unwearied kindness, civility,
and generosity of the friends I encountered during my stay
in Lancashire.

It was a question which would be the best place to settle
in—Liverpool or Manchester. I had seen striking evidences
of the natural aptitude of Lancashire workmen for every
sort of mechanical employment, and had observed their un-
sparing energy while at work. I compared them with the
workmen whom I had seen in London, and found them
superior. They were men of greater character; they struck
harder on the anvil; their minds were more capacious;
their ingenuity was more inventive. I felt assured that in
either Liverpool or Manchester—the centres of commercial
and manipulative energy—I could settle down with my
limited capital and tools, and in course of time contrive to
get on, helped by energy, self-reliance, and determination.
I also found that the demand for machine-making tools was
considerable, and that their production would soon become
an important department of business. It might be carried
on with little expenditure of capital, as the risks were small
and the returns were quick. I resolved to cultivate that
moderate and safe class of mechanical business, at all events
at the outset.

I first went to Liverpool. I presented my letter of
introduction to Mr. Roscoe, head of the Mersey Steel and
Iron Company. He received me with great kindness, and
gave me much good advice. I called upon Edward Berry,
engineer, and also upon William Fawcett, who had received
me with so much kindness on my former visit. I cannot
omit mentioning also the friendly reception which I received
from Dr. Sillar. He had been a medical student at Edinburgh, and had during that time met with some kindness
from my father. He expressed his remembrance of it to

me with grateful effusion ; and added his personal introduc-
tion, with that of my letters, to some of the leading men in
Liverpool.   I may mention that Dr. Sillar was the son of
Burns's "Brother Poet" Davie, to whom the well-known
"Epistle" was addressed.

Among the other well known men to whom I was introduced at Liverpool was John Cragg, a most intelligent and
enterprising ironfounder.   He was an extensive manufacturer
of the large sugar-boiling pans used in the West Indies.   He
had also given his attention to the introduction of iron into
buildings of different sorts.   Being a man of artistic taste
he had even introduced cast-iron into Gothic architecture.
In order to exhibit, in an impressive form, the uses of his
favourite metal, he erected at his own cost a very elegant
church in the northern part of Liverpool.   Cast-iron was
introduced, not only in the material parts of the structure,
but into the Gothic columns and Gothic tracery of the
windows, as well as into the lofty and elegant spire.   Iron
was also employed in the external ornamental details, where
delicate yet effective decoration was desirable.   The famous
architect, Edward Blore, was the designer of the church ;
and the whole details of the building—of which cast-iron
formed the principal material—were executed to his entire
satisfaction.[1]

My introduction to Mr. Cragg led to an acquaintance,
and then to a friendship.   When the ice was broken—
which was very soon—he told me that he was desirous of
retiring from the more active part of his business.   Whether
he liked my looks or not I do not know ; but, quite unexpectedly, he made me a very tempting offer to enter his
works as his successor.   He had already amassed a fortune,

---

[1] So far as I can recollect the name of the church was St. James's.   It
exhibited a very early introduction of iron as an important element in architectural construction.   Iron was afterwards largely introduced into mills, mill
gearing, and buildings generally.

and I might do the same.   I could only thank him most
sincerely for his kindness.   But, on carefully thinking the
matter over, I declined the proposal.   My principal reason
was, that the special nature of his foundry work did not
quite harmonise with my desire to follow the more strictly
mechanical part of the iron business.   Besides, I thought I
had a brighter prospect of success before me; though I
knew that I had many difficulties to contend against.   Did
I throw away my chances in declining the liberal proposal
of Mr. Cragg ?   The reader will be able to judge from the
following pages.   But to the last I continued a most friendly
intercourse with my intended patron, while he on his part
took an almost paternal interest in my progress.[1]

After my visit to Liverpool I passed on to Manchester.
I was fortunate in having introductions to some of the
leading men there, such as John Kennedy, William Fairbairn, the Grant Brothers, and lastly, that most admirable
man, Benjamin Hick, engineer, of Bolton.   To narrate in
detail all the instances of warm and hospitable kindnesses
which I received from men in Lancashire, even from the outset of my career there, would fill a volume.

I first went to see my friend Edward Tootal, who had
given me so kind a reception in 1830.   I was again cordially received ; he now promised to befriend me, which he
did most effectually.   I next visited John Chippendale, of
the firm of Thomson, Chippendale, and Company, calico
printers.   I had met him at a friend's house in London,
where he had offered, if I ever visited Manchester, to introduce me to some of the best men there.   I accordingly
called upon him at his counting-house.   It happened to be
Tuesday, the market day, when all the heads of manufac-
facturing establishments in and around Manchester met
together at the Exchange between 12 and 1 ; and thus all
were brought to a focus in a very convenient manner.

---

[1] Mr. Cragg died in 1853, aged 84.

Mr. Chippendale first introduced me to Mr. John Kennedy, one of the most distinguished men in Manchester. I had a special letter of introduction to him from Buchanan of Catrine, and his partner Smith of Deanstone. I explained to him the object of my visit to Manchester, and he cordially entered into my views. He left his occupation at the time, and went with me to see a place which he thought might be suitable for my workshop. The building was near at hand—in Dale Street, Piccadilly. It had been used as a cotton mill, but was abandoned by the owner in favour of more suitable and extensive premises. It was now let out in flats for manufacturing purposes. Power was supplied to each flat from a shaft connected with a large mill up the street, the owner of which had power to spare. The flat shown to me was 130 feet long by 27 feet wide, and the rent was only £50 a year. I thought the premises very suitable, but I took a night to sleep over it. I thanked Mr. Kennedy very much for his kindness, and for the trouble which he had taken on behalf of an unknown stranger.

On this memorable day I had another introduction, through the kindness of Mr. Chippendale, which proved of great service to me. It was to the Messrs. Grant, the famous "Brothers Cherryble" of Dickens. I was taken to their counting-house in Cannon Street, where I was introduced to Daniel Grant. Although business was at its full height he gave me a cordial reception. But, to save time, he invited me to come after the Exchange was over and take "tiffin" with him at his hospitable mansion in Mosely Street. There, he said, I should meet some of the most enterprising men in Lancashire. I was most happy, of course, to avail myself of his invitation. I went thither accordingly, and the first thing that Daniel did was to present me in the most cordial manner to "his noble brother William," as he always affectionately called him. William was the head of the firm, and he, too, gave me a warm and hearty welcome. He asked me to sit beside him at the head of the table.

During dinner—for indeed it was such, being the survival of the old-fashioned one o'clock dinner of a departing age —William entered into conversation with me. He took occasion to inquire into the object of my visit to Manchester. I told him, as briefly as I could, that I intended to begin the business of a mechanical engineer on a very moderate scale, and that I had been looking out for premises wherein to commence operations. He seemed interested, and asked more questions. I related to him my little history, and told him of my desires, hopes, and aspirations. "What was my age?" "Twenty-six." "That is a very young age at which to begin business on your own account." "Yes; but I have plenty of work in me, and I am very economical." Then he pressed his questions home. "But what is your capital?" I told him that my capital in cash was £63. "What!" he said, "that will do very little for you when Saturday nights come round." "That's true," I answered; "but as there will be only myself and Archy Torry to provide for, I think I can manage to get along very well until profitable work comes in."

He whispered to me to "keep my heart up!" With such views, he said, I was sure to do well. And if, he added, on any Saturday night I wanted money to pay wages or other expenses, I would find a credit for £500 at 3 per cent at his office in Cannon Street, "*and no security.*" These were his very words. What could have been more generous? I could only whisper my earnest thanks for his warm-hearted kindness. He gave me a kindly squeeze of the hand in return, which set me in a glow of gladness. He also gave me a sort of wink that I shall never forget—a most knowing wink. In looking at me he seemed to turn his eye round and brought his eyebrows down upon it in a sudden and extraordinary manner. I thought it was a mere confirmation of his kind

advice to "keep my heart up!" It was not until two
years after that I found, from a mutual friend, that the eye
in question was *made of glass!* Sometimes the glass eye
got slightly out of its place, and Mr. Grant had to force it
in again by this odd contortion of his eyebrows, which I
translated into all manner of kind intentions.

As soon as the party broke up I went to Wren and
Bennett, the agents for the flat of the old mill which I had

MY FACTORY FLAT AT MANCHESTER.

seen in Dale Street. I inspected it again, and found that
it was in all respects suitable for my purpose. I may
mention in passing that the flat below mine was in the
occupation of a glass cutter, whose glass-cutting lathes and
grindstones were supplied with power from the same upright
shaft that was to serve me in the same manner on the flat
above. Encouraged by the support of William Grant, I
immediately entered into a contract for my premises as a
yearly tenant. Nothing could have been more happily
arranged for my entering into business as a mechanical
engineer and machine tool maker. The situation of the

premises was excellent, being in the heart of Manchester. There was a powerful crab crane, or hoisting apparatus, in the upper story, and the main chains came down in front of the wide doors of my workshop, so that heavy castings or cases of machinery might be lifted up or let down with the utmost ease and convenience. At the same time I was relieved from looking after the moving power and its natural accompaniment of trouble and expense in the way of fuel and attendance.

When I had settled the contract for taking the place, I wrote down to Edinburgh by that night's post to tell my father of the happy results of my visit to Manchester, and also to inform my right hand man, Archy Torry, that I should soon be with him. He was to prepare for packing up my lathes, planing machines, drilling machines, and other smaller tools, not forgetting my father's foot lathe, of which I had made such effective use.[1] I soon followed up my letter. I was in Edinburgh in a few days' time, and had all my tools packed up. In the course of about ten days I returned to Manchester, and was followed by Archy Torry and the ponderous cases of machinery and engineer's tools. They were all duly delivered, hoisted to my flat, and put in their proper places. I was then ready for work.

The very first order I received was from my friend Edward Tootal. It was a new metallic piston for the small steam-engine that gave motion to his silk-winding machinery. It was necessary that it should be done over night, in order that his factory should be at work as usual in the morning. My faithful Archy and I set to work accordingly. We removed the old defective piston, and replaced it by a new

---

[1] I have still this foot-lathe in full and perfect and almost daily action. I continue to work with it now, after sixty-three years of almost constant use. It is a lathe that I duly prize and venerate, not only because it was my father's, but also because it was, in practical fact, the progenitor, more or less directly, of all the mechanical productions of my long and active life.

and improved one, made according to my own ideas of how so important a part of a steam-engine should be constructed. We conveyed it to Mr. Tootal's factory over night, and by five o'clock in the morning gave it a preliminary trial to see that everything was in order. The "hands" came in at six, and everything was set to work. It was no doubt a very small order, but the piston was executed perfectly and satisfactorily. The result of its easier action, through reduced friction, was soon observable in the smaller consumption of coal. Mr. Tootal and his brother were highly pleased at my prompt and careful attention to their little order, and it was the forerunner of better things to come.

Orders soon came in. My planing machine was soon fully occupied. When not engaged in executing other work it was employed in planing the flat cast-iron inking tables for printing machines. These were made in considerable numbers by Messrs. Wren and Bennett (my landlords) under the personal superintendence of Ebenezer Cowper, brother of the inventor, who, in conjunction with Mr. Applegarth, was the first to produce a really effective newspaper printing machine. I had many small subsidiary jobs sent to me to execute. They not only served to keep my machine tools properly employed, but tended in the most effective way to make my work known to some of the best firms in Manchester, who in course of time became my employers.

In order to keep pace with the influx of work I had to take on fresh hands. I established a smithy down in the cellar flat of the old mill in Dale Street, so that all forge work in iron and steel might be promptly and economically produced on the premises. There was a small iron foundry belonging to a Mr. Heath, about three minutes' walk from my workshop, where I had all my castings of iron and brass done with promptness, and of excellent quality. Mr. Heath very much wanted a more powerful steam-engine to drive his cupola blowing fan. I had made a steam-engine in Edin-

burgh and brought it with me.   There it lay in my workshop,
where it remained unused, for I was sufficiently supplied
with power from the rotating shaft.   Mr. Heath offered to
buy it.   The engine was accordingly removed to his iron
foundry, and I received my full quota of value in castings.

Week by week my orders grew, and the flat of the old
mill soon assumed a very busy aspect.   By occasionally
adding to the number of my lathes, drilling machines, and
other engineers' tools, I attracted the attention of employers.   When seen in action they not only facilitated and
economised the production of my own work, but became
my best advertisements.   Each new tool that I constructed
had some feature of novelty about it.   I always tried for
simplicity and perfectness of workmanship.   I was punctual
in all my engagements.   The business proved safe and profitable.   The returns were quick.   Sometimes one-third of
the money was paid in advance on receipt of the order, and
the balance was paid on delivery at my own premises.   All
risk of bad debts was avoided.   Thus I was enabled to carry
on my business with a very moderate amount of capital.

My crowded workshop and the active scene it presented,
together with the satisfaction my work gave to my employers,
induced several persons to offer to enter into partnership
with me.   Sometimes it was on their own account, or for
a son or relation for whom they desired an opening.   But I
fought shy of such proposals.   It was a very riskful affair
to admit as partners young men whose character for ability
might be very doubtful.   I was therefore satisfied to go on
as before.   Besides, I had the kind and disinterested offer
of the Brothers Grant, which was always available, though,
indeed, I did not need to make use of it.   I had also the
good fortune to be honoured by the friendship of Edward
Lloyd, the head of the firm of Jones, Lloyd, and Co.   I had
some moderate financial transactions with the bank.   Mr.
Lloyd had, no doubt, heard something of my industry and

economy. I never asked him for any accommodation ; but on one occasion he invited me into his parlour, not to sweat me, but to give me some most kindly hints and advice as to the conduct of my financial affairs. He volunteered an offer which I could not but feel proud of. He said that I should have a credit of £1000 at my service, at the usual bank rate. He added, " As soon as you can, lay by a little capital of your own, and baste it with its own gravy!" A receipt which I have carefully followed through life, and I am thankful to say with satisfactory results.

Before I conclude this chapter, let me add something more about my kind friends the Brothers Grant. It is well that their history should be remembered, as the men who personally knew them will soon be defunct. The three brothers, William, Daniel, and John Grant, were the sons of a herdsman or cattle-dealer, whose occupation consisted in driving cattle from the far north of Scotland to the rich pastures of Cheshire and Lancashire. The father was generally accompanied by his three sons, who marched barefoot, as was the custom of the north country lads in those days. Being shrewd fellows, they observed with interest the thriving looks and well-fed condition of the Lancashire folks. They were attracted by the print works and cotton mills which lay by the Irwell, as it crept along in its bright and rural valley towards Manchester. When passing the works of Sir Robert Peel at Nuttal, near Bury, they admired the beauty of the situation. The thought possessed them that they would like to obtain some employment in the neighbourhood. They went together in search of a situation. It is said that when they reached the crown of the hill near Walmsley, from which a beautiful prospect is to be seen, they were in doubt as to the line of road which they should pursue. To decide their course, a stick was put up, and they agreed to follow the direction in which it should fall. The stick fell in the direction of Ramsbotham, then a little village in the bottom of the valley, on the river Irwell. There they went, and found employment.

They were thrifty, economical, and hard-working; and they soon saved money. Their savings became capital, and they invested it in a little print work. Their capital grew, and they went on investing it in print works and cotton mills. They became great capitalists and manufacturers; and by their industry, ability, and integrity, were regarded as among the best men in Lancashire. As a memorial of the event which enabled them to take up their happy home at Ramsbotham, they caused to be erected at the top of Walmsley Hill, a lofty tower, overlooking the valley, as a kind of public thank-offering for the prosperity and success which they had achieved in their new home. Their well-directed diligence made the valley teem with industry, activity, health, joy, and opulence. They never forgot the working-class from which they had sprung, and as their labours had contributed to their wealth, they spared no expense in providing for the moral, intellectual, and physical interests of their work-people. Whenever a worthy object was to be achieved, the Brothers Grant were always ready with their hearty and substantial help. They contributed to found schools, churches, and public buildings, and many a deserving man did they aid with their magnanimous bounty.

I may also mention that they never forgot their first impression of the splendid position of the first Sir Robert Peel's works at Nuttal. In course of time Sir Robert had, by his skill and enterprise, acquired a large fortune, and desired to retire from business. By this time the Grant Brothers had succeeded so well that they were enabled to purchase the whole of his works and property in the neighbourhood. They proceeded to introduce every improvement in the way of machinery and calico printing, and thus greatly added to the quality of their productions. Their name became associated with everything that was admirable. They abounded

in hospitality and generosity. In the course of many long
years of industry, enterprise, and benevolence, they earned
the goodwill of thousands, the gratitude of many, and the
respect of all who knew them. I was only one of many
who had cause to remember them with gratefulness. How
could I acknowledge their kindness? There was one way;
it was a very small way, but I will relate it.

Soon after my introduction to the Grants, and before I
had brought my tools to Manchester, William invited me to
join a gathering of his friends at Ramsbotham. The church
built at his cost had just been finished, and it was to be
opened with great eclat on the following Sunday. He
asked me to be his guest, and I accepted his invitation with
pleasure. As it was a very fine day at the end of May, I
walked out to Ramsbotham, and enjoyed the scenery of the
district. Here was the scene of the Grant Brothers' industry
and prosperity. I met many enterprising and intelligent
men, to whom William Grant introduced me. I was greatly
pleased with the ceremonies connected with the opening of
the church.

On the Monday morning, William Grant, having seen
some specimens of my father's artistic skill as a landscape
painter, requested me to convey to him his desire that he
should paint two pictures—one of Castle Grant, the residence of the chief of the Clan Grant, and the other of Elgin
Cathedral. These places were intimately associated with
his early recollections. The brothers had been born in the
village adjoining Castle Grant; and Elgin Cathedral was one
of the principal old buildings of the north. My father
replied, saying that he would be delighted to execute the
pictures for a gentleman who had given me so kindly a reception, but that he had no authentic data—no drawings, no
engravings—from which to paint them; and that he was
now too old to visit the places. I therefore resolved to do
what I could to help him to paint the pictures.

o

As it was necessary that I should go to London before returning to Edinburgh to pack up my machine tools, I went thither, and after doing my business, I embarked for Dundee by the usual steamer. I made my way from there, *via* Perth and Dunkeld, to Inverness, and from thence I proceeded to Elgin. I made most careful drawings of the remains of that noble cathedral. I endeavoured to include all that was most beautiful in the building and its surrounding scenery. I then went on to Castle Grant, through a picturesque and romantic country. I found the castle amidst its deep forests of pine, larch, elm, and chestnut. The building consists of a high quadrangular pile of many stories, projecting backwards at each end, and pierced with windows of all shapes and sizes. I did my best to carry away a graphic sketch of the old castle and its surroundings; and then, with my stock of drawings, I prepared to return to Inverness on foot.

The scenery was grand and beautiful. The weather was fine, although after mid-day it became very hot. A thunderstorm was evidently approaching. The sun was obscured by a thunder-cloud; the sky flashed with lightning, and the rain began to pour down. I was then high up on a wild-looking moor, covered with heather and vast boulders. There was no shelter to be had, for not a house was in sight. I did not so much mind for my clothes, but I feared very much for my sketches. Taking advantage of the solitude, I stripped myself, put my sketches under my clothes, and thrust them into a hollow underneath a huge boulder. I sat myself down on the top of it, and there I had a magnificent shower-bath of warm rain. I never enjoyed a bath under such romantic circumstances. The thunder-clouds soon passed over my head, and the sun broke out again cheerily. When the rain had ceased I took out my clothes and drawings from the hollow, and found them perfectly dry. I set out again on my long

walk to Inverness; and reached it just in time to catch the
Caledonian Canal steamer.   While passing down Loch Ness
I visited the romantic Fall of Foyers; then through Loch
Lochy, past Ben Nevis to Loch Linnhe, Oban, and the Kyles
of Bute, to Glasgow, and from thence to Edinburgh.

I had the pleasure of placing in my father's hands the
sketches I had made.   He was greatly delighted with them.

AN EXTEMPORISED SHOWER BATH.

They enabled him to set to work with his usual zeal, and
in the course of a short time he was able to execute, *con
amore*, the commission of the Brothers Grant.   So soon as
I had completed my sketches I wrote to Daniel Grant and
informed him of the result of my journey.   He afterwards
expressed himself most warmly as to my prompt zeal in
obtaining for him authentic pictures of places so dear to the
brothers, and so much associated with their earliest and most
cherished recollections.

I have already referred to the Brothers Cowper. They
were among my most attached friends at Manchester.
Many of my most pleasant associations are connected with
them. Edward Cowper was one of the most successful
mechanics in bringing the printing machine to a state of
practical utility. He was afterwards connected with Mr.
Applegarth of London, the mechanical engineer of the *Times*
newspaper.[1] He invented for the proprietors a machine that
threw off from 4500 to 5000 impressions in the hour. In
course of time the Brothers Cowper removed the manufacture
of their printing machines from London to Manchester.
There they found skilled and energetic workmen, ready to
carry their plans into effect. They secured excellent premises, supplied with the best modern machine tools, in the
buildings of Wren and Bennett, about two minutes' walk
from my workshop, which I rented from the same landlords.

I had much friendly intercourse with the Cowpers,
especially with Ebenezer the younger brother, who took up
his residence at Manchester for the purpose of specially
superintending the manufacture of the printing machines.
These were soon in large demand, not only for the printing
of books but of newspapers. One of the first booksellers
who availed himself of the benefits of the machine was Mr.
Charles Knight, who projected the *Penny Magazine* of 1832,
and sold it to the extent of about 180,000 copies weekly.
It was also adopted by the Messrs. Chambers of Edinburgh,
and the proprietors of the *Magasin Pittoresque* of Paris.
The Universities of Oxford and Cambridge also used Cowper's machine in printing vast numbers of bibles and prayer-

---

[1] Mr. Kœnig's machines, first used at the *Times* office, were patented in
1814. They were too complicated and expensive, and the inking was too
imperfect for general adoption. They were superseded by Mr. Edward Cowper's machine, which he invented and patented in 1816. He afterwards
added the inking roller and table to the common press. The effect of Mr.
Cowper's invention was to improve the quality and speed of printing, and to
render literature accessible to millions of readers.

books, thereby reducing their price to one-third of the former cost. There was scarcely a newspaper of any importance in the country that was not printed with a Cowper's machine.

As I possessed some self-acting tools that were specially suited to execute some of the most refined and important parts of the printing machine, the Messrs. Cowper transferred their execution to me. This was a great advantage to both. They were relieved of the technical workmanship; while I kept my men and machine tools fully employed at times when they might otherwise have been standing idle. Besides, I derived another advantage from my connection with the Brothers Cowper, by having frequent orders to supply my small steam-engines, which were found to be so suitable for giving motion to the printing machines. At first the machines were turned by hand, and very exhausting work it was; but the small steam-engine soon relieved the labourer from his heavy work.

Edward frequently visited Manchester to arrange with his brother as to the increasing manufacture of the printing machines, and also to introduce such improvements in the minor details as the experience and special requirements of the printing trade suggested. It was on these occasions that I had the happy opportunity of becoming intimately acquainted with him; and this resulted in a firm friendship which continued until the close of his admirable life. The clear and masterly way in which, by some happy special faculty, he could catch up the essential principles and details of any mechanical combination, however novel the subject might be, was remarkable; and the quaint and humorous manner in which he treated all such subjects, in no small degree caused his shrewd and intelligent remarks to take a lasting hold of the memory.

On many occasions Edward Cowper gave Friday evening lectures on technical subjects at the Royal Institution, London. Next to Faraday, no one held the attention of a delighted audience in so charming a manner as he did. Like Faraday, he possessed the power of clearly unveiling his subject, and stripping it of all its complicated perplexities. His illustrations were simple, clear, and understandable. Technical words were avoided as much as possible. He threw the ordinary run of lecturers far into the shade. Intelligent boys and girls could understand him. Next to Faraday no one filled the theatre of the Institution with such eager and crowded audiences as he did. His choice of subjects, as well as his masterly treatment, always rendered his lectures instructive and attractive. He was one of the most kind-hearted of men, and the cheerful way in which he laid aside his ordinary business to give instruction and pleasure to others endeared him to a very wide circle of devoted friends.

CHAPTER XI.

BRIDGEWATER FOUNDRY——PARTNERSHIP.

My business went on prosperously. I had plenty of orders, and did my best to execute them satisfactorily. Shortly after the opening of the Liverpool and Manchester Railway there was a largely increased demand for machine-making tools. The success of that line led to the construction of other lines, concentrating in Manchester; and every branch of manufacture shared in the prosperity of the time.

There was a great demand for skilled, and even for unskilled labour. The demand was greater than the supply. Employers were subjected to exorbitant demands for increased rates of wages. The workmen struck, and their wages were raised. But the results were not always satisfactory. Except in the cases of the old skilled hands, the work was executed more carelessly than before. The workmen attended less regularly; and sometimes, when they ought to have been at work on Monday mornings, they did not appear until Wednesday. Their higher wages had been of no use to them, but the reverse. Their time had been spent for the most part in two days' extra drinking.

The irregularity and carelessness of the workmen naturally proved very annoying to the employers. But it gave an increased stimulus to the demand for self-acting machine tools, by which the untrustworthy efforts of hand labour

might be avoided. The machines never got drunk; their
hands never shook from excess; they were never absent
from work ; they did not strike for wages; they were unfailing in their accuracy and regularity, while producing the
most delicate or ponderous portions of mechanical structures.

It so happened that the demand for machine tools, con-
sequent upon the increasing difficulties with the workmen,
took place at the time that I began business in Manchester,
and I had my fair share of the increased demand.   Most of
my own machine tools were self-acting—planing machines,
slide lathes, drilling, boring, slotting machines, and so on.
When set up in my workshop they distinguished themselves
by their respective merits and efficiency.   They were, in fact,
their own best advertisements.   The consequence was that
orders for similar machines poured in upon me, and the floor
of my flat became completely loaded with the work in hand.

The tenant below me, it will be remembered, was a glass-
cutter. He observed, with alarm, the bits of plaster from
the roof coming down among his cut glasses and decanters.
He thought that the rafters overhead were giving way, and
that the whole of my machinery and engines would come
tumbling down upon him some day and involve him in ruin.
He probably exaggerated the danger ; still there was some
cause for fear.

The massive castings on my floor were moved about from
one part to another, when the floor quivered and trembled
under the pressure.   The glass-cutter complained to the
landlord ; and the landlord expostulated with me.   I did
all that I could to equalise the pressure, and prevent vibra-
tion as much as possible.   But at length, in spite of all my
care, an accident occurred which compelled me to take
measures to remove my machinery to other premises.   As
this removal was followed by consequences of much importance to myself, I must endeavour to state the circumstances
under which it occurred.

My kind friend, John Kennedy, continued to take the greatest interest in my welfare.   He called in upon me occasionally; he admired the quality of my work, and the beauty of my self-acting machinery.   More than that, he recommended me to his friends.   It was through his influence that I obtained an order for a high-pressure steam-engine of twenty horse-power to drive the machinery connected with a distillery at Londonderry, in Ireland.   I was afraid at first that I could not undertake the job.   The size of the engine was somewhat above the height of my flat, and it would probably occupy too much space in my already overcrowded workshop.   At the same time I was most anxious not to let such an order pass me.   I wished to please my friend Mr. Kennedy ; and besides, the execution of the engine might lead to further business.

At length, after consideration, I undertook to execute the order.   Instead of constructing the engine perpendicularly, I constructed it lying upon its side.   There was a little extra difficulty, but I managed to complete it in the best style.   It had next to be taken to pieces for the purpose of being conveyed to Londonderry.   It was then that the accident happened.   My men had the misfortune to allow the end of the engine beam to crash through the floor !   There was a terrible scattering of lath and plaster and dust.   The glass-cutter was in a dreadful state.   He rushed forthwith to the landlord, and called upon him to come at once and *judge for himself !*

Mr. Wren *did* come, and *did* judge for himself.   He looked in at the glass shop, and saw the damage that had been done amongst the tumblers and decanters.   There was the hole in the roof, through which the end of the engine beam had come and scattered the lath and plaster.   The landlord then came to me.   The whole flat was filled with machinery, including the steam-engine on its side, now being taken to pieces for the purpose of shipment to Ireland.

Mr. Wren, in the kindest manner, begged me to remove from the premises as soon as I could, otherwise the whole building might be brought to the ground with the weight of my machinery. "Besides," he argued, "you must have more convenient premises for your rapidly extending business." It was quite true. I must leave the place and establish myself elsewhere.

The reader may remember that while on my journey on foot from Liverpool to Manchester in 1830, I had rested myself for a little on the parapet of the bridge overlooking the canal near Patricroft, and gazed longingly upon a plot of land situated along the canal side. On the afternoon of the day on which the engine beam crashed through the glass-cutter's roof, I went out again to look at that favourite piece of land. There it was, unoccupied, just as I had seen it some years before. I went to it and took note of its dimensions. It consisted of about six acres. It was covered with turf, and as flat and neat as a bowling-green. It was bounded on one side by the Bridgewater Canal, edged by a neat stone margin 1050 feet long, on another side by the Liverpool and Manchester Railway, while on a third side it was bounded by a good road, accessible from all sides. The plot was splendidly situated. I wondered that it had not been secured before. It was evidently waiting for me!

I did not allow the grass to grow beneath my feet. That very night I ascertained that the proprietor of this most beautiful plot was Squire Trafford, one of the largest landed proprietors in the district. Next morning I proceeded to Trafford Hall for the purpose of interviewing the Squire. He received me most cordially. After I had stated my object in calling upon him, he said he would be exceedingly pleased to have me as one of his tenants. He gave me a letter of introduction to his agent, Mr. Thomas Lee, of Princes Street, Manchester, with whom I was to arrange as to the terms. I was offered a lease of the six-acre plot for

999 years, at an annual rental of 1¾d. per square yard.
This proposal was most favourable, as I obtained the advantage of a fee-simple purchase without having to sink capital
in the land.   All that I had to provide for was the annual
rent.

My next step in this important affair was to submit the
proposal to the judgment of my excellent friend Edward
Lloyd, the banker.   He advised me to close the matter as
soon as possible, for he considered the terms most favourable.   He personally took me to his solicitors, Dennison,
Humphreys, and Cunliffe, and introduced me to them.   Mr.
Humphreys took the matter in hand.   We went together to
Mr. Lee, and within a few days the lease was signed, and I
was put into possession of the land upon which the Bridgewater Foundry was afterwards erected.[1]

I may mention briefly the advantages of the site.   The
Bridgewater Canal, which lay along one side of the foundry,
communicated with every water-way and port in England,
whilst the railway alongside enabled a communication to
be kept up by rail with every part of the country.   The
Worsley coal-boats came alongside the wharf, and a cheap
and abundant supply of fuel was thus insured.   The railway station was near at hand, and afforded every opportunity for travelling to and from the works, while I was
at the same time placed within twenty minutes of Manchester.

Another important point has to be mentioned.   A fine
bed of brick-clay lay below the surface of the ground, which
supplied the material for bricks.   Thus the entire works
may be truly said to have " risen out of the ground;" for
the whole of the buildings rested upon the land from
which the clay below was dug and burned into bricks.

_____

[1] I called the place the Bridgewater Foundry as an appropriate and humble
tribute to the memory of the first great canal maker in Britain—the noble
Duke of Bridgewater.

Then, below the clay lay a bed of New Red Sandstone rock, which yielded a solid foundation for any superstructure, however lofty or ponderous.

As soon as the preliminary arrangements for the lease of the six-acre plot had been made, I proceeded to make working drawings of a temporary timber workshop; as I was anxious to unload the floor of my flat in Dale Street, and to get as much of my machinery as possible speedily removed to Patricroft. For the purpose of providing the temporary accommodation, I went to Liverpool and purchased a number of logs of New Brunswick pine. The logs were cut up into planks, battens, and roof-timbers, and were delivered in a few days at the canal wharf in front of my plot. The building of the workshops rapidly proceeded. By the aid of some handy active carpenters, superintended by my energetic foreman, Archy Torry, several convenient well-lighted workshops were soon ready for the reception of my machinery. I had a four horse-power engine, which I had made at Edinburgh, ready to be placed in position, together with the boiler. This was the first power I employed in starting my new works.

I must return for a moment to the twenty horse-power engine, which had been the proximate cause of my removal from Dale Street. It was taken to pieces, packed, and sent off to Londonderry. When I was informed that it was erected and ready for work I proceeded to Ireland to see it begin its operations.

I may briefly say that the engine gave every satisfaction, and I believe that it continues working to this day. I had the pleasure of bringing back with me an order for a condensing engine of forty horse-power, required by Mr. John Munn for giving motion to his new flax mill, then under construction. I mention this order because the engine was the first important piece of work executed at the Bridgewater Foundry.

This was my first visit to Ireland. Being so near the
Giant's Causeway, I took the opportunity, on my way homewards, of visiting that object of high geologic interest, together with the magnificent basaltic promontory of Fairhead.
I spent a day in clambering up the terrible-looking crags.
In a stratum of red hematite clay, underneath a solid
basaltic crag of some sixty feet or more in thickness, I
found the charred branches of trees—the remains of some

BRIDGEWATER FOUNDRY. BY ALEXANDER NASMYTH.

forest that had, at some inconceivably remote period, been
destroyed by a vast out-belching flow of molten lava from
a deep-seated volcanic store underneath.

I returned to Patricroft, and found the wooden workshops nearly finished. The machine tools were, for the
most part, fixed and ready for use. In August 1836 the
Bridgewater Foundry was in complete and efficient action.
The engine ordered at Londonderry was at once put in
hand, and the concern was fairly started in its long career
of prosperity. The wooden workshops had been erected
upon the grass. But the sward soon disappeared. The

hum of the driving belts, the whirl of the machinery, the sound of the hammer upon the anvil, gave the place an air of busy activity. As work increased, workmen increased. The workshops were enlarged. Wood gave place to brick. Cottages for the accommodation of the work-people sprang up in the neighbourhood ; and what had once been a quiet grassy field became the centre of a busy population.

It was a source of vast enjoyment to me, while engaged in the anxious business connected with the establishment of the foundry, to be surrounded with so many objects of rural beauty. The site of the works being on the west side of Manchester, we had the benefit of breathing pure air during the greater part of the year. The scenery round about was very attractive. Exercise was a source of health to the mind as well as the body. As it was necessary that I should reside as near as possible to the works, I had plenty of opportunities for enjoying the rural scenery of the neighbourhood. I had the good fortune to become the tenant of a small cottage in the ancient village of Barton, in Cheshire, at the very moderate rental of £15 a year. The cottage was situated on the banks of the river Irwell, and was only about six minutes' walk from the works at Patricroft. It suited my moderate domestic arrangements admirably.

The village was surrounded by apple orchards and gardens, and situated in the midst of tranquil rural scenery. It was a great treat to me, after a long and busy day at the foundry, especially in summer time, to take my leisure walks through the green lanes, and past the many picturesque old farmhouses and cottages which at that time presented subjects of the most tempting kind for the pencil. Such quiet summer evening strolls afforded me the opportunity for tranquil thought. Each day's transactions furnished abundant subjects for consideration. It was a happy period in my life. I was hopeful for the future, and everything had so far prospered with me.

When I had got comfortably settled in my cozy little
cottage, my dear sister Margaret came from Edinburgh to
take charge of my domestic arrangements. By her bright
and cheerful disposition she made the cottage a very happy
home. Although I had neither the means nor the disposition to see much company, I frequently had visits from
some of my kind friends in Manchester. I valued them all
the more for my sister's sake, inasmuch as she had come
from a bright household in Edinburgh, full of cheerfulness,
part of which she transferred to my cottage.

At the same time, it becomes me to say a word or two
about the great kindness which I received from my friends
and well-wishers at Manchester and the neighbourhood.
Amongst these were the three brothers Grant, Benjamin
Hick of Bolton, Edward Lloyd the banker, John Kennedy,
and William Fairbairn. I had not much leisure during the
week days, but occasionally on Sunday afternoons my sister
and myself enjoyed their cordial hospitality. In this way I
was brought into friendly intercourse with the most intelligent and cultivated persons in Lancashire. The remembrance of the delightful evenings I spent in their society
will ever continue one of the cherished recollections of my
early days in Manchester.

I may mention that one of the principal advantages of
the site of my works was its connection with the Liverpool
and Manchester Railway, as well as with the Bridgewater
Canal. There was a stone-edged roadway along the latter,
where the canal barges might receive and deliver traffic in
the most convenient manner. As the wharfage boundary
was the property of the trustees of the Bridgewater Canal,
it was necessary to agree with them as to the rates to
be charged for the requisite accommodation. Their agent
deferred naming the rent until I had finally settled with
Squire Trafford as to the lease of his land, and then, after
he supposed he had got us into a cleft stick, he proposed

so extravagant a rate, that we refused to use the wharf upon his terms.

It happened, fortunately for us, that this agent had involved himself in a Chancery suit with the trustees, which eventually led to his retirement. The property then merged into the hands of Lord Francis Egerton, heir to the Bridge-water Estates. The canal was placed under the manage-ment of that excellent gentleman, James Loch, M.P. Lord Francis Egerton, on his next visit to Worsley Hall, called upon me at the foundry. He expressed his great pleasure at having us as his near neighbours, and as likely to prove such excellent customers of the canal trustees. Be-cause of this latter circumstance, he offered us the use of the wharf free of rent. This was quite in accordance with his generous disposition in all matters. But as we desired the agreement to be put in a regular business-like form, we arranged with Mr. Loch to pay 5s. per an-num as a formal acknowledgment, and an agreement to this effect was accordingly drawn up and signed by both parties.

Lord Francis Egerton was soon after created Earl of Ellesmere. He became one of the most constant visitors at the foundry, in which he always took a lively interest. He delighted to go through the workshops, and enjoy the sight of the active machinery and the work in progress. When he had any specially intelligent visitors at Worsley Hall, which was frequently the case, he was sure to bring them down to the foundry in his beautiful private barge, and lead them through the various departments of the establish-ment. One of his favourite sights was the pouring out of the molten iron into the moulds for the larger class of castings; when some twelve or sixteen tons, by the aid of my screw safety ladle, were decanted with as much neatness and exactness as the pouring out of a glass of wine from a decanter. When this work was performed towards dark, Lord Ellesmere's poetic fancy and artistic eye enabled him to enjoy the sight exceedingly.[1]

I must here say a few words as to my Screw Safety Ladle. I had observed the great danger occasioned to workmen by the method of emptying the molten iron into the casting moulds. The white-hot fluid was run from the melting furnace into a large ladle with one or two cross handles

OLD FOUNDRY LADLE.

and levers, worked by a dozen or fifteen men. The ladle contained many tons of molten iron, and was transferred by a crane to the moulds. To do this required the greatest caution and steadiness. If a stumble took place, and the

---

[1] I had the happiness to receive the kindest and most hospitable attention from Lord Ellesmere and his family. His death, which occurred in 1857, at the early age of fifty-seven, deprived me of one of my warmest friends. The Countess of Ellesmere continued the friendship until her death, which occurred several years later. The same kindly feelings still exist in the children of the lamented pair, all of whom evince the admirable qualities which so peculiarly distinguished their parents, and made them universally beloved by all classes, rich and poor.

ladle was in the slightest degree upset, there was a splash of hot metal on the floor, which, in the recoil, flew against the men's clothes, set them on fire, or occasioned frightful scalds and burns.

To prevent these accidents I invented my Safety Foundry Ladle. I applied a screw wheel, keyed to the trunnion of the ladle, which was acted on by an endless screw attached to the sling of the ladle; and by this means one man could

SAFETY FOUNDRY LADLE.

move the largest ladle on its axis, and pour out its molten contents with the most perfect ease and safety. Not only was all risk of accident thus removed, but the perfection of the casting was secured by the steady continuous flow of the white-hot metal into the mould. The nervous anxiety and confusion that usually attended the pouring of the metal required for the larger class of castings was thus entirely obviated.

At the same time I introduced another improvement in connection with these foundry ladles which, although of minor importance, has in no small degree contributed to the perfection of large castings. This consisted in hanging " the skimmer " to the edge of the ladle, so as to keep back the *scoriæ* that invariably floats on the surface of the melted metal. This was formerly done by hand, and many accidents were the consequence. But now the clear flow of pure metal into the moulds was secured, while the *scoriæ* was mechanically held back. All that the attendant has to do is to regulate the inclination of the Skimmer so as to keep its lower edge sufficiently under the surface of the outflowing metal. The preceding illustrations will enable the reader to understand these simple but important technical improvements.

These inventions were made in 1838. I might have patented them, but preferred to make them over to the public. I sent drawings and descriptions of the Safety Foundry Ladle to all the principal founders both at home and abroad ; and I was soon after much gratified by their cordial expression of its practical value. The ladle is now universally adopted. The Society of Arts of Scotland, to whom I sent drawings and descriptions, did me the honour to present me with their large silver medal in acknowledgment of the invention.

In order to carry on my business with effectiveness it was necessary that I should have some special personal assistance. I could carry on the whole " mechanical " department as regards organisation, designing, and construction ; but there was the " financial " business to be attended to,—the counting-house, the correspondence, and the arrangement of money affairs. I wanted some help with respect to these outer matters.

When I proceeded to take my plot of land at Patricroft some of my friends thought it a very bold stroke, especially for a young man who had been only about three years in business. Nevertheless, there were others who watched my progress with special interest, and were willing to join in my adventure—though adventure it was not.   They were ready to take a financial interest in my affairs.   They did me the compliment of thinking me a *good investment*, by offering to place their capital in my concern as sleeping partners.

But I was already beyond the "sleeping partner" state of affairs.   Whoever joined me must work as energetically as I did, and must give the faculties of his mind to the prosperity of the concern.   I communicated the offers I had received to my highly judicious friend Edward Lloyd. He was always willing to advise me, though I took care never to encroach upon his kindness.   He concurred with my views, and advised me to fight shy of sleeping partners. I therefore continued to look out for a working partner.   In the end I was fortunate.   My friend, Mr. Thomas Jeavons, of Liverpool, having been informed of my desire, made inquiries, and found the man likely to suit me.   He furnished him with a letter of introduction to me, which he presented one day at the works.

The young man became my worthy partner, Holbrook Gaskell.   He had served his time with Yates and Cox, iron merchants, of Liverpool.   Having obtained considerable experience in the commercial details of that business, and being possessed of a moderate amount of capital, he was desirous of joining me, and embarking his fortune with mine.   He was to take charge of the counting-house department, and conduct such portion of the correspondence as did not require any special technical knowledge of mechanical engineering.   The latter must necessarily remain in my hands, because I found that the "off-hand" sketches which I introduced in my letters as explanatory of mechanical designs and suggestions were much more intelligible than any amount of written words.

I was much pleased with the frank and friendly manner

of Mr. Gaskell, and I believe that the feeling between us
was mutual.   With the usual straightforwardness that
prevails in Lancashire, the articles of partnership were at
once drawn up and signed, and the firm of Nasmyth and
Gaskell began.   We continued working together with
hearty zeal for a period of sixteen successive years; and I
believe Mr. Gaskell had no reason to regret his connection
with the Bridgewater Foundry.

The reason of Mr. Gaskell leaving the concern was the
state of his health.   After his long partnership with me, he
was attacked by a serious illness, when his medical adviser
earnestly recommended him to retire from all business
affairs.   This was the cause of his reluctant retirement.
In course of time the alarming symptoms departed, and he
recovered his former health.   He then embarked in an
extensive soda manufactory, in conjunction with one of our
pupils, whose taste for chemistry was more attractive to
him than engine making.   A prosperous business was
established, and at the time I write these lines Mr. Gaskell
continues a hale and healthy man, the possessor of a large
fortune, accumulated by the skilful manner in which he has
conducted his extensive affairs.

CHAPTER XII.

FREE TRADE IN ABILITY——THE STRIKE——DEATH OF MY FATHER.

I HAD no difficulty in obtaining abundance of skilled workmen in South Lancashire and Cheshire. I was in the neighbourhood of Manchester, which forms the centre of a population gifted with mechanical instinct. From an early period the finest sort of mechanical work has been turned out in that part of England. Much of the talent is inherited. It descends from father to son, and develops itself from generation to generation. I may mention one curious circumstance connected with the pedigree of Manchester: that much of the mechanical excellence of its workmen descends from the Norman smiths and armourers introduced into the neighbourhood at the Norman Conquest by Hugo de Lupus, the chief armourer of William the Conqueror, after the battle of Hastings, in 1060.

I was first informed of this circumstance by William Stubbs of Warrington, then maker of the celebrated "Lancashire files." The "P. S.," or Peter Stubbs's files, were so vastly superior to other files, both in the superiority of the steel and in the perfection of the cutting, which long retained its efficiency, that every workman gloried in the possession and use of such durable tools. Being naturally interested in everything connected with tools and mechanics, I was exceedingly anxious to visit the factory where these admirable files were made. I obtained an introduction to William Stubbs, then head of the firm, and was received by him with much cordiality. When I asked him if I might be favoured with a sight of his factory, he replied that he had no factory, as such; and that all he had to do in supplying his large warehouse was to serve out the requisite quantities of the pure cast steel as rods and bars to the workmen; and that they, on their part, forged the metal into files of every description at their own cottage workshops, principally situated in the neighbouring counties of Cheshire and Lancashire.

This information surprised as well as pleased me. Mr. Stubbs proceeded to give me an account of the origin of this peculiar system of cottage manufacture in his neighbourhood. It appears that Hugo de Lupus, William the Conqueror's Master of Arms, the first Earl of Chester, settled in North Cheshire shortly after the Conquest. He occupied Halton Castle, and his workmen resided in Warrington and the adjacent villages of Appleton, Widnes, Prescot, and Cuerdley. There they produced coats of steel, mail armour, and steel and iron weapons, under the direct superintendence of their chief.

The manufacture thus founded continued for many centuries. Although the use of armour was discontinued, these workers in steel and iron still continued famous. The skill that had formerly been employed in forging chain armour and war instruments was devoted to more peaceful purposes. The cottage workmen made the best of files, and steel tools of other kinds. Their talents became hereditary, and the manufacture of wire in all its forms is almost peculiar to Warrington and the neighbourhood. Mr. Stubbs also informed me that most of the workmen's peculiar names for tools and implements were traceable to old Norman-French words. He also stated that at Prescot a peculiar class of workmen has long been established, cele-

brated for their great skill in clock and watchmaking; and
that, in his opinion, they were the direct descendants of a
swarm of workmen from Hugo de Lupus's original Norman
hive of refined metal-workers, dating from the time of the
Conquest.

To return to my narrative. In the midst of such a
habitually industrious population, it will be obvious that
there was no difficulty in finding a sufficient supply of able
workmen. It was for the most part the most steady,
respectable, and well-conducted classes of mechanics who
sought my employment—not only for the good wages they
received, but for the sake of their own health and that of
their families; for it will be remembered that the foundry
and the workmen's dwellings were surrounded by the fresh,
free, open country. In the course of a few years the
locality became a thriving colony of skilled mechanics. In
order to add to the accommodation of the increasing numbers, an additional portion of land, amounting to eight acres,
was leased from Squire Trafford on the same terms as
before. On this land suitable houses and cottages for the
foremen and workmen were erected. At the same time
substantial brick workshops were built in accordance with
my original general plan, to meet the requirements of our
rapidly expanding business, until at length a large and
commodious factory was erected, as shown in the annexed
engraving.

The village of Worsley, the headquarters of the Bridgewater Canal, supplied us with a valuable set of workmen.
They were, in the first place, labourers; but, like all Lancashire men, they were naturally possessed of a quick aptitude for mechanical occupations connected with machinery.
Our chief employment of these so-called labourers was in
transporting heavy castings and parts of machinery from
one place to another. To do this properly required great
care and judgment, in order that the parts might not be

BRIDGEWATER FOUNDRY, PATRICROFT.   FROM A PAINTING BY ALEXANDER NASMYTH.

disturbed, and that the workmen might proceed towards
their completion without any unnecessary delay. None but
those who have had practical acquaintance with the importance of having skilful labourers to perform these apparently
humble, but in reality very important functions, can form
an adequate idea of the value of such services.

All the requisite qualities we required were found in the
Worsley labourers. They had been accustomed to the
heaviest class of work in connection with the Bridgewater
Canal. They had been thoroughly trained in the handling
of all manner of ponderous objects. They performed their
work with energy and willingness. It was quite a treat to
me to look on and observe their rapid and skilful operations
in lifting and transporting ponderous portions of machinery,
in which a vast amount of costly work had been embodied.
After the machines or engines had been finished, it was the
business of the same workmen to remove them from the
workshops to the railway siding alongside the foundry, or
to the boats at the canal wharf. In all these matters the
Worsley men could be thoroughly depended upon.

Where they showed the possession, in any special degree,
of a true mechanical faculty, I was enabled to select from
the working labourers the most effective men to take charge
of the largest and most powerful machine tools—such as
planing machines, lathes, and boring machines. The ease
and rapidity with which they caught up all the technical
arts and manipulations connected with the effective working
of these machines was extraordinary. The results were
entirely satisfactory to myself, and they proved equally
satisfactory to the men themselves by the substantial rise in
their wages which followed their advancement to higher
grades of labour. Thus I had no difficulty in manning my
machine tools by drawing my recruits from this zealous and
energetic class of Worsley labourers. It is by this "selection of the fittest" that the true source of the prosperity of

every large manufacturing establishment depends. I believe that *Free Trade in Ability* has a much closer relation to national prosperity than even Free Trade in Commodities.

But here I came into collision with another class of workmen—those who are of opinion that employers should select for promotion, not those who are the fittest and most skilful, but those who have served a seven years' apprenticeship and are members of a Trades' Union. It seemed to me that this interference with the free selection and promotion of the fittest was at variance with free choice of the best men, and that it was calculated, if carried out, to strike at the root of the chief source of our prosperity. If every workman of the same class went in the same rut, and were paid the same uniform rate of wages, irrespective of his natural or acquired ability, such a system would destroy the emulative spirit which forms the chief basis of manipulative efficiency and practical skill, and on which, in my opinion, the prosperity of our manufacturing establishments mainly depends. But before I proceed to refer to the strike of Unionists, which for a time threatened to destroy, or at all events to impede the spirit of enterprise and the free choice of skilful workmen, in which I desired to conduct the Bridgewater Foundry, I desire to say a few words about the excellent helpers, in the shape of foremen engineers, who zealously helped me in my undertaking from beginning to end.

I must place my most worthy, zealous, and faithful Archy Torry at the top of the list. He rose from being my only workman when I first started in Manchester, to be my chief general foreman. The energy and devotion which he brought to bear upon my interests set a high example to all in my employment. Although he was in some respects deficient in his knowledge of the higher principles of engineering and mechanical construction, I was always ready to supply that defect. His hearty zeal and cheerful temper,

and his energetic movement when among the men, had a
sympathetic influence upon all about him. His voice had
the same sort of influence upon them as the drum and fife
on a soldier's march: it quickened their movements. We
were often called in by our neighbour manufacturers to
repair a breakdown of their engines. That was always a
sad disaster, as all hands were idle until the repair was
effected. Archy was in his glory on such occasions. By
his ready zeal and energy he soon got over the difficulty,
repaired the engines, and set the people to work again. He
became quite famous in these cases of extreme urgency.
He never spared himself, and his example had an excellent
effect upon every workman under him.

Another of my favourite workshop lieutenants was James
Hutton. He had been leading foreman to my worthy friend
George Douglass, of Old Broughton, Edinburgh. He was
fully ten years my senior, and when working at Douglass's I
looked up to him as a man of authority. I had obtained
from him many a valuable wrinkle in mechanical and
technical construction. After I left Edinburgh he had
emigrated to the United States for the purpose of bettering
his condition. But he promised me that if disappointed in
his hopes of settling there, he should be glad to come into
my service if I was in a position to give him employment.
Shortly after my removal to Patricroft, and when everything had been got into full working order, I received a
letter from him in which he said that he was anxious to
return to England, and asking if there was any vacancy in
our establishment that he might be employed to fill up. It
so happened that the foremanship of turners was then
vacant. I informed Hutton of the post; and on his return
to England he was duly enrolled in our staff.

The situation was a very important one, and Hutton
filled it admirably. He was a sound practical man, and
thoroughly knew every department of engineering mechan-

ism.   As I had provided small separate rooms or offices for
every department of the establishment for the use of the
foremen, where they kept their memoranda and special tools,
I had often the pleasure of conferring with Hutton as to
some point of interest, or when I wished to pass my ideas
and designs through the ordeal of his judgment, in order
that I might find out any lurking defect in some proposed
mechanical arrangement.   Before he gave an opinion, Hutton
always took a pinch of snuff to stimulate his intellect, or
rather to give him a little time for consideration.   He would
turn the subject over in his mind.   But I knew that I could
trust his keenness of insight.   He would give his verdict
carefully, shrewdly, and truthfully.   Hutton remained a
faithful and valued servant in the concern for nearly thirty
years, and died at a ripe old age.   Notwithstanding his
mechanical intelligence, Hutton was of too cautious a temperament to have acted as *a general* foreman or manager,
otherwise he would have been elevated to that position.   A
man may be admirable in details, but be wanting in width,
breadth, and largeness of temperament and intellect.   The
man who possesses the latter gifts becomes great in organisation; he soon ceases to be a " hand," and becomes a " head,"
and such men generally rise from the employed to the
employer.

Another of my excellent assistants was John Clerk.   He
had been for a long time in the service of Fairbairn and
Lillie; but having had a serious difference with one of the
foremen, he left their service with excellent recommendations.   I soon after engaged him as foreman of the pattern-
making department.   He was a most able man in some of
the more important branches of mechanical engineering.
He had, besides, an excellent knowledge of building operations.   I found him of great use in superintending the
erection of the additional workshops which were required
in proportion as our business extended.   He made out full-

sized chalk-line drawings from my original pencil sketches, on the large floor of the pattern store, and from these were formed the working drawings for the new buildings. He had a wonderful power of rapidity and clearness in apprehending new subjects, and the way in which he depicted them in large drawings was quite masterly. John Clerk and I spent many an hour on our knees together on the pattern store floor, and the result of our deliberations usually was some substantial addition to the workshops of the foundry, or some extra large and powerful machine tool. This worthy man left our service to become a partner in an engineering concern in Ireland ; and though he richly deserved his promotion, he left us to our very great regret.

The last of our foremen to whom I shall refer was worthy Thomas Crewdson. He entered our service as a smith, in which pursuit he displayed great skill. We soon noted the high order of his natural ability ; promoted him from the ranks, and made him foreman of the smith's and forge-work department. In this he displayed every quality of excellence, not only in seeing to the turning out of the forge work in the highest state of perfection, but in managing the men under his charge with such kind discretion as to maintain the most perfect harmony in the workshops. This is always a matter of great importance—that the foreman should inspire the workmen with his own spirit, and keep up their harmony and activity to the most productive point. Crewdson was so systematic in his use of time that we found that he was able also to undertake the foremanship of the boiler-making department, in addition to that of the smith work ; and to this he was afterwards appointed, with highly satisfactory results to all concerned.

So strongly and clearly impressed is my mind with the recollection of the valuable assistance which I received during my engineering life from those vice-regents of practical management at Patricroft, that I feel that I cannot proceed

further in my narrative without thus placing the merits of
these worthy men upon record.   It was a source of great
good fortune to me to be associated with them, and I consider them to have been among the most important elements
in the prosperity of the Bridgewater Foundry.   There were
many others, in comparatively humble positions, whom I
have also reason to remember with gratitude.   In all well-
conducted concerns the law of "selection of the fittest"
sooner or later comes into happy action, when a loyal and
attached set of men work together harmoniously for their
own advantage as well as for that of their employers.

It was not, however, without some difficulty that we
were allowed to carry out our views as to *Free Trade in
Ability.*   As the buildings were increased more men were
taken on—from Manchester, Bolton, Liverpool, as well as
from more distant places.   We were soon made to feel that
our idea of promoting workmen according to their merits,
and advancing them to improved positions and higher wages
in proportion to their skill, ability, industry, and natural
intelligence, was quite contrary to the views of many of our
new employées.   They took advantage of a large access of
orders for machinery, which they knew had come into the
foundry, to wait upon us suddenly, and to lay down their
Trade Union law for our observance.

The men who waited upon us were deputed by the
Engineer Mechanics Trades' Union to inform us that there
were men in our employment who were not, as they termed
it, "legally entitled to the trade;" that is, they had
never served a regular seven years' apprenticeship.   "These
men," said the delegates, "are filling up the places, and
keeping out of work, the legal hands."   We were accordingly requested to discharge the workmen whom we had
promoted, in order to make room for members of the Trades'
Union.

To have complied with this request would have altered

the whole principles and practice on which we desired to
conduct our business. I wished, and my partner agreed
with me, to stimulate men to steadfast and skilful work by
the hope of promotion. It was thus that I had taken
several of the Worsley men from the rank of labourers, and
raised them to the class of mechanics with correspondingly
higher wages. We were perfectly satisfied with the conduct
of these workmen, and with the productive results of their
labour. We thought it fair to them as well as to ourselves
to resist the order to discharge them, and we consequently
firmly refused to submit to the dictation of the Unionists.

The delegates left us with a distinct intimation that if
we continued to retain the illegal men in our employment
they would call out the Union men, and strike until " the
grievance " was redressed. The Unionists, no doubt, fixed
upon the right time to place their case before us. We
wanted more workmen to execute the advantageous orders
which had come in; and they thought that the strike would
put an entire stop to our operations. On engaging the
workmen we had never up to this time concerned ourselves
with the question of whether they belonged to the Trades'
Union or not. The only proof we required of a man was
his ability. If, after a week's experience, he proved himself
an efficient workman, we engaged him.

The strike took place. All the Union men were " called
out," and left the works. Many of them expressed their
great regret at leaving us, as they were perfectly satisfied
with their employment as well as with their remuneration.
But they were nevertheless compelled to obey the mandate
of the Council. The result was that more than half our
men left us. Those who remained were very zealous.
Nothing could exceed their activity and workfulness. We
appealed to our employers. They were most considerate in
not pressing us for the speedy execution of the work we
had in hand. We made applications in the neighbourhood

for other mechanics in lieu of those who had left us. But the men on strike, under orders from the Union, established pickets round the works, who were only too efficient in preventing those desirous of obtaining employment from getting access to the foundry.

Our position for a time seemed to be hopeless. We could not find workmen enough to fill our shops or to execute our orders. What were we to do under the circumstances ? We could not find mechanics in the neighbourhood ; but might they not be found elsewhere ? Why not bring them from a distance ? We determined to try. Advertisements were inserted in the Scotch newspapers, announcing our want of mechanics, smiths, and foundrymen. We appointed an agent in Edinburgh, to whom applications were to be made. We were soon in receipt of the welcome intelligence that numbers of the best class of mechanics had applied, and that the agent's principal difficulty lay in making the proper selection from amongst them.

A selection was, however, made of over sixty men, who appeared in every respect likely to suit us. With true Scotch caution they deputed two of their number to visit our works and satisfy themselves as to the state of the case. We had great pleasure in receiving these two clear-headed cautious pioneers. We showed them over the workshops, and pointed out the habitations in the neighbourhood with their attractive surroundings. The men returned to their constituents, and gave such a glowing account of their mission that we had no difficulty in obtaining the men we required. Indeed, we might easily have obtained three times the number of efficient mechanics. Sixty-four of the most likely men were eventually selected, men in the zenith of their physical powers. We made arrangements for their conveyance to Glasgow, from whence they started for Liverpool by steamer. They landed in a body at the latter port, many of them accompanied by their wives and children, and eight-day clocks ! A special train was engaged for the conveyance of the whole—men, women, and children, bag and baggage—from Liverpool to Patricroft, where suitable accommodation had been provided for them.

The arrival of so powerful a body of men made a great sensation in the neighbourhood. The men were strong, respectable looking, and well dressed. The pickets were " dumbfoundered." They were brushed to one side by the fresh arrivals. They felt that their game was up, and they suddenly departed. The men were taken over the workshops, with which they appeared quite delighted. They were told to be ready to start next morning at six, after which they departed to their lodgings. The morning arrived and the gallant sixty-four were all present. After allotting to each his special work, they gave three hearty cheers, and dispersed throughout the workshops.

We had no reason to regret the alterations which had been accomplished through the Strike ordered by the Trades' Union. The new men worked with a will. They were energetic, zealous, and skilful. They soon gave evidence of their general handiness and efficiency in all the departments of work in which they were engaged. We were thus enabled to carry out our practice of free trade in ability in our own way, and we were no longer interfered with in our promotion of the workmen who served us the best. In short, we had Scotched the strike ; we conquered the union in their wily attempt to get us under their withering control; and the Bridgewater Foundry resumed its wonted activity in every department.

It was afterwards a great source of happiness to me to walk through the various workshops and observe the cheerful and intelligent countenances of the new men, and to note the energetic skill with which they used their tools in the advancement of their work. General handiness is one of the many valuable results that issues from the practice of handling the variety of materials which are more or less employed in mechanical structures. At the time that I refer to, the skilful workmen employed in the engineering establishments of Scotland (which were then comparatively small in size) were accustomed to use all manner of mechanical tools. They could handle with equally good effect the saw, the plane, the file, and the chisel; and, as occasion required, they could exhibit their skill at the smith's forge with the hammer and the anvil. This was the kind of workmen with which I had reinforced the foundry. The men had been bred to various branches of mechanics. Some had been blacksmiths, others carpenters, stone masons, brass or iron founders; but all of them were *handy* men. They merely adopted the occupation of machine and steam-engine makers because it offered a wider field for the exercise of their skill and energy.

I may here be allowed to remark that we owe the greatest advances in mechanical invention to Free Trade in Ability. If we look carefully into the narratives of the lives of the most remarkable engineers, we shall find that they owed very little to the seven years' rut in which they were trained. They owed everything to innate industry, energy, skill, and opportunity. Thus, Brindley advanced from the position of a millwright to that of a canal engineer; Smeaton and Watt, from being mechanical instrument makers, advanced to higher positions,—the one to be the inventor of the modern lighthouse, the other to be the inventor of the condensing steam-engine. Some of the most celebrated mechanical and civil engineers—such as Rennie, Cubitt, and Fairbairn—were originally millwrights. All these men were many-handed. They had many sides to their intellect. They were resourceful men. They afford the best illustrations of the result of Free Trade in Ability.

The persistent aim at an indolent equality which Union

men aim at, is one of the greatest hindrances to industrial progress. When the Union Delegates called upon me to insist that none but men who had served seven years' apprenticeship should be employed in the works, I told them that I preferred employing a man who had acquired the requisite mechanical skill in two years rather than another who was so stupid as to require seven years' teaching. The delegates regarded this statement as preposterous and heretical. In fact, it was utter high treason. But in the long run we carried our point.

It is true, we had some indenture-bound apprentices. These were pupils who paid premiums. In some cases we could not very well refuse to take them. And yet they caused a great deal of annoyance and disturbance. They were irregular in their attendance, consequently they could not be depended upon for the regular operations of the foundry. They were careless in their work, and set a bad example to the unbound. We endeavoured to check this disturbing element by agreeing that the premium should be payable in six months' portions, and that each party should be free to terminate the connection at the end of each succeeding six months, or at a month's notice from any time. By this means we secured more care and regularity on the part of the pupil apprentices.

But the arrangement which we greatly preferred was to employ intelligent well-conducted young lads, the sons of labourers or mechanics, and advance them by degrees according to their merits. They took charge of the smaller machine tools, by which the minor details of the machines in progress were brought into exact form without having recourse to the untrustworthy and costly process of chipping and filing. A spirit of emulation was excited amongst them. They vied with each other in executing their work with precision. Those who excelled were paid an extra weekly wage. In course of time they took pride, not only in the quantity but in the *quality* of their work ; and in the long run they became skilful mechanics. We were always most prompt to recognise their skill in a substantial manner. There was the most perfect freedom between employer and employed. Every one of these lads was at liberty to leave at the end of each day's work. This arrangement acted as an ever-present check upon master as well as apprentice. The only bond of union between us was mutual interest. The best of them remained in our service because they knew our work and were pleased with the surroundings ; while we on our part were always desirous of retaining the men we had trained, because we knew we could depend upon them. Nothing could have been more satisfactory than the manner in which this system worked.

In May 1835 I had the great happiness of receiving a visit from my dear father. I was then in Dale Street, Manchester, where my floor was overloaded with the work in progress. My father continued to take a great interest in mechanical undertakings, and he was pleased with the prosperity which had followed my settlement in this great manufacturing centre. He could still see his own lathe, driven by steam power, in full operation for the benefit of his son. His fame as an artist was well known in Manchester, for many of his works were possessed by the best men of the town. I had the pleasure of introducing him to the Brothers Grant, John Kennedy, Edward Lloyd, George Murray, James Frazer, William Fairbairn, and Hugh and Joseph Birley, all of whom gave him a most cordial welcome, and invited him to enjoy their hospitality.

In 1838 he visited me again. I had removed to Patricroft, and the Bridgewater Foundry was in full operation. My father was then in his eightieth year. He was still full of life and intellect. He was vastly delighted to witness the rapid progress which I had made since his first visit. He took his daily walk through the busy workshops, where many processes were going on which greatly interested him. He was sufficiently acquainted with the technical details of mechanical work to enjoy the sight, especially when self-acting tools were employed. It was a great source of pleasure to him to have " a crack " with the most intelligent foremen and mechanics. These, on their part, treated him with the most kind and respectful attention. The Scotch workmen regarded him with special veneration. They knew that he had been an intimate friend of Robert Burns, their own best-beloved poet, whose verses shed a charm upon their homes, and were recited by the fireside, in the fields, or at the workman's bench.

They also knew that he had painted the only authentic portrait of their national bard. This fact invested my father with additional interest in their eyes. Their respect for him culminated in a rather extraordinary demonstration. On the last day of his visit the leading Scotch workmen procured "on the sly" an arm-chair, which they fastened to two strong bearing poles. When my father left the works at the bell-ringing at mid-day, he was approached by the workmen, and respectfully requested to "take the chair." He refused; but it was of no use. He was led to the chair, and took it. He was then raised and carried in triumph to my house. He was carefully set down at the little garden-gate, where the men affectionately took leave of him, and ended their cordial good wishes for his safe return home with three hearty cheers. I need scarcely say that my father was greatly affected by this kind demonstration on the part of the workmen.

His life was fast drawing to a close. He had borne the heat and burden of the day; and was about to be taken home like a shock of corn in full season. After a long and happy life, blessed and cheered by a most affectionate wife, he laid down his brushes and went to rest. In his later years he rejoiced in the prosperity of his children, which

was all the more agreeable as it was the result of the
example of industry and perseverance which he had ever
set before them.   My father untiringly continued his professional occupations until 1840, when he had attained the
age of eighty-two.   His later works may be found wanting

ALEXANDER NASMYTH.   AFTER A CAMEO BY SAMUEL JOSEPH.

in that degree of minute finish which characterised his
earlier productions; but in regard to their quality there
was no falling off, even to the last picture which he painted.
The delicate finish was amply compensated by the increase
in general breadth and effectiveness, so that his later works
were even more esteemed by his brother-artists.

The last picture he painted was finished eight days before his death. It was a small work. The subject was a landscape with an autumnal evening effect. There was a picturesque cottage in the middle distance, a rustic bridge over a brook in the foreground, and an old labouring man, followed by his dog, wearily passing over it on his way towards his home. From the chimney of his cottage a thin streak of blue smoke passed upwards through the tranquil evening air. All these incidents suggested the idea, which no doubt he desired to convey, of the tranquil conclusion of his own long and active life, which was then, too evidently, drawing to a close. The shades of evening had come on when he could no longer see to work, and he was obliged to lay down his pencil. My mother was at work with her needle close by him; and when he had finished, he asked her what he should call the picture. Not being ready with an answer, he leant back in his chair, feeling rather faint, and said, " Well, I think I had better call it *Going Home.*" And so it was called.

MONUMENT TO ALEXANDER NASMYTH.

Next morning his strength had so failed him that he could not get up. He remained there for eight days, and then he painlessly and tranquilly passed away. While on his deathbed he expressed the desire that his remains should be placed beside those of a favourite son who had died in early youth. "Let me lie," he said, "beside my dear Alick." His desire was gratified. He was buried beside his son in St. Cuthbert's churchyard, under the grandest portion of the great basaltic rock on which Edinburgh Castle stands. His grave is marked by a fine Runic Cross, admirably sculptured by Rhind of Edinburgh.

One of the kindest letters my mother received after her great loss was one from Sir David Wilkie. It was dated 18th April 1840. "I hasten," he said, "to assure you of my most sincere condolence on your severe affliction, feeling that I can sympathise in the privation you suffer from losing one who was my earliest professional friend, whose art I at all times admired, and whose society and conversation was perhaps the most agreeable that I ever met with.

"He was the founder of the Landscape Painting School of Scotland, and by his taste and talent has for many years taken a lead in the patriotic aim of enriching his native land with the representations of her romantic scenery; and, as the friend and contemporary of Ramsay, of Gavin Hamilton, and the Runcimans, may be said to have been the last remaining link that unites the present with the early dawn of the Scottish School of Art."

I may add that my mother died six years later, in 1846, at the same age as my father, namely eighty-two.

CHAPTER XIII.

MY MARRIAGE——THE STEAM HAMMER.

BEFORE I proceed to narrate the later events of my industrial life, it is necessary to mention, incidentally, an important subject. As it has been the source of my greatest happiness in life, I cannot avoid referring to it.

I may first mention that my earnest and unremitting pursuit of all subjects and occupations, such as I conceived were essential to the acquirement of a sound practical knowledge of my profession, rendered me averse to mixing much in general society. I had accordingly few opportunities of enjoying the society of young ladies. Nevertheless, occasions now and then occurred when bright beings moved before me like meteors. They left impressions on my memory, which in no small degree increased the earnestness of my exertions to press forward in my endeavours to establish myself in business, and thereby to acquire the means of forming a Home of my own.

Many circumstances, however, conspired to delay the ardently longed for condition of my means, such as should induce me to solicit some dear one to complete my existence by her sweet companionship, and enter with me into the most sacred of all the partnerships of life. In course of time I was rewarded with that success which, for the most part, ensues upon all honourable and unremitting business efforts. This cheered me on; although there were still many causes for anxiety, which made me feel that I must not yet solicit some dear heart to forsake the comforts of an affluent home to share with me what I knew must for some years to come be an anxious and trying struggle for comfort and comparative independence. I had reached my thirtieth year before I could venture to think that I had securely entered upon such a course of prosperity as would justify me in taking this the most important step in life.

It may be a trite but not the less true remark that some of the most important events originate in apparently chance occurrences and circumstances, which lead up to results that materially influence and even determine the subsequent course of our lives. I had occasion to make a business journey to Sheffield on the 2d of March 1838, and also to attend to some affairs of a similar character at York. As soon as I had completed my engagement at Sheffield, I had to wait for more than two dreary hours in momentary expectation of the arrival of the coach that was to take me on to York. The coach had been delayed by a deep fall of snow, and was consequently late. When it arrived, I found that there was only one outside place vacant; so I mounted to my seat. It was a very dreary afternoon, and the snow was constantly falling.

As we approached Barnsley I observed, in the remaining murky light of the evening, the blaze of some ironwork furnaces near at hand. On inquiring whose works they were, I was informed that they belonged to Earl Fitzwilliam, and that they were under the management of a Mr. Hartop. The mention of this name, coupled with the sight of the ironworks, brought to my recollection a kind invitation which Mr. Hartop had given me while visiting my workshop in Manchester to order some machine tools, that if I ever happened to be in his neighbourhood, he would be most happy to show me anything that was interesting about the ironworks and colliery machinery under his management.

I at once decided to terminate my dreary ride on the top of the coach. I descended, and, with my small valise in hand I trudged over some trackless snow-covered fields, and made my way by the shortest cut towards the blazing iron furnaces. On reaching them I was informed that Mr. Hartop had just gone to his house, which was about a mile distant. I accordingly made my way thither the best that I could through the deep snow. I met with a cordial welcome, and with the hospitable request that I should take up my quarters there for the night, and have a round of the ironworks and the machinery on the following day. I cheerfully acceded to the kind invitation. I was then introduced to his wife and daughter in a cosy room, where I spent a most pleasant evening. As Mr. Hartop was an enthusiast in all matters relating to mechanism and mechanical engineering subjects generally, we found plenty to converse about; while his wife and daughter, at their needlework, listened to our discussions with earnest and intelligent attention.

On the following day I was taken a round of the ironworks, and inspected their machinery, as well as that of the collieries, in the details of which Mr. Hartop had introduced many common-sense and most effective improvements. All of these interested me, and gave me much pleasure. In the evening we resumed our " cracks " on many subjects of mutual interest. The daughter joined in our conversation with most intelligent remarks; for, although only in her twenty-first year, she had evidently made good use of her time, aided by her clear natural faculties of shrewd observation. Mr. Hartop having met with some serious reverse of fortune, owing to the very unsatisfactory conduct of a partner, had in a manner to begin business life again on his own account; and although he had to reduce his domestic

establishment considerably in consequence, there was in all
its arrangements a degree of neatness and perfect systematic
order, combined with many evidences of elegant taste and
good sense which pervaded the whole, that enhanced in no
small degree the attractiveness of the household.   The chief
of these, however, was to me their daughter Anne!   I soon
perceived in her, most happily and attractively combined,
all the conditions that I could hope for and desire to meet
with in the dear partner of my existence.

As I had soon to proceed on my journey, I took the
opportunity of telling her what I felt and thought, and so
ardently desired in regard to our future intercourse.   What
little I did was to this great purpose ; and, so far as I could
judge, all that I said was received in the best spirit that I
could desire.   I then communicated my hopes and wishes
to the parents.   I explained to them my circumstances,
which happily were then beginning to assume an encouraging prospect, and realising, in a substantial form, a return
for the earnest exertions which I had made towards establishing a home of my own.   They expressed their concur-
rence in the kindest manner ; and it was arranged that
if business continued to progress as favourably as I hoped,
our union should take place in about two years from that
time.

Everything went on hopefully and prosperously.   The
two years that intervened looked very long in some respects,
and very short in others ; for I was always fully occupied,
and labour shortens time.   At length the two years came
to an end.   My betrothed and myself continued of the same
mind.   The happy " chance " event of our meeting on the
evening of the 2d of March 1838 culminated in our marriage
at the village church of Wentworth on the 16th of June
1840—a day of happy memory !   From that day to this
the course of our united hearts and lives has continued to
run on with steady uninterrupted harmony and mutual

happiness. Forty-two years of our married life finds us the same affectionate and devoted " cronies" that we were at the beginning; and there is every prospect that, under God's blessing, we shall continue to be so to the end.

I was present at the opening of the Liverpool and Manchester Railway, on the 15th of September 1830. Every one knows the success of the undertaking. Railways became the rage. They were projected in every possible direction; and when made, locomotives were required to work them. When George Stephenson was engaged in building his first locomotive at Killingworth, he was greatly hampered not only by the want of handy mechanics, but by the want of efficient tools. But he did the best that he could. His genius overcame difficulties. It was immensely to his credit that he should have so successfully completed his engines for the Stockton and Darlington, and afterwards for the Liverpool and Manchester Railway.

Only a few years had passed, and self-acting tools were now enabled to complete, with precision and uniformity, machines that before had been deemed almost impracticable. In proportion to the rapid extension of railways the demand for locomotives became very great. As our machine tools were peculiarly adapted for turning out a large amount of first class work, we directed our attention to this class of business. In the course of about ten years after the opening of the Liverpool and Manchester Railway, we executed considerable orders for locomotives for the London and Southampton, the Manchester and Leeds, and the Gloucester railway companies.

The Great Western Railway Company invited us to tender for twenty of their very ponderous engines. They proposed a very tempting condition of the contract. It was, that if, after a month's trial of the locomotives, their working proved satisfactory, a premium of £100 was to be added to the price of each engine and tender. The locomotives were made and delivered; they ran the stipulated number of test miles between London and Bristol in a perfectly satisfactory manner; and we not only received the premium, but, what was much more encouraging, we received a special letter from the Board of Directors, stating their entire satisfaction with the performance of our engines, and desiring us to refer other contractors to them with respect to the excellence of our workmanship. This testimonial was altogether spontaneous, and proved extremely valuable in other quarters.

I may mention that, in order to effect the prompt and perfect execution of this order, I contrived several special machine tools, which assisted us most materially. These tools for the most part rendered us more independent of mere manual strength and dexterity, while at the same time they increased the accuracy and perfection of the work. They afterwards assisted us in the means of perfecting the production of other classes of work. At the same time they had the important effect of diminishing the cost of production, as was made sufficiently apparent by the balance-sheet prepared at the end of each year.

My connection with the Great Western Company shortly led to a most important event in connection with my own personal history. It appears that the *Great Western* steamship had been very successful in her voyages between Bristol and New York; so much so, indeed, that the directors of the Company ordered the construction of another vessel of much greater magnitude—the *Great Britain*. Mr. Francis Humphries, their engineer, came to Patricroft to consult with me as to the machine tools, of unusual size and power, which were required for the construction of the immense engines of the proposed ship, which were to be made on the vertical trunk principle. Very complete works were erected

at Bristol for the accommodation of the requisite machinery.
The tools were made according to Mr. Humphries's order ;
they were delivered and fitted to his entire approval, and
the construction of the gigantic engines was soon in full
progress.

An unexpected difficulty, however, was encountered with
respect to the enormous wrought-iron intermediate paddle-
shaft.   It was required to be of a size and diameter the like
of which had never been forged.   Mr. Humphries applied to
the largest firms throughout the country for tenders of the
price at which they would execute this important part of
the work, but to his surprise and dismay he found that not
one of them would undertake so large a forging.   In this
dilemma he wrote a letter to me, which I received on the
24th of November 1839, informing me of the unlooked-for
difficulty.   " I find," he said, " that there is not a forge
hammer in England or Scotland powerful enough to forge
the intermediate paddle-shaft of the engines for the *Great
Britain !*   What am I to do ?   Do you think I might dare
to use cast-iron ? "

This letter immediately set me a-thinking.   How was
it that the existing hammers were incapable of forging a
wrought-iron shaft of thirty inches diameter?   Simply
because of their want of compass, of range and fall, as well
as of their want of power of blow.   A few moments' rapid
thought satisfied me that it was by our rigidly adhering
to the old traditional form of a smith's hand hammer—of
which the forge and tilt hammer, although driven by water
or steam power, were mere enlarged modifications—that the
difficulty had arisen; as, whenever the largest forge hammer
was tilted up to its full height, its range was so small that
when a piece of work of considerable size was placed on
the anvil, the hammer became " gagged ; " so that, when
the forging required the most powerful blow, it received
next to no blow at all, as the clear space for the fall of

the hammer was almost entirely occupied by the work on the anvil.

The obvious remedy was to contrive some method by which a ponderous block of iron should be lifted to a sufficient height above the object on which it was desired to strike a blow, and then to let the block full down upon the forging, guiding it in its descent by such simple means as should give the required precision in the percussive action of the falling mass.   Following up this idea, I got out my " Scheme Book," on the pages of which I generally *thought out*, with the aid of pen and pencil, such mechanical adaptations as I had conceived in my mind, and was thereby enabled to render them visible.   I then rapidly sketched out my Steam Hammer, having it all clearly before me in my mind's eye.   In little more than half an hour after receiving Mr. Humphries's letter narrating his unlooked-for difficulty, I had the whole contrivance, in all its executant details, before me in a page of my Scheme Book, a reduced photographed copy of which I append to this description.   The date of this first drawing was the 24th November, 1839.

My Steam Hammer, as thus first sketched, consisted of, first, a massive anvil on which to rest the work ; second, a block of iron constituting the hammer or blow-giving portion ; and, third, an inverted steam cylinder to whose piston-rod the hammer-block was attached.   All that was then required to produce a most effective hammer was simply to admit steam of sufficient pressure into the cylinder, so as to act on the under-side of the piston, and thus to raise the hammer-block attached to the end of the piston-rod.   By a very simple arrangement of a slide valve, under the control of an attendant, the steam was allowed to escape, and thus permit the massive block of iron rapidly to descend by its own gravity upon the work then upon the anvil.

Thus, by the more or less rapid manner in which the

attendant allowed the steam to enter or escape from the
cylinder, any required number or any intensity of blows

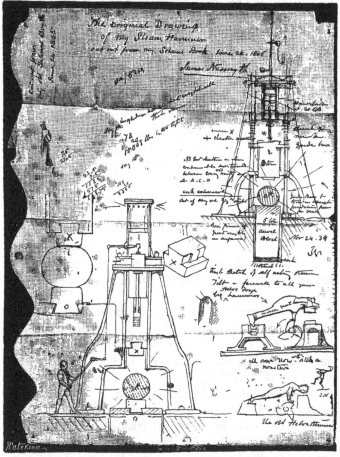

FIRST DRAWING OF STEAM HAMMER, 24TH NOV. 1839.

could be delivered.   Their succession might be modified in
an instant.   The hammer might be arrested and suspended
according to the requirements of the work.   The workman

R

might thus, as it were, *think in blows*. He might deal
them out on to the ponderous glowing mass, and mould or
knead it into the desired form as if it were a lump of clay;
or pat it with gentle taps according to his will, or at the
desire of the forgeman.

Rude and rapidly sketched out as it was, this, my first
delineation of the steam hammer, will be found to comprise
all the essential elements of the invention. Every detail
of the drawing retains to this day the form and arrangement which I gave to it forty-three years ago. I believed
that the steam hammer would prove practically successful;
and I looked forward to its general employment in the
forging of heavy masses of iron. It is no small gratification to me now, when I look over my rude and
hasty first sketch, to find that I hit the mark so exactly,
not only in the general structure but in the details; and
that the invention as I then conceived it and put it into
shape, still retains its form and arrangements intact in
the thousands of steam hammers that are now doing
good service in the mechanical arts throughout the civilised
world.

But to return to my correspondence with the Great
Western Company. I wrote at once to Mr. Humphries,
and sent him a sketch of my proposed steam hammer. I
told him that I felt assured he would now be able to
overcome his difficulty, and that the paddle-shaft of
the *Great Britain* might now be forged. Mr. Humphries
was delighted with my design. He submitted it to Mr.
Brunel, engineer-in-chief of the steamship; to Mr.
Guppy, the managing director; and to other persons interested in the undertaking,—by all of whom it was
heartily approved. I accordingly gave the Company permission to communicate my design to such forge proprietors as might feel disposed to erect the steam hammer,
the only condition that I made being, that in the event of

its being adopted I was to be allowed to supply it in accordance with my design.

But the paddle-shaft of the *Great Britain* was never forged. About that time the substitution of the Screw for the paddle-wheel as a means of propulsion was attracting much attention. The performances of the *Archimedes*, contrived by Mr. Francis P. Smith, were so satisfactory that Mr. Brunel, after he had made an excursion in that vessel, recommended the directors to adopt the new propelling power. After much discussion, they yielded to his strongly-urged advice. The consequence was, that the great engines which Mr. Humphries had so elaborately designed, and which were far advanced in construction, were given up, to his inexpressible regret and mortification, as he had pinned his highest hopes as a practical engineer on the results of their performance. And, to crown his distress, he was ordered to produce fresh designs of engines specially suited for screw propulsion. Mr. Humphries was a man of the most sensitive and sanguine constitution of mind. The labour and the anxiety which he had already undergone, and perhaps the disappointment of his hopes, proved too much for him; and a brain fever carried him off after a few days' illness. There was thus, for a time, an end of the steam hammer required for forging the paddle-shaft of the *Great Britain*.

Very bad times for the iron trade, and for all mechanical undertakings, set in about this time. A wide-spread depression affected all conditions of industry. Although I wrote to the heads of all the great firms, urging the importance of my invention, and forwarding designs of my steam hammer, I was unable to obtain a single order. It is true, they cordially approved of my plan, and were greatly struck by its simplicity, unity, and apparent power. But the substance of their replies was, that they had not sufficient orders to keep the forge hammers they already possessed in work.

They promised, however, that in the event of trade recovering from its depression, they would probably adopt the new power.[1]

In the meantime my invention was taken up in an entirely new and unexpected quarter. I had for some years ,been supplying foreign customers with self-acting machine tools. The principals of continental manufacturing establishments were accustomed to make frequent visits to England for the purpose of purchasing various machine tools required for the production of the ponderous as well as the lighter parts of their machinery. We gave our foreign visitors every facility and opportunity for seeing our own tools at work, and they were often so much pleased that, when they came to order one special tool, they ended by ordering many,—the machine tools in full activity thus acting as their most effective advertisements.[2]

In like manner I freely opened my Scheme Book to any foreign visitors. There I let them see the mechanical thoughts that were passing through my mind, reduced to pen and ink drawings. I did not hesitate to advocate the advantage of my steam hammer over every other method of forging heavy masses of iron; and I pointed out the draw-

---

[1] Among the heads of firms who sent me cordial congratulations on my design, were Benjamin Hick, of the Soho Ironworks, Bolton, a man whose judgment in all matters connected with engineering and mechanical construction was held in the very highest regard; Messrs. Rushton and Eccersley, Bolton Ironworks ; Messrs. Howard and Ravenhill, Rotherhithe Ironworks, London; Messrs. Hawkes, Crashaw, and Company, Newcastle-upon-Tyne ; George Thorneycroft, Wolverhampton; and others.

[2] Some establishments in the same line of business were jealous of the visits of foreigners ; but to our views, restriction in the communication of new ideas on mechanical subjects to foreigners of intelligence and enterprising spirit served no good purpose, as the foreign engineer was certain to obtain all the information he was in quest of from the drawings in the Patent Office, or from the admirable engravings contained in the engineering publications of the day. It was better to derive the advantage of supplying them with the machines they were in quest of, than to wait until the demand was supplied by foreigners themselves.

ing in my Scheme Book in confirmation of my views.   The
book was kept in the office to be handy for such occasions ;
and in many cases it was the means of suggesting ideas of
machine tools to our customers, and thus led to orders which
might not have been obtained without this effective method
of prompting them.   Amongst our foreign visitors was M.
Schneider, proprietor of the great ironworks at Creuzot, in
France.   We had supplied him with various machine tools,
and he was so pleased with their action that the next time
he came to England he called at our office at Patricroft.
M. Bourdon, his mechanical manager, accompanied him.

I happened to be absent on a journey at the time ; but
my partner, Mr. Gaskell, was present.   After showing them
over the works, as an act of courtesy he brought them my
Scheme Book and allowed them to examine it.   He pointed
out the drawing of my Steam Hammer, and told them the
purpose for which it was intended.   They were impressed
with its simplicity and apparent practical utility,—so much
so, that M. Bourdon took careful notes and sketches of the
constructive details of the hammer.

I was informed on my return of the visit of MM.
Schneider and Bourdon, but the circumstance of their having
inspected the designs in my Scheme Book, and especially
my original design of the steam hammer, was regarded
by my partner as too ordinary and trivial an incident of
their visit to be mentioned to me.   The exhibition of my
mechanical designs to visitors at the Foundry was a matter of
almost daily occurrence.   I was, therefore, in entire ignorance
of the fact that these foreign visitors had taken with them to
France a copy of the plan and details of my steam hammer.

It was not until my visit to France in April 1842 that
the upshot of their visit was brought under my notice in an
extraordinary manner.   I was requested by M. Bouchier,
Minister of Marine, to visit the French dockyards and
arsenals for the purpose of conferring with the director of

each with reference to the supply of various machine tools
for the proper equipment of the marine engine factories in
connection with the Royal Dockyards.   In order to render
this journey more effective and instructive, I visited most
of the French engineering establishments which had been
supplied with machine tools by our firm.   Amongst these
was of course the famous firm of Schneider, whose works
at Creuzot lay not far out of the way of my return journey.[1]
I accordingly made my way thither, and found M. Bourdon
at his post, though M. Schneider was absent.

M. Bourdon received me with much cordiality.   As he
spoke English with fluency I was fortunate in finding him
present, in order to show me over the works ; on entering
which, one of the things that particularly struck me was the
excellence of a large wrought-iron marine engine single crank,
forged with a remarkable degree of exactness in its general
form.   I observed also that the large eye of the crank had
been punched and drifted with extraordinary smoothness and
truth.   I inquired of M. Bourdon "how that crank had been
forged ?"   His immediate reply was, "*It was forged by your
steam hammer !*"

Great was my surprise and pleasure at hearing this statement.   I asked him how he had come to be acquainted
with my steam hammer ?   He then narrated the circumstance of his visit to the Bridgewater Foundry during my
absence.   He told me of my partner having exhibited to
him the original design, and how much he was struck by
its simplicity and probable efficiency ; that he had taken
careful notes and sketches on the spot ; that among the
first things he did after his return to Creuzot was to put in
hand the necessary work for the erection of a steam hammer ;
and that the results had in all respects realised the high
expectations he had formed of it.

M. Bourdon conducted me to the forge department of

---

[1] The particulars of this journey are referred to in a future Chapter.

the works, that I might, as he said, "*see my own child;*" and there it was, in truth—a thumping child of my brain. Until then it had only existed in my Scheme Book; and yet it had often and often been before my mind's eye in full action.   On inspecting the steam hammer I found that Bourdon had omitted some important details, which had led to a few mishaps, especially with respect to the frequent breaking of the piston-rod at its junction with the hammer block.   He had effected this, in the usual way, by means of a cutter wedge through the rod; but he told me that it often broke through the severe jar during the action of the hammer.   I sketched for him, then and there, in full size on a board, the elastic packing under the end of the piston-rod, which acted, as I told him, like the cartilage between the bones of the vertebræ, preventing the destructive effects of violent jars.   I also communicated to him a few other important details, which he had missed in his hasty inspection of my design.   Indeed, I felt great pleasure in doing so, as I found Bourdon to be a most intelligent mechanic, and thoroughly able to appreciate the practical value of the information I communicated to him.   He expressed his obligation to me in the warmest terms, and the alterations which he shortly afterwards effected in the steam hammer, in accordance with my plans, enabled it to accomplish everything that he could desire.

I had not yet taken out a patent for the Steam Hammer. The reason was this.   The cost of a patent, at the time I invented it, was little short of £500, all expenses included. My partner was unwilling to lay out so large a sum upon an invention for which there seemed to be so little demand at that time; and I myself had the whole of my capital embarked in the concern.   Besides, the general depression still continued in the iron trade; and we had use for every farthing of money we possessed.   I had been warned of the risk I ran by freely exhibiting my original design, as well as by sending drawings of it to those who I thought were most likely to bring the invention into use. But nothing had as yet been done in England. It was left for France, as I have described, to embody my invention in an actual steam hammer.

I now became alarmed, and feared lest I should lose the benefits of my invention. As my partner declined to help me, I applied to my brother-in-law, William Bennett. He was a practical engineer, and had expressed himself as highly satisfied with its value. He had also many times cautioned me against "publishing" its advantages so widely, without having first protected it by a patent. He was therefore quite ready to come to my assistance. He helped me with the necessary money, and the invention was placed in a position of safety so far as my interests were concerned. In return for his kindness I stipulated that the reimbursement of his loan should be a first charge upon any profits arising from the manufacture of the steam hammer; and also that he should have a share in the profits during the period of the patent rights. Mr. Bennett lived for many years, rejoicing in the results of his kindness to me in the time of my difficulty. I may add that the patent was secured in June 1842, or less than two months after my return from France.

Soon after this, the iron trade recovered from its depression. The tide of financial prosperity of the Bridgewater Foundry soon set in, and my partner's sanguine confidence in my ability to raise it to the condition of a thriving and prosperous concern was justified in a most substantial manner. In order to make the most effective demonstration of the powers and capabilities of my steam hammer, I constructed one of 30 cwt. of hammer block, with a clear four feet range of fall. I soon had it set to work; and its energetic services helped us greatly in our smith and forge work. It was admired by all observers.

People came from a distance to see it. Mechanics and iron-founders wondered at the new power which had been born. The precision and beauty of its action seemed marvellous. The attendant could, by means of the steam slide-valve lever in his hand, transmit his will to the action of the hammer, and thus *think* in blows. The machine combined great power with gentleness. The hammer could be made to give so gentle a blow as to crack the end of an egg placed in a wine glass on the anvil; whilst the next blow would shake the parish,[1] or be instantly arrested in its descent midway.

Hand-gear was the original system introduced in working the hammer. A method of self-acting was afterwards added. In 1843, I admitted steam above the piston, to aid gravitation. This was an important improvement. The self-acting arrangement was eventually done away with, and hand-gear again became all but universal. Sir John Anderson, in his admirable *Report on the Vienna Exhibition of* 1873, says: "The most remarkable features of the Nasmyth hammers were the almost entire abandonment of the old self-acting motion of the early hammers and the substitution of new devices, and in the use of hand-gear only in all attempts to show off the working. There is no real saving, as a general rule, by the self-acting arrangement, because one attendant is required in either case, and on the other hand there is frequently a positive loss in the effect of the blow. By hand-working, with steam on top of piston, the full force can be more readily maintained until the blow is fully delivered; it is thus more of a 'dead blow' than was formerly the case with the other system."

---

[1] This is no mere figure of speech. I have heard the teacups rattle in the cupboard in my house a quarter of a mile from the place where the hammer was at work. I was afterwards informed that the blows of my great steam hammer at Woolwich Arsenal were sensibly felt at Greenwich Observatory, about two miles distant.

There was no want of orders when the valuable qualities
of the steam hammer came to be seen and experienced. The
first order came from Rushton and Eckersley of Bolton, who,
by the way, had seen the first copy of my original design a
few years before. The steam hammer I made for them was
more powerful than my own. The hammer block was of
five tons weight, and had a clear fall of five feet. It gave
every satisfaction, and the fame of its performances went
abroad amongst the ironworkers. The Lowmoor Ironworks
Company followed suit with an order for one of the same
size and power; and another came from Hawkes and Co. of
Newcastle-upon-Tyne.

One of·the most important uses of the steam hammer
was in forging anchors. Under the old system, anchors—
upon the soundness of which the safety of ships so often
depends—were forged upon the " bit by bit " system. The
various pieces of an anchor were welded together, but at the
parts where the different pieces of iron were welded together,
flaws often occurred; the parts would break off—blades from
the stock, or flukes from the blades—and leave the vessel,
which relied upon the security of its anchor, to the risk of
the winds and the waves. By means of the steam hammer
these·risks were averted. The slag was driven out during
the hammering process. The anchor was sound throughout
because it was welded as a whole.

Those who are technically acquainted with smith work
as it used to be practised, by what I term the " bit by bit "
system—that is, of building up from many separate parts
of iron, afterwards welded together into the required form
—can appreciate the vast practical value of the Die method
brought into general use by the controllable but immense
power of the steam hammer. At a very early period of
my employment of the steam hammer, I introduced the
system of stamping masses of welding hot iron as if it
had been clay, and forcing it into suitable moulds or dies

placed upon the anvil. This practice had been in use on a small scale in the Birmingham gun trade. The ironwork of fire-arms was thus stamped into exact form. But, until we possessed the wide range and perfectly controllable powers of the steam hammer, the stamping system was confined to comparatively small portions of forge work. The new power enabled the die and stamp system to be applied to the largest class of forge work; and another era in the working of ponderous masses of smith and forge work commenced, and has rapidly extended until the present time. Without entering into further details, the steam hammer has advanced the mechanical arts, especially with relation to machinery of the larger class, to an extent that is of incalculable importance.

Soon after my steam hammer had exhibited its merits as a powerful and docile agent in percussive force, and shown its applicability to some of the most important branches of iron manufacture, I had the opportunity of securing a patent for it in the United States. This was through the kind agency of my excellent friend and solicitor, the late George Humphries of Manchester. Mr. Humphries was a native of Philadelphia, and the intimate friend of Samuel Vaughan Merrick, founder of the eminent engineering firm of that city. Through his instrumentality I forwarded to Mr. Merrick all the requisite documents to enable a patent to be secured at the United States Patent Office at Washington. I transferred the patent to Mr. Merrick in order that it might be worked to our mutual advantage. My invention was thus introduced into America under the most favourable auspices. The steam hammer soon found its way into the principal ironworks of the country. The admirable straightforward manner in which our American agent conducted the business from first to last will ever command my grateful remembrance.

CHAPTER XIV.

TRAVELS IN FRANCE AND ITALY.

I HAVE already referred to my visit to Creuzot, in France. I must explain how it was that I was induced to travel abroad. The French Government had ordered from our firm some powerful machine tools, which were manufactured, delivered, and found to give every satisfaction. Shortly after, I received a letter from M. Bouchier, the Minister of Marine, inviting me to make a personal visit to the French naval arsenals for the purpose of conferring with the directing officials as to the mechanical equipments of their respective workshops.

I accordingly proceeded to Paris, and was received most cordially by the Minister of Marine. After conferring with him, I was furnished with letters of introduction to the directing officers at Cherbourg, Brest, Rochefort, Indret, and Toulon. While in Paris I visited some of the principal manufacturing establishments, the proprietors of which had done business with our firm. I also visited Arago at the Observatory, and saw his fine array of astronomical instruments. The magnificent collections of antiquities at the Louvre and Hotel Cluny occupied two days out of the four I spent in Paris; after which I proceeded on my mission. Rouen lay in my way, and I could not fail to stay there and indulge my love for Gothic architecture. I visited the magni-

ficent Cathedral and the Church of St. Ouen, so exquisite
in its beauty, together with the refined Gothic architectural
remains scattered about in that interesting and picturesque
city.   I was delighted beyond measure with all that I saw.
With an eye to business, however, I paid a visit to the works
which had been established by the late Joseph Locke in the
neighbourhood of Rouen for the supply of locomotives to the
Havre, Rouen, and Paris Railway.   The works were then
under the direction of Mr. Buddicom.

I went onward through Caen to Bayeux.   There I rested
for a few hours for the purpose of visiting the superb Norman Cathedral, and also to inspect the celebrated Bayeux
tapestry.   I saw the needlework of Queen Matilda and her
handmaidens, which so graphically commemorates the history
of the Norman Conquest.   In the evening I reached Cherbourg.   I was cordially received by the directing officer of
the dockyard, which is of very large extent and surrounded
by fortifications.   My business was with the smithy or *atélier
des forges*, and the workshops or *atéliers des machines*.   There
I recognised many of the machine-tools manufactured at the
Bridgewater Foundry, doing excellent work.

My next visit was to Brest, the chief naval arsenal of
France.   It combines a dockyard, arsenal, and fortress of the
first class.   Everything has been done to make the place
impregnable.   The harbour is situated on the north side of
one of the finest havens in the world, and is almost land-
locked.   Around the harbour run quays of great extent,
alongside of which the largest ships can lie—five artificial
basins being excavated out of the solid rock.   The whole
of the harbour is defended by tier above tier of batteries.
Foreigners are not permitted to enter the dockyard without
special permission; but as I was armed with my letter of
introduction from the Minister of Marine, I was admitted
and cordially received, as at Cherbourg.   I went through the
Government foundry and steam-factory, for which I had

supplied many of my machine tools. I found the establishment to be the largest and most complete that I had seen.

From Brest I went to Rochefort, an excellent naval arsenal, though much smaller than those at Cherbourg and Brest. Next, to Indret on the Loire. Here is the large factory where marine engines are made for the royal steamers. The works were superintended by M. Rosine, a most able man. I was so much pleased with him that I spent two days in his society. I have rarely met with a more perfect union of the sound practical mechanic, of strong common sense, and yet with a vivid imagination, which threw a light upon every subject that he touched. It was delightful to see the perfect manner in which he had arranged all the details of the engine factory under his superintendence, and to observe the pride which he took in the accuracy of the work turned out by his excellent machinery. It was a treat to see the magnificent and intricate iron castings produced there.

As M. Rosine spoke English fluently, we had discussions on a vast variety of topics, not only relating to technical subjects, but on other matters relating to art and mechanical drawing. He was one of the few men I have met who had in perfection the happy accomplishment of sketching with true artistic spirit any object that he desired to bring before you. His pencil far outstripped language in conveying distinct ideas on constructive and material objects. The time that I spent in the company of this most interesting man will ever remain vivid in my memory. It grieved me greatly to hear of his premature death about two years after the date of my visit. He must have been a sad loss to his deeply attached friends,[1] as well as to the nation whom he so faithfully served.

---

[1] The only man I ever met, to whom I might compare Rosine, was my lamented friend Francis Humphries, engineer of the Great Western Steamship Company. Both were men of the same type, though Rosine was several octaves higher in the compass and vividness of his intellect.

On my way to Toulon I passed through Bordeaux, and by Avignon to Nismes. At the latter city I was delighted by the sight of the exquisite Roman temple, the *Maison Carrée*. It is almost perfect. But the most interesting of the Roman remains at Nismes is the magnificent Amphitheatre. In viewing this grand specimen of architecture, as well as the old temples, cathedrals, and castles, I felt that we moderns are comparative pigmies. Our architecture wants breadth, grandeur, sublimity.

It appears to me that one of the chief causes of the inferiority and defects of Modern Architecture is, that our designers are so anxious to display their taste in ornamentation. They first design the exterior, and then fit the interiors of their buildings into it. The *purpose* of the building is thus regarded as a secondary consideration. In short, they *utilise ornament* instead of *ornamenting utility*—a total inversion, as it appears to me, of the fundamental principle which ought to govern all classes of architectural structures. This is, unfortunately, too evident in most of our public buildings.

One thing I was especially struck with at Nismes—the ease with which some thousands of people might issue, without hindrance, from the Amphitheatre. The wedge-shaped passages radiate from the centre, and, widening outwards, would facilitate the egress of an immense crowd. Contrast this with the difficulty of getting out of any modern theatre or church in case of alarm or fire. Another thing is remarkable—the care with which the huge blocks of magnesian limestone[1] have been selected. Some of the stone slabs are eighteen feet long; they roof over the corridors; yet they still retain the marks of the Roman chisel. Every individual chip is as crisp as on the day on which it was made; even

---

[1] I believe *Dolomite* is the proper geological term. This fine material abounds in this part of France, and has materially contributed to the durability of the Roman mason work.

the delicate " scribe " marks, by which the mason, some.1900
years ago lined out his work on the blocks of stone he was
about to chip into its required form, are still perfectly distinct.

This wonderfully durable stone is of the same material as
that employed by lithographers.   Though magnesian, it is of
a different quality from that employed in building our Houses
of Parliament.   As this was carefully selected, the latter
was carelessly *un*selected.   Most probably it was the result
of a job.   It was quarried at random, in the most ignorant
way; some of it proved little better than chalk; and though
all sorts of nostrums have been tried, nothing will cure the
radical defect.   This, however, is a wide digression from my
subject of the admirable mason work, and the wonderful
skill and forethought employed in erecting that superb arena
and the other Roman buildings at Nismes.

I proceeded to Marseilles, where I had some business to
transact with Philip Taylor and Company, the engineering
firm.   They were most kind and attentive to me while there,
and greatly added to the enjoyment of my visit to that remarkable city.   From Marseilles I proceeded to Toulon, the last
of the marine dockyards I had to visit.   There was no railway between the places at that time, and it was accordingly
necessary that I should drive along the usual road.   In the
course of my journey to Toulon I went through the Pass
of Col d'Ollioulles.   It was awfully impressive.   The Pass
appeared to consist of a mighty cleft between two mountains;
made during some convulsion of Nature.   There was only
room for the carriage road to pass between the cliffs.   The
ruins of a Saracenic castle stood on the heights to guard the
passage.   It was certainly the most romantic scene I had
ever beheld.

Looking down into the deep cleft below me, at the
bottom of which ran a turbulent stream, I saw the narrow
road along which our carriage was to pass.   And then suddenly I emerged in full sight of the Mediterranean, with

the calm blue heavens resting over the deep blue sea. There were palms, cactuses, and orange trees, mixed with olive groves. The fields were full of tulips and narcissuses, and the rocks by the roadside were covered with boxwood and lavender. Everything gave evidence of the sunny South. I had got a glimpse of the Mediterranean a few days before; but now I saw it in its glory.

I arrived in due time at Toulon. The town is not very striking in itself. It is surrounded by an amphitheatre of mountains of hard magnesian limestone. These are almost devoid of vegetation. This it is which gives so peculiar an arid aspect to this part of the coast. Facing the south, the sun's rays reflected from the bare surface of the rocks, places one at mid-day as if in the focus of a great burning mirror, and sends every one in quest of shade. This intense temperature has its due effect upon the workers in the dockyard. I found the place far inferior to the others which I had visited. The heat seemed to engender a sort of listlessness over the entire place. The people seemed to be falling asleep. Though we complain of cold in our northern hemisphere, it is a great incentive to work. Even our east wind is an invigorator; it braces us up, and strengthens our nerves and muscles.

It is quite possible that the workmen of the Toulon dockyard might fire up and work with energy provided an occasion arose to call forth their dormant energy. But without the aid of an almost universal introduction of self-acting machinery in this sleepy establishment, to break, with the busy hum of active working machinery, the spell of indolence that seemed to pervade it, there appeared to me no hope of anything like continuous and effective industry or useful results. The docks looked like one vast knacker's yard of broken-down obsolete ships and wretched old paraphernalia—unfortunately a characteristic of other establishments nearer home than Toulon.

s

After transacting my business with the directing officers
of this vast dockyard ,I returned to Marseilles.   There I
found letters requiring me to proceed to Naples, in order to
complete some business arrangements in that city.   I was
exceedingly rejoiced to have an opportunity of visiting the
south of Italy.   I set out at once.   A fine new steamer of
the Messageries Impériales, the *Ercolana*, was ready to
sail from the harbour.   I took my place on board.   I found
that the engines had been made by Maudsley Sons and
Field; they were of their latest improved double-cylinder
construction.   When I went down into the engine-room I
felt myself in a sense at home; for the style of the engines
brought to my mind many a pleasant remembrance of the
days gone by.

We steamed out of the harbour, and passed in succession
the beautiful little islands which gem the bay of Marseilles.
Amongst others, the isle of If, crowned by its castle, once a
State prison, and the Chateau d'If, immortalised by Dumas.
Then Pomègne, Ratoneau, and other islands.   We were now
on the deep blue Mediterranean, watching the graceful curves
of the coast as we steamed along.   Soon after, we came in
sight of the snow-capped maritime Alps behind Nice.   The
evening was calm and clear, and a bright moon shone overhead.   Next morning I awoke in the harbour of Genoa,
with a splendid panoramic view of the city before me.   I
shall never forget the glorious sight of that clear bright
morning as long as I live.

As the steamer was to remain in the harbour until two
o'clock, I landed with the other passengers and saw the
wonders of the city.   I felt as if I were in a new world.
On every side and all around me were objects of art lighted
up by glorious sunshine.   The picturesque narrow streets,
with the blue sky overhead and the bright sunshine lighting
up the beautiful architecture of the palatial houses, relieved
by masses of clear shade, together with the picturesque

dresses of the people, and the baskets of oranges and lemons
with the leaves on the boughs on which they had been born
and reared, the brilliant greenery of the inner courts into
which you peeped while passing along the Strada Nuova,
literally a street of palaces, threw me into a fervency of
delight. Here, indeed, was architecture to be proud of—
grand, imposing, and massive—chastely yet gloriously ornamented. There was nothing of the gingerbread order here

The plan of these palaces is admirable. They are open
to the street, so that all the inner arrangements may be
seen. There is the court, surrounded by arcades, the arches
of which rest upon columns ; the flights of marble steps on
each side, leading to the great hall or to the principal apartments ; and inside the court, the pink daphnes and Tan-
gerine orange trees, surrounded by greenery, with which
the splendour of the marble admirably contrasts ; — the
whole producing a magnificent effect. I remembered that
Genoa *la superba* was one of my father's pet subjects when
talking of his first visit to Italy ; and now I could confirm
all that he had said about the splendour of its palaces.

I do not know of anything more delightful than to grope
one's way through a foreign city, especially such a city as
Genoa, and come unexpectedly upon some building that
one has heard of—that has dimly lived in the mind like a
dream—and now to see it realised in fact. It suddenly
starts into life, as it were, surrounded by its natural associations. I hate your professional guides and their constant
chatter. Much better to come with a mind prepared with
some history to fall back upon, and thus be enabled to
compare the present with the past, the living with the
dead.

I climbed up some of the hills surrounding Genoa—for
it is a city of ups and downs. I wandered about the
terraced palaces surrounded by orange groves, and surveyed
the fortified heights by which the place is surrounded.

What exquisite bits of scenery there were to sketch; what a rich combination of nature and art! And what a world of colour, with the clear blue sea in the distance! Altogether, that one day at Genoa—though but a succession of glimpses —formed a bright spot in my life, that neither time nor distance can dim or tarnish.

I returned to the harbour two hours before the steamer was to leave. To commemorate my visit, I mounted the top of the paddle-box, took out my sketch book, and made a panoramic view of Genoa as seen from the harbour. I did it in pencil at the time, and afterwards filled it up with ink. When the pages of the sketch book had been joined together the panoramic view extended to about eight feet long. The accuracy of the detail, as well as the speed with which the drawing was done, were perhaps rather creditable to the draughtsman—at least so my artistic friends were pleased to tell me. Indeed, many years after, a friend at court desired to submit it to the highest Lady in the land, and, being herself an artist, she expressed herself as highly gratified with the performance.

The next station the steamer touched at was Leghorn. As the vessel was not to start until next day, there was sufficient time for me to run up to Pisa. There I spent a delightful day, principally in wandering about that glorious group of buildings situated so near to each other—the Cathedral, the Baptistery, the Campo Santo, and the Campanile or Leaning Tower. What interested me most at the Cathedral was the fine bronze lamp suspended at the end of the nave, which initiated in the mind of Galileo the invention of the pendulum. Thousands had seen the lamp swinging before him, but he alone *would* know "the reason why." Then followed the discovery which paved the way for Newton's law of gravitation—one of the grandest laws of the universe. Some of the finest works of Andrea del Sarto, son of the Tailor, are found here. Indeed, the

works of that great painter are little known out of Pisa and
Florence. I was reluctant to tear myself away from Pisa;
but the *Ercolina* could not wait, and I was back in good
time, and soon under weigh.

The next port we touched at was Civita Vecchia, one of
the most dreary places that can be
imagined, though at one time an
Etruscan city, and afterwards the
port of Trajan. I did not land,
as there were some difficulties in
the way of passports. We steamed
on; and next morning when I
awoke we were passing the coast
of Ischia. We could scarcely see
the island, for a thick mist had
overspread the sea. Naples was
still hidden from our sight, but
over the mist I could observe the
summit of Vesuvius vomiting forth
dense clouds of white smoke. The
black summit of the crater appeared floating in the clear blue
sky. But the heat of the sun
shortly warmed the mist, and it
floated away like a curtain.

A MONK ON BOARD.

A grand panorama then lay
before us. Naples looked bright and magnificent under the
sunlight. The sea was so smooth that the buildings and
towers and convents and spires were reflected in the water.
On our left lay the Bay of Baiæ, with its castles and temples
and baths, dating from the days of the Roman Republic.
To the right lay Castellamare, Sorrento, and the island of
Capri. But the most prominent object was Vesuvius in
front, with its expanding cloud of white smoke over the
landscape.

On landing, I took up my quarters at the Hotel Victoria.
I sallied forth to take my first hasty view of the Chiaia, the
streets, and the principal buildings.   But, in accordance with
my motto of " Duty *first*, pleasure *second*," I proceeded to
attend to the business respecting which I had visited Naples.

DISTANT VIEW OF VESUVIUS.

That, however, was soon disposed of.   In a few days I was
able to attend to pleasure.   I made my way to the Museo
Borbonico, now called the National Museum.   I found it a
rich mine of precious treasures, consisting of Greek, Etruscan,
and Roman antiquities of every description.   Not the least
interesting part of the Museum is the collection of marbles,

pictures, and articles of daily use, dug up from the ruins of
the buried city of Pompeii. Every spare hour that I could
command was occupied in visiting and revisiting this wonderful Museum.

Herculaneum and Pompeii were also visited, but, more
than all, the crater of Vesuvius. During my visit the mountain was in its normal state. I mounted the volcanic ashes
with which it is strewn, and got to the top. There I could
look down into the pit from which the clouds of steam are
vomited forth. I went down to the very edge of the crater,
stood close to its mouth, and watched the intermittent up-
rushing of the blasts of vapour and sulphureous gases. To
keep clear of these I stood to the windward side, and was
thus out of harm's way.

What struck me most was the wonderfully brilliant
colours of the rugged lava rocks forming the precipitous
cliffs of the interior walls of the crater. These brilliant
colours were the result of the sublimation and condensation
on their surfaces of the combinations of sulphur and chloride
of iron, quite as bright as if they had been painted with bright
red, chrome, and all the most brilliant tints. Columns of
all manner of chemical vapours ascended from the clefts and
deep cracks, at the bottom of which I clearly saw the bright
hot lava.

I rolled as big a mass of cool lava as I could, to the
edge of the crater and heaved it down; but I heard no
sound. Doubtless the depth was vast, or it might probably
have fallen into the molten lava, and thus make no noise.
On leaving this horrible pit edge, I tied the card of the
Bridgewater Foundry to a bit of lava and threw it in, as a
token of respectful civility to Vulcan, the head of our craft.

I had considerably more difficulty in clambering up to
the top edge of the crater than I had in coming down. Once
or twice, indeed, I was half choked by the swirls of sulphureous and muriatic acid vapour that environed me before I

could reach the upper edge.   I sat down in a nook, though
it was a very hot one, and made a sketch or two of the
appearance of the crater, which may perhaps interest my
readers.   But I feel that it is quite beyond my power
either by pen or pencil, to convey an idea of the weird
unearthly aspect which the funnel-shaped crater of Vesuvius
presented at that time.   An eruption of unusual violence
had occurred shortly before I saw it.   Great rounded blocks
of lava had been thrown high into the air again and again,

SKETCH OF THE CRATER.

and had fallen back into the terrible focus of volcanic
violence.   Vast portions of the rugged and precipitous sides
of the crater had fallen in, and were left in a state of the
wildest confusion.   When I visited the place the eruption
had comparatively subsided.   The throat of the crater was
a rugged opening of more than forty feet diameter, leading
down to—Where ?   Echo answers, " Where ? "
    And yet there is no doubt but that the great mass of
materials which lay around me as I made my sketch, had
been shot up from inconceivable depths beneath the solid
crust of the earth.   There still remains an enormous mass
of molten materials that has been shut up beneath that

crust, since the surface of the globe assumed its present condition. The mineral matter had converged towards its centre of gravity, and the arrestment of the momentum of the coalescing particles resulted in intense heat, and the molten lava of the volcano.

This seems to me to be the true origin of volcanic heat. It has played a great part in the physical history of the globe. Volcanic action has been, as it were, the universal plough! It has given us mountains, hills, and valleys. It has given us picturesque scenery, gorges, precipices, waterfalls. The upheaving agent has displayed the mineral treasures of the earth, and enabled man, by intelligent industry, to use them as mines of material blessings. This is indeed a great and sublime subject.

I had remained near the mouth of the crater for about five hours. Evening was approaching. My drawings were finished, and I prepared to leave. My descent from the summit of the crater edge was comparatively rapid, though every footstep went down some fifteen inches through the volcanic ashes. I descended by the eastern side, and was soon at the base of the great cone. I made my way by tortuous walking round the erupted masses of lava, and also by portions of the lava streams, which, on losing their original fluidity, had become piled up and contorted into gigantic masses.

At the extreme edges of the flow, where the lava had become viscid, these folds and contortions were very remarkable. They were piled fold over fold,—the result of the mighty pressure from behind. It was sad to see so many olive gardens burnt and destroyed ; the trees were as black as charcoal. It is singular to see the numbers of orange and olive growers who choose to live so near to the " fiery element." But the heat presses forward the growth of vegetation. To be there is like living in a hothouse ; and the soil is extraordinarily fertile. Hence the number of vine-

yards quite close to the base of Vesuvius. The cultivators
endeavour to enclose their gardens with hard masses of lava,
so as to turn off the flow of the molten streams in other
directions; but the lava bursts through the walls again and
again, and the gardens are often utterly burnt up and
ruined.   Almost every field at the base of Vesuvius
contains a neat little oratory, with a statue of the Virgin
and Child, to which the cultivators repair in times of peril
and calamity.   But chapel, statue, and gardens are alike
swept away by the tremendous descent of the molten lava.

As the night was growing dark, I made my way from
these riskful farms to Rosina, a little village on the way back
to Naples.   As I had had nothing to eat or drink during this
thirst-producing journey, I went into a wine shop and asked
for some refreshment.   The wine shop was a sort of vault,
with a door like that of a coach-house, but with a bench
and narrow table.   The good woman brought me a great
green glass bottle like a vitriol carboy!   It contained more
than six gallons of wine, and she left me with a big glass to
satisfy my wants.   The wine was the veritable *Lachryma
Christi*—a delightful light claret—for producing which the
vineyards at the base of Vesuvius are famous.   After some
most glorious swigs from this generous and jovial carboy,
accompanied with some delightful fresh-made bread, I felt
myself up to anything.   After washing down the dust that
I had swallowed during the day, I settled with my liberal
landlady (indeed she was mightily pleased with only ten-
pence), and started for Naples.

I had still an eight-mile walk before me, but that was
nothing to my vigorous powers at that time.   The moon had
risen during my stay in the wine house, and it shone with a
bright clear light.   After a few miles walking I felt a little
tired, for the day's exercise had been rather toilsome.   A
fine carriage passed me on the road with a most tempting
platform behind.   I hailed the driver, and was allowed to

mount.   I was soon bowling along the lava-paved road, and in a short time I arrived at Naples.   I made another excursion to the crater of Vesuvius before I left, as well as visits to Herculaneum and Pompeii, which exceedingly interested me.   But these I need not attempt to relate.   I refer my readers to Murray's *Guide Book*, where both are admirably described.

After completing my business affairs at Naples, and sowing the seeds of several orders, which afterwards bore substantial results, I left the city by the same line of steamers.   I passed again Civita Vecchia, Leghorn, Genoa, and Marseilles.   On passing through the south of France I visited the works of several of our employers, and carried back with me many orders.   It was when at Creuzot that I saw the child of my own brain, the steam hammer, in full and efficient work.   But this I have referred to in a previous chapter.

CHAPTER XV.

IN 1840 I furnished Sir Edward Parry with a drawing of my steam hammer, in the hope that I might induce him to recommend its adoption in the Royal Dockyards. Sir Edward was at that time the head director of the steam marine of England. That was after the celebrity he had acquired through his Arctic voyages. I was of opinion that the hammer might prove exceedingly useful in forging anchors and large iron work in those great establishments. Sir Edward appeared to be much struck with the simplicity and probable efficiency of the invention. But the Admiralty Board were very averse to introducing new methods of manufacturing into the dockyards. Accordingly, my interview with Sir Edward Parry, notwithstanding his good opinion, proved fruitless.

Time passed by. I had furnished steam hammers to the principal foundries in England. I had sent them abroad, even to Russia. At length it became known to the Lords of the Admiralty that a new power in forging had been introduced. This was in 1843, three years after I had submitted my design to Sir Edward Parry. The result was that my Lords appointed a deputation of intelligent officers to visit my foundry at Patricroft to see the new invention. It consisted of Captain Denison (brother of the late Speaker), and Captain Burgman, Resident Engineer at Devonport Dockyard. They were well able to understand the powerful agency of the steam hammer for marine forge work. I gave them every opportunity for observing its action. They were much pleased, and I may add astonished, at its range, power, and docility.

Besides showing them my own steam hammer, I took the deputation to the extensive works of Messrs. Rushton and Eccersley, where they saw one of my five ton hammer-block steam hammers in full action. It was hammering out some wrought-iron forgings of the largest class, as well as working upon smaller forgings. By exhibiting the wide range of power of the steam hammer the gentlemen were entirely satisfied of its fitness for all classes of forgings for the naval service. They reported to the Admiralty accordingly, and in a few days we received an official letter, with an order for a steam hammer having a 50 cwt. hammer-block, together with the appropriate boiler, crane, and forge furnace, so as to equip a complete forge shop at Devonport Dockyard. This was my first order from the Government for a steam hammer.

When everything was ready I set out for Devonport to see the hammer and the other portions of the machinery carefully erected. In about a fortnight it was ready for its first stroke. As good luck would have it, the Lords of the Admiralty were making their annual visit of inspection to the dockyard that day. They arrived too late in the afternoon for a general inspection of the establishment; but they asked the superintending admiral if there was anything of importance which they might see before the day closed. The admiral told them that the most interesting novelty in the dockyard was the starting of Nasmyth's steam hammer. " Very well," they said, " let us go and see that."

I was there with the two mechanics I had brought with me from Patricroft, to erect the steam hammer. I took

share and share alike in the work.   The Lords were introduced to me, and I proceeded to show them the hammer.
I passed it through its paces.   I made it break an eggshell
in a wine-glass without injuring the glass.   It was as neatly
effected by the two-and-a-half ton hammer as if it had been
done by an egg-spoon.   Then I had a great mass of white-
hot iron swung out of the furnace by a crane and placed
upon the anvil block.   Down came the hammer on it with
ponderous blows.   My Lords scattered, and flew to the extremities of the workshop, for the splashes and sparks of hot
metal flew about.   I went on with the hurtling blows of the
hammer, and kneaded the mass of iron as if it had been
clay.

After finishing off the forging, my Lords gathered round
the hammer again, when I explained to them the rationale
of its working, and the details of its construction.   They
were greatly interested, especially Mr. Sidney Herbert (afterwards Earl of Pembroke), then Secretary to the Admiralty,
and Sir George Colborn, a fine specimen of the old admiral.
Indeed, all the members of the Board were more or less
remarkable men.   They honoured me with their careful
attention, and expressed their admiration at the hammer's
wonderful range of power and delicacy of touch, and the
controllable application of the force of steam.

This afternoon was a most important one for me in more
ways than one, although I cannot venture to trouble my
readers with the details.   It was followed, however, by an
order to supply all the Royal Dockyard forge departments
with a complete equipment of steam hammers, with all the
requisite accessories.   These were supplied in due time, and
gave in every case the highest satisfaction.   The forgings
were found to be greatly better, and almost absurdly cheaper
than those done by the old building-up process.   The danger
of flaws was entirely done away with; and, in the case of
anchors, this was a consideration of life and death to the

JAMES NASMYTH

STEAM HAMMER IN FULL WORK.   FROM A PAINTING BY JAMES NASMYTH.

seamen who depend for their security upon the soundness of the forging.

Besides my introduction to that admirable man, Mr. Sidney Herbert, I had the happiness of being introduced to Captain Brandreth, Director of Naval Works. The whole of the buildings on shore, including the dockyards, were under his control. One of the most important affairs that the Lords of the Admiralty had to attend to on their visit to Devonport was to conclude the contract for constructing the great docks at Keyham. This was a large extension of the Devonport Docks, intended for the accommodation of the great steamships of the Royal Navy, as well as for an increase of the graving docks and workshops for their repair. An immense portion of the shore of the Hamoaze had to be walled in so as to exclude the tide and enable the space to be utilised for the above purposes. To effect this a vast amount of pile driving was rendered necessary, in order to form a firm foundation for the great outer dock wall, about a mile and a quarter in length.

Messrs. Baker and Sons were the contractors for this work. They were present at the first start of my steam hammer at Devonport. They were, like the others, much impressed by its vast power and manageableness. They had an interview with me as to its applicability for driving piles for the immense dock, this being an important part of their contract. Happily, I had already given some attention to this application of the powers of the steam hammer. In fact, I had secured a patent for it. I had the drawings for the steam hammer pile-driving machine with me. I submitted them to Mr. Baker, and he saw its importance in a moment. "That," he said, "is the very thing that I want to enable me to complete my contract satisfactorily." Thousands of enormous piles had to be driven down into the deep silt of the shore; and to have driven them down by the old system of pile driving would have occupied a long time, and would also have been very costly.

The drawings were of course submitted to Captain Brandreth. He was delighted with my design. The steam pile driver would be, in his opinion, the prime agent for effecting the commencement of the great work originated by himself. At first the feat of damming out such a high tide as that of the Hamoaze seemed very doubtful, because the stiff slate silt was a treacherous and difficult material to penetrate. But now, he thought, the driving would be rendered comparatively easy. With Captain Brandreth's consent the contractors ordered of me two of my steam hammer pile-drivers. They were to be capable of driving 18-inch square piles of 70 feet in length into the silt of the Hamoaze.

This first order for my pile driver was a source of great pleasure to me. I had long contemplated this application of the power of the steam hammer. The machine had long been in full action in my "mind's eye," and now I was to see it in actual reality. I wrote down to my partner by that night's post informing him of the happy circumstance. The order was for two grand steam hammer pile drivers, each with four-ton hammer-blocks. The wrought iron guide case and the steam cylinder were to weigh in all seven tons. All this weight was to rest on the shoulders of the pile. The blows were to be about eighty in the minute. This, I thought, would prove thoroughly effective in driving the piles down into the earth.

I have said that the steam pile driver was in my mind's eye long before I saw it in action. It is one of the most delightful results of the possession of the constructive faculty, that one can build up in the mind mechanical structures and set them to work in imagination, and observe beforehand the various details performing their respective functions, as if they were in absolute material form and action. Unless this happy faculty exists *ab initio* in the brain of the mechanical engineer, he will have a hard and disappointing life before him. It is the early cultivation of the imagination which gives the right flexibility to the thinking faculties. Thus business, commerce, and mechanics are all the better for a little *healthy* imagination.

SPACE TO BE ENCLOSED AT THE HAMOAZE.

So soon as I had returned home, I set to work and prepared the working drawings of the steam pile drivers They were soon completed, conveyed to Devonport, and erected on the spot where they were to be used. They were ready on the 3d of July 1845. Some preliminary pile driving had been done in the usual way, in order to make a stage or elevated way for my pile driver to travel along the space where the permanent piles were to be driven. I arranged my machines so that they might travel

T

by their own locomotive powers along the whole length of
the coffer dam, and also that they should hoist up the great

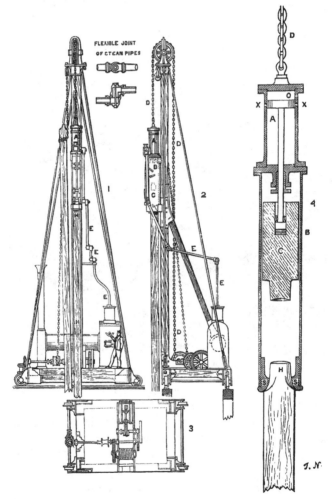

logs of Baltic timber which formed the piles into their
proper places before being driven.

The entire apparatus of the machine was erected on a strong timber platform, and was placed on wheels, so that it might move along the rails laid down upon the timber way. The same boiler that supplied the steam hammer part of the apparatus served to work the small steam-engine fixed to the platform for its locomotion, and also to perform the duty of rearing the next pile which had to be driven. The steam was conveyed to the hammer cylinder by the jointed pipe seen in the annexed engraving. The pipe accommodated itself to any elevation or descent of the hammer. The whole weight of the cylinder, hammer-block, and guide box, supported by the shoulders of the pile, amounting to seven tons in all, rested upon the shoulders of the pile as a " persuader ;" and the eighty blows per minute of the four-ton hammer came down with tremendous energy upon the top of the pile head.[1] No soil, that piles could penetrate, could resist such effective agencies.

There was a great deal of curiosity in the dockyard as to the action of the new machine. The pile-driving machine-

---

[1] EXPLANATION OF THE DIAGRAM OF THE STEAM PILE DRIVER.—The chief feature of novelty of the pile-driving machine consists in the employment of the direct action of the Steam Hammer as the blow-giving agent, and also in the manner in which the dead weight of the entire apparatus, consisting of the hammer-block C, the steam cylinder A, and its guide case B, is employed to importantly aid the effect of the rapid and energetic blows of the steam hammer. These ponderous parts rest on the shoulders of the pile H all the while it is being driven, the pile in this respect being *the only* support of the apparatus A B C. So that, besides the eighty blows per minute that the four-ton steam hammer energetically deals out on to the head of the pile from a four foot fall, the dead weight of the apparatus constantly acts as a most effective " predisposer " to the sinking of the pile into the ground ; the hoisting chain D being let slack the while, so as to allow A B C to " follow down " the pile H, while the eighty blows per minute are incessantly showered on its head. The upward stroke of the piston, with its attached hammer-block C, is arrested at the proper height not only by allowing the steam that raised it to escape, but as soon as the piston passes the escape holes X X, the confined air above the piston at O rebounds, and so aids most effectively in increasing the energy of the fall of the hammer-block C on the pile head.

men gave me a good-natured challenge to vie with them in
driving down a pile. They adopted the old method, while
I adopted the new one. The resident managers sought out
two great pile logs of equal size and length—70 feet long
and 18 inches square. At a given signal we started together. I let in the steam, and the hammer at once began
to work. The four-ton block showered down blows at the
rate of eighty a minute; and in the course of *four and a
half minutes* my pile was driven down to its required depth.
The men working at the ordinary machine had only begun to
drive. It took them upwards of *twelve hours* to complete
the driving of their pile !

Such a saving of time in the performance of similar work
—by steam *versus* manual labour—had never before been witnessed. The energetic action of the steam hammer, sitting
on the shoulders of the pile high up aloft, and following it
suddenly down, the rapidly hammered blows keeping time
with the flashing out of the waste steam at the end of
each stroke, was indeed a remarkable sight. When my pile
was driven, the hammer-block and guide case were speedily
re-hoisted by the small engine that did all the labouring and
locomotive work of the machine; the steam hammer portion
of which was then lowered on to the shoulders of the next
pile in succession. Again it set to work. At this the
spectators, crowding about in boats, pronounced their approval in the usual British style of "three cheers!" My
new pile-driver was thus acknowledged as another triumphant
proof of the power of steam.

The whole of the piles for this great work were speedily
driven in. The wall was constructed, and the docks
were completed in an unusually short time. The success
of my pile-driver was followed by numerous orders. It
was used for driving the immense piles required for the
High Level Bridge at Newcastle, the great Border Bridge
at Berwick-upon-Tweed, the Docks at Tynemouth, the Docks

at Birkenhead, the Docks at Grimsby, the new Westminster Bridge, the great bridge at Kief in Russia, the bridge at Petersburg, the forts at Cronstadt, the Embarrage of the Nile, at Yokohama in Japan, and at other places. It enabled a solid foundation to be laid for the enormous superstructures erected over them, and thus contributed to the permanence of many important undertakings.

The mechanical principles on which the efficiency of the steam pile-driver chiefly depends are as simple as I believe they are entirely novel and original. The shoulder of the pile acts as the sole supporter of the ponderous mass of the hammer-block, cylinder, and guide box. This heavy weight acts as a predisposing agency to drive the pile down, while the momentum given by the repeated fall of the hammer, at eighty blows the minute, brings the constant dead weight into full action. I am not aware of any other machine in which such a combination of mechanical forces is employed.

Another very effective detail consisted in employing the waste steam in the upper part of the cylinder for the purpose of acting as a buffer to resist any undue length of the upward stroke of the piston. But for this the cylinder covers might have been knocked off. The elastic buffer of waste steam also acted as a help to the downward blow of the hammer-block. The simplicity and effectiveness of these arrangements forms—if I may be allowed to say so —a happy illustration of my "Definition of Engineering," *the application of common sense to the use of materials.*

The folding-up steam pipe with which the steam was conveyed from the boiler to the cylinder at all heights, and the way in which the folding joints accommodated themselves to the varying height of the cylinder, was another of my happy thoughts. In fact, this invention, like most others, was the result of a succession of happy thoughts. The machine in its entirety, was the result of a number of common-sense contrivances, such as I generally delight in.

At all events, this most effective and novel machine was a special favourite with me.

I may mention, before concluding this branch of my subject, that pile-driving had before been conducted on what I might term the artillery or cannon-ball principle. A small mass of iron was drawn slowly up, and suddenly let down on the head of the pile at a high velocity. This was *destructive*, not *impulsive* action. Sometimes the pile was shivered into splinters, without driving it into the soil; in many cases the head of the pile was shattered into matches, and this in spite of a hoop of iron about it to keep the layers of wood together. Yet the whole was soon beat into a sort of brush. Indeed, a great portion of the men's time was consumed in reheading the piles. On the contrary, I employed great mass and moderate velocity. The fall of the steam hammer block was only three or four feet, but it went on at eighty blows the minute, and the soil into which the pile was driven never had time to grip or thrust it up—an impediment well known to ordinary pile drivers. At the end of the driving by my steam hammer, the top of the pile was always found neat and smooth, indeed more so than when the driving began.

I may again revert to my interview with the Lords of the Admiralty on the occasion of my first meeting them at Devonport. I was living at the hotel where they usually took up their quarters while making their annual visitation of the dockyard. I was honoured with an invitation to confer with Sir George Colburn, Mr. Sydney Herbert, and Captain Brandreth on a subject of considerable importance; namely, the proving of chain cables and anchors required for the Royal Navy. The question was mooted as to whether or not some permanent injury was done to both by the test strains to which they were submitted before being put on board ship. This was a subject of vital importance. The members of the Board requested me to be one of a

committee to inquire into the subject. I felt much gratified
by the invitation, and gladly accepted it.

On discussing the subject with these gentlemen that
evening, I found that Sir George Colburn entertained an
ingenious theory in support of his apprehensions as the
effect of " over-proof" straining of cables and anchors. It
was that they were originally in the condition of a strong
man who had to lift some heavy weight, requiring him to
exert his muscular strength *to the utmost;* and, although he
might perform the feat, it was at the cost of a permanent
injury, and he might never be able to lift the same weight
again.   This, however true it might be with regard to flesh
and bone structures, was scarcely true with respect to mechanical agencies.   I proposed a simple experiment with
chain cables, which, it occurred to me, would show quite a
different result—namely, that the capability of resisting the
severest proof-strain would *rise* rather than *fall* at each successive proof of the same chain cable.

To test the correctness of my supposition, we had a first
class chain cable put into the proof machine, and subjected it
to such a strain as to break it again and again, until at last
it was divided almost into single links.   As I expected, the
proof or breaking strain kept rising and rising as each
successive remaining portion of the cable was torn asunder,
thus showing that no injury to the *natural tenacity* of the
chain had resulted from the increased proofs to which it had
been subjected, and that the last broken chains had been
much more resisting than the first.   The same class of de-
monstrative experiments was made with anchors, and other
wrought-iron work used in the service.   The Admiralty
officers were much gratified with the result, as removing
a groundless but very natural apprehension, heightened, no
doubt, by the suggestions that had been made to the
Admiralty, that their standard proof strain was not only
too high in itself, but produced permanent damage to what

at the outset was of the toughest iron.   My system of *continued* proof-straining was, in fact, another exemplification of the " Survival of the Fittest " !

A very interesting truth came out in the course of our experiments.   It was that the chief cause of failure in the links of chain cables arose, not so much from their want of tenacity, or from the quality of the iron, but from some defective welding in the making of the links.   To get at this truth, many excellent cables as received from the contractors, as well as veteran ones that had held great ships riding at anchor in terrible gales, were pulled asunder link by link by an intentional destructive strain by the proving machine.   An exact account was taken of the nature of the fracture of each.   The result was that in eight cases out of ten, the fracture was found to result from a defectively welded part of the chain-link.   The practically trained eye could see the scoriæ which indicates the defective welding. Though long unseen, it was betrayed at once when the link was torn open by the proof strain.

My services on this committee proved a source of great enjoyment to me.   I had frequent occasion to visit the dockyards and workshops, accompanied by Captain Brandreth, surveyor-general of the Admiralty landworks, Mr. Thomas Lloyd, engineer-in-chief of the Admiralty, and Mr. Jeremiah Owen, chief of the metal material required in the equipment of the navy.   I was requested to suggest any improvement in the workshops that I thought would add to the efficiency of the department; and I trust that my recommendations proved of practical good to the service.   At the same time, I have reason to know that many of the recommendations of the committee, though cordially acknowledged by the higher powers, were by a sort of passive resistance practically shelved.

I was much amused, when I first went to Devonport dockyard, to notice the punctilious observance of forms and

ceremonies with respect to the various positions of officials
—from the admiral-superintendent down the official grades
of dignity, to the foremen of departments, and so on. I did
not care for all this *panjandrum* of punctiliousness, but was,
I hope, civil and chatty with everybody. I had a good
word for the man as well as for the foreman. I received
some kind and good-natured hints as to the relative official
superiority that prevailed in the departments, and made out
a scale or list of the various *strata* accordingly. This gamut
of eminence was of use to me in my dealings with dockyard
officials. I was enabled to mind my p's and q's in communicating with them.

The first Sunday that I spent at Devonport I went to
the dockyard church—the church appointed for officials and
men employed by the Government. The seats were appointed
in the order of rank, employments, and rate of pay. The
rows of seats were all marked with the class of employers
that were expected to sit in them. Labourers were near
the door. The others were in successive rows forward, until
the pew of the " Admiral Superintendent," next the Altar
rails, was reached. I took my seat among the " artificers,"
being of that order. On coming out of church the master-
attendant, next in dignity to the admiral-superintendent,
came up to me to say how distressed he was to see me
" among the artificers," and begged me in future to use his
seat. No doubt this was kindly intended, and I thanked
him for his courtesy. Nevertheless I kept to my class of
artificers. I did not like the " breest o' the laft "[1] principle.
No doubt the love of distinction, within reasonable limits,
is a great social prime mover; but at Devonport, with the

---

[1] "The breest o' the laft" is the seat of dignity. The best places in
churches are occupied by "superior" people. In Scotland the chief men—
the Provosts, Bailies, and Councillors—have a seat appropriated to them in
the front part of the gallery, generally opposite the minister. This is "*the
breest o' the laft.*" The same principle pervades society generally.

splitting up into ranks and dignities even amongst the workmen, I found it simply amusing.

I afterwards met with several veterans in the service of
the Admiralty, who are well served by such conscientious
and well-selected men. It is the schemers and the satellites who haunt the contractors that are the vermin of
dockyards. I gave them all a very wide berth. But worst
of all are the men who get their employment through par-
liamentary influence. They are a detestable set. They
always have some "grievance" to pester people about. I
hope things are better now.

I may add, with respect to the steam hammer pile-
driving machines, that I received an order for two of them
from Mohammed Ali, the Pasha of Egypt. These were
required for driving the piles in that great work—the bar-
rage of the Nile near Cairo. The good services of these
machines so pleased the Pasha that he requested us to receive
three selected Arab men into our works. He asked that
they should have the opportunity of observing the machinery
processes and the system of management of an English
engineering factory. The object of the Pasha was that the
men should return to Egypt and there establish an engine
manufactory, so as to render him in a measure independent
of foreign help. For British workmen, when imported into
Egypt, had a great tendency to deteriorate when removed
from the wholesome stimulus to exertion in competition with
their fellows.

My firm had no objection to the introduction of the Arab
workmen. Accordingly, one day we received a visit from
an excellent Egyptian officer, Edim Bey, accompanied by his
secretary Rushdi Effendi, who spoke English fluently. He
thus made our interview with the Bey easy and agreeable.
He conveyed to us, in the most courteous manner, the wishes
of the Pasha; and the three workmen were at once received.
Every opportunity was given them to observe and under-

stand the works going forward. They were intelligent-looking young men, about twenty-five years of age. One of them was especially bright looking, quick in the expression of his eyes, and active in his manner. His name was Affiffi Lalli; the names of the others I forget.

These young men were placed under charge of the foremen of the departments that each fancied to be most to his taste. Affiffi was placed in the fitting department, in which skilful manipulation was required. He exhibited remarkable aptitude, and was soon able to hold his own alongside of our best workmen. Another was set to the turning department, and did fairly well. The third was placed in the foundry, where he soon became efficient in moulding and casting brass and iron work. He lent a hand all round, and picked up a real practical knowledge of the various work of his department. During their sojourn in our works they became friendly with their colleagues; and in fact became quite favourites with the men, who were always willing to help them. But Affiffi Lalli was regarded as the genius of the trio. He showed a marked and intelligent aptitude for acquiring technical skill in all the branches of our business.

After remaining with us for about four years they were ready to return to Cairo, and show what they had learned in practical and technical mechanical knowledge during their stay in England. The three Arab workmen were placed in their suitable departments in the Pasha's workshops. But such was the natural energy of Affiffi, that when he was set to work beside the slow, dilatory, and stupid native workmen, he became greatly irritated. The contrast between the active energetic movements which he had seen at the Bridgewater Foundry, and the ineffective, blundering, and untechnical work of his fellows was such that he could not stand it any longer. So one fine day he disappeared from the works, took refuge on board a British steamer, and at the

risk of his neck made his way back to the Bridgewater
Foundry !

As we were reluctant to take back a man who had
escaped from the Pasha's employment—excellent workman
though he was—we declined to employ him.   But I gave
Affiffi a note of introduction to Boulton and Watt of Soho,
Birmingham, and there he was employed.   He afterwards
passed into other firms, and having employed his skill in
making some needle machinery at Redditch, he settled down
there.   He married a Warwickshire lass, and had a family
—half Arab, half English—and has now a thriving foundry
and engineer workshop of his own.   This little narrative
shows that the Arab has still much of the wonderful energy
and skill that once made the Moors masters of a large part
of South-western Europe.

We had many visitors at the foundry—from London, from
the manufacturing districts, and from foreign countries.   One
day a young gentleman presented a letter from Michael Faraday, dated " Royal Institution, 29th May 1847," requesting
me to pay him some attention and show him round the works.
I did so with all my heart, and wrote to Mr. Faraday intimat-
ing how much pleasure it gave me to serve him in any respect.
I cannot refrain from giving his answer.   He said :—

" MY DEAR SIR—That you should both show kindness to the bearer
of my letter, and prove that you did so with pleasure by writing me a
letter in return, was indeed more than I ought or could have expected ;
but it was very gratifying and pleasant to my mind.   I only wish that
the circumstances of my life were such as to enable me to take advantage of such goodwill on your part, and to be more in your company
and conversation than is at present possible.

" I could imagine great pleasure from such a condition of things ;
but though our desires, and even our hopes at times, spread out beforehand over a large extent, it is wonderful how, as the future becomes
the present, the circumstances that surround us limit the sphere to
which our real life is circumscribed.   If ever I come your way I hope
to see your face ; and the hope is pleasant, though the reality may
never arrive.

"You tell me of the glorious work of your pile-driver, and it must be indeed a great pleasure to witness the result. Is it not Shakespeare who says, 'The pleasure we delight in physics pain'? In all your fatigue and labour you must have this pleasure in abundance, and a most delightful and healthy enjoyment it is. I shall rejoice to see some day a blow of the driver and a tap of the hammer.

"You speak of some experiments on tempering in which we can help you. I hope when you do come to town you will let us have the pleasure of doing so. Our apparatus, such as it is, shall be entirely at your service. I made, a long while ago, a few such experiments on steel wire, but could eliminate no distinct or peculiar results. You will know how to look at things, and at your hand I should expect much.

"Here we are just lecturing away, and I am too tired to attempt anything, much less to do anything just now; but the goodwill of such men as you is a great stimulus, and will, I trust even with me, produce something else praiseworthy.—Ever, my dear Nasmyth, yours most truly,   M. FARADAY."

CHAPTER XVI.

In the autumn of 1842 I had occasion to make a journey to Nuremberg in company with my partner Mr. Gaskell. We had been invited to a conference with the directors of the Nuremberg and Munich Railroad as to the supply of locomotives for working their line. As this was rather an important and extensive transaction, we thought it better not to trust to correspondence, but to see the directors on the spot. We found that there were several riskful conditions attached to the proposed contract, which we considered it imprudent to agree to. We had afterwards good reason to feel satisfied that we had not yielded to the very tempting commercial blandishments that were offered to us, but that we refrained from undertaking an order that required so many important modifications.

Nevertheless, I was exceedingly delighted with the appearance of the city of Nuremberg. It carries one back to the mediæval times! The architecture, even of the ordinary houses, is excellent. St. Lawrence, St. Sebald's, and the Frauenkirche, are splendid specimens of Gothic design. The city is surrounded by old walls and turrets, by ramparts and bastions, enclosed by a ditch faced with masonry. Very few cities have so well escaped the storm of war and sieges in the Middle Ages, and even in modern times.

Everything has been carefully preserved, and many of the best houses are still inhabited by the families whose forefathers originally constructed them. But "progress" is beginning to affect Nuremberg. It is the centre of railways; buildings are extending in all directions ; tram-cars are running in the streets ; and before long the ditch will be filled up, the surrounding walls and towers demolished, and the city thrown open to the surrounding country.

I visited the house of Albert Durer, one of the greatest artists who ever lived. He was a man of universal genius —a painter, sculptor, engraver, mathematician, and engineer. He was to Germany what Leonardo da Vinci was to Italy. His house is wonderfully preserved. You see his entrance hall, his exhibition room, his bedroom, his studio, and the opening into which his wife—that veritable Xantippe— thrust the food that was to sustain him during his solitary hours of labour. I saw his grave, too, in the old churchyard beyond the Thiergarten gate. I saw the bronze plate commemorating the day of his death. "*Emigravit* 8 *idus Aprilis* 1528." "Emigravit" only, for the true artist never dies. Hans Sachs's grave is there too—the great Reformation poet of Luther's time.

Adam Krafft must have been a great sculptor, though his name is little known out of Nuremberg. Perhaps his finest work is in St. Lawrence Cathedral—the *Sacramentshäuslein*, or the repository for the sacred wafer—a graceful tapering stone spire of florid Gothic open work, more than sixty feet high, which stands at the opening of the right transept. Its construction and decoration occupied the sculptor and his two apprentices no less than five years ; and all that he received for his hard labour and skilful work was 770 gulden, or about £80 sterling. No wonder that he died in the deepest distress. St. Sebald's and the Frauenkirche also contain numerous specimens of his admirable work.

In the course of the following year (1843), it was necessary for me to make a journey to St. Petersburg. My object was to endeavour to obtain an order for a portion of the locomotives required for working the line between that city and Moscow. The railway had been constructed under the engineership of Major Whistler, father of the well-known artist; and it was shortly about to be opened. It appeared that the Emperor Nicholas was desirous of securing a home supply of locomotives, and that, like a wise monarch, he wished to employ his own subjects rather than foreigners in producing them. No one could object to this.

The English locomotive manufacturers were not aware of the Emperor's intention. When I arrived in the city I expected an order for locomotives. The representatives of the principal English firms were there like myself; they, too, expected a share of the order. It so happened that at the table d'hôte dinner, I sat near a very intelligent American, with whom I soon became intimate. He told me that he was very well acquainted with Major Whistler, and offered to introduce me to him. By all means! There is nothing like friendly feelings in matters of business.

The Major gave me a frank and cordial reception, and informed me of the position of affairs. The Emperor, he said, was desirous of training a class of Russian mechanics to supply not only the locomotives but to keep them constantly in repair. He could not solely depend upon foreign artisans for the latter purpose. The locomotives must be made in Russia. The Emperor had given up the extensive premises of the Imperial China Manufactory, which were to be devoted to the manufacture of engines.

The Major appointed Messrs. Eastwick, Harrison, and Wynants, with the approval of the Government, to supply the entire mechanical plant of the railway. I saw that it would be of no use to apply for any order for locomotives; but I offered to do all that I could to supply the necessary

materials. In the course of a few days I was introduced to
Joseph Harrison, the chief mechanic of the firm ; and I
then entered into a friendship which proved long and lasting. He gave me a very large order for boilers, and for
other detail parts of the Moscow engines,—all of which helped
him forward in the completion of the locomotives. We
also supplied many of our special machine tools, without
which engines could not then be very satisfactorily made or
kept in repair.

The enjoyment of my visit to St. Petersburg was much
enhanced by frequent visits to my much valued friend
General Alexander Wilson. He was a native of Edinburgh, and delighted to enjoy cracks with me upon subjects of mutual interest. His sister, who kept house for
him, joined in our conversation. She had been married to
the Emperor Paul's physician, who was also a Scotsman,
and was able to narrate many terrible events in relation to
Russian Court affairs. The General had worked his way
upwards, like the rest of us. During the principal part of
his life he had superintended the great mechanical establishments at Alexandrosky and Colpenha, where about 3000
operatives were employed. These establishments were originally founded by the Empress Catherine for the purpose
of creating a native manufacturing population capable of
carrying on textile and mechanical works of all kinds. The
sail-cloth for the Russian navy was manufactured at Alexandrosky by excellent machinery. Cotton fabrics were also
manufactured, as well as playing cards, which were a Crown
monopoly. The great establishment at Colpenha consisted
of a foundry, a machine manufactory, and a mint—where
the copper money of the empire was coined. General
Wilson was the directing chief officer of all these establishments.

Through him I had the happiness of being introduced
to General Greg, son of the great admiral who shed such

U

honour on the Russian flag during the reign of the Empress
Catherine.   He was then well advanced in years, but full
of keen intelligence and devoted to astronomical pursuits.
He was in a great measure the founder of the Imperial
Observatory at Pulkowa, situated on an appropriate eminence
about eight miles from St. Petersburg.   The observatory
was furnished under his directions with the most magnificent
astronomical instruments.   I had the honour to be introduced by him to the elder Struve, whose astronomical
labours procured him a well-earned reputation throughout
Europe.   I had the rare happiness of spending some nights
with Struve, when he showed me the wonderful capabilities
of his fine instruments.   The observatory is quite imperial
in its arrangement and management, and was supported in
the most liberal manner by the Emperor Nicholas.   Indeed, it is a perfect example of what so noble an establishment should be.

Struve most kindly invited me to come whenever the
state of the weather permitted him to show forth the
wonderful perfection of his instruments, — a rare chance,
which I seized many opportunities of enjoying.   It was quite
a picture to see the great pleasure, but intense enjoyment,
with which the profound astronomer would seat himself at
his instrument and pick out some exquisite test objects,
such as the double stars in Virgo, Cygnus, or Ursa Major.
The beautiful order and neatness with which the instruments were kept in their magnificent appropriate apartments, each having its appropriate observer proceeding
quietly with his allotted special work, with nothing to
break the silence but the "tick, tack!" of the sidereal
clock—this was indeed a most impressive sight!   And the
kindly companionable manner of the great master of the
establishment was in all respects in harmony with the
astronomical work which he conducted in this great Temple
of the Universe!

Through my friendship with General Wilson I was enabled to extend my acquaintance with many of my countrymen who had been long settled at St. Petersburg in connection with commercial affairs. I enjoyed their kind hospitality, and soon found myself quite at home amongst them. I remained in the city for about two months. During that time I was constantly about. The shops, the streets, the houses, the museums, were objects of great interest. The view of the magnificent buildings along the sides of the quay is very imposing. Looking from the front of the statue of Peter the Great you observe the long façade of the Admiralty, the column of Alexander, the Winter Palace, and other public buildings. The Neva flows in front of them in a massive volume of pure water. On an island opposite stands the citadel. The whole presents a *coup d'œil* of unexampled architectural magnificence.

I was much interested by the shops and their signboards. The latter were fixed all over the fronts of the shops, and contained a delineation of the goods sold within. There was no necessity for reading. The pictorial portraits told their own tale. They were admirable specimens of what is called still-life pictures ; not only as regards the drawing and colouring of each object, but with respect to the grouping, which was in most cases artistic and natural. Two reasons were given me for this style of artistic sign-painting : one was that many of the people could not read the written words defining the articles sold within ; and the other was that the severe and long-continued frosts of the St. Petersburg winter rendered *large* shop windows impossible for the proper display of the goods. Hence the small shop-windows to keep out the cold, and the large painted sign-boards to display the articles sold inside.

I was also greatly pleased with the manner in which the Russians employ ivy in screening their windows during summer. Ivy is a beautiful plant, and is capable of forming a most elegant window-screen.   Nothing can be more beautiful than to look through green leaves.   Nearly every window of the ground flat of the houses in St. Petersburg is thus screened.   The neat manner in which the ivy plants are trained over ornamental forms of cane is quite a study in its way.   And though the ivy is very common, yet a common thing, being a thing of beauty, may be a "joy for ever."   In the finer and most important mansions, the sides of the flight of wide steps that lead up to the reception rooms were beautifully decorated by oleander plants, growing in great vigour, with their fine flowers as fresh as if in a carefully-kept conservatory.   Other plants of an ornamental kind were mixed with the oleander, but the latter appeared to be the favourite.[1]

About the end of my visit I was about to call upon one of my customers with reference to my machine tools ; for though I pursued pleasure at occasional times, I never lost sight of business.   It was a very dull day, and the streets about the Winter Palace were almost deserted.   I was sitting in my drosky with my roll of drawings resting on my thigh—somewhat in the style of a commander-in-chief as represented in the old pictures—when I noticed a drosky coming out of the gates of the Winter Palace.   I observed that it contained a noble-looking officer in a blue military cloak sitting behind his drosky driver.   My driver instantly took off his hat, and I, quickly following his example, took off my hat and bowed gracefully, keeping my extended hand

---

[1] While passing through Lubeck on my way out to St. Petersburg I was much struck with the taste for flower-plants displayed by the people of that old-world city.   The inner side of the lower house windows were all beautifully decorated with flowers, which were evidently well cared for.   Some of the windows were almost made up with flowers.   Perhaps the long-continued winter of these parts has caused the people to study and practise within-door culture with such marked success.   It is a most elegant pursuit, and should be cultivated everywhere.   It is thoroughly compatible with the exquisite cleanliness and tidiness of the houses at Lubeck.

on the level of my head—a real royal salute. The person
was no other than the Emperor Nicholas! He fixed his
peculiarly fine eyes upon me, and gave me one of the
grandest military salutes, accompanied, as I thought, with
a kindly smile from his magnificent eyes as he passed close
by me.

As I had been lunching with a Dutch engineer about
half an hour before, and had a glass or two of champagne,
this may have had something to do with my daring to give
the Emperor, in his own capital, what I was afterwards told
was not a bow but a brotherly recognition between poten-
tates, and only by royal usage allowed to be so given,—
namely, swaying off the hat at arm's length level with the
head, so as to infer royal equality, or something of that
sort. When I narrated to some Russian friends what I
had done, they told me that I need not be surprised if I
received a visit from the chief of police next morning for
my daring to salute the Emperor in such a style. But the
Emperor was doubtless more amused than offended, and I
never received the expected visit.

To anticipate a little. Soon afterwards the Emperor
sent me a present of a magnificent diamond ring through his
ambassador in England—Baron Brunow. It was also accompanied, as the Baron informed me, with the Emperor's most
gracious thanks for the manner in which my steam hammer
had driven the piles for his new forts at Cronstadt, which
he had seen with his own eyes. The steam-hammer pile-
driver had also been used for driving the piles of the great
bridge at Kieff. I next received an order for one of my
largest steam hammers for the Imperial Arsenal, and it was
followed by many more. It is a singular fact, as showing
the readiness of the Russian and other foreign Governments
to adopt at an early date any mechanical improvement of
ascertained utility, that I supplied steam hammers to the
Russian Government twelve months before our Admiralty

availed themselves of its energetic action.   But Athelstane
the Unready has always been found dreadfully slow—in
peace, as well as in war.

Before I leave this part of my subject, I must not omit
to mention my friend Mr. Francis Baird, the zealous son of
Sir Charles Baird.   The latter was among the first to establish iron foundries and engine works at St. Petersburg.   At
the time of my visit he was far advanced in years, and
unable to attend personally to the very large business which
he had established.   But he was nevertheless full of geniality.   He greatly enjoyed the long conversations which he
had with me about his friends in Scotland, many of whom I
knew.   He also told me about the persons in his employment.   He said that the workmen were all serfs, or the sons
of serfs.   The Empress Catherine had given them to him
for the purpose of being trained in his engine foundry, and
in his sugar refinery, which was another part of his business.
I had rarely seen a more faithful and zealous set of workmen than these Russian serfs.   They were able and skilful,
and attached to their employers by some deeper and stronger
tie than that of mere money wages.   Indeed, they were
treated by Sir Charles Baird and his son with the kindest
and most paternal care, and they duly repaid their attachment by their zeal in his service and the excellent quality
of their work.

The most important business in hand at the time of my
visit to the foundry was the moulding and casting of the
magnificent bronze capitals of the grand portico of the
Izak Church.   This building is one of the finest in St.
Petersburg.   It is of grand proportions,—simple, noble,
and massive.   It is built upon a forest of piles.   The walls
of the interior are covered with marble.   The malachite
columns for the screen are fifty feet high, and exceed everything that has yet been done in that beautiful fabric.   The
great dome is of iron overlaid with gold.   This, as well as

the Corinthian capitals of bronze, was manufactured at the foundry of the Bairds. The tympanum of the four great porticos consisted of colossal groups of alto-relievo figures, many of which were all but entirely detached from the background of the subject. It was a kind of foundry work of the highest order, all the details and processes requiring the greatest care. To my surprise every one engaged in this gigantic and refined metal work was a serf. The full-sized plaster models which they used in moulding were executed by a resident French artist. He was a true artist, and of the highest order. But to see the skilful manner in which these native workmen, drawn from the staff of the Bairds' ordinary foundry workers, performed their duties, was truly surprising. It would make our best bronze statuary founders wince to be asked to execute such work. Judging from what I saw of the Russian workmen in this instance, I should say that Russia has a grand future before it.

Having satisfactorily completed all my business arrangements in St. Petersburg, I prepared to set out homewards. But as I had some business to transact at Stockholm and Copenhagen I resolved to visit those cities. I left St. Petersburg for Stockholm by a small steamer, which touched at Helsingfors and Abo, both in Finland. The weather was beautiful. Clear blue sky and bright sunshine by day, and the light prolonged far into the night. Even in September the duration of the sunshine is so great and the night so short that the air has scarcely time to cool till it gets heated again by the bright morning rays. Even at twelve at night the sun dips but a little beneath the bright horizon on the north. The night is so bright in the Abo latitude that one can read the smallest print.

Nothing can be more beautiful than the charming scenery we passed through in our tortuous voyage to Stockholm. We threaded along and past the granite islands which crowd the shores of the Baltic. They are covered with pines,

which descend to the water's edge. We swept them with
our paddle-boxes, and dipped their bright green fronds into
the perfectly clear sea. For about two days our course lay
through those beautiful small islands. It seemed like a
voyage through fairyland. And it continued in this exquisite tranquil way until we reached that crowning feature of
all—the magnificent city of Stockholm, sleeping, as it were,
on the waters of the Mälar Lake, and surrounded by noble
mountains clad with pines. With the exception of Edinburgh, Genoa, and Naples, I had never beheld so noble a
city with such magnificent surroundings.

I spent but a short time in Stockholm, but quite
sufficient to enable me to see much that was grandly beautiful
in its neighbourhood. Lakes, rocks, and noble trees abounded,
and exquisite residences peeped out through the woods,
giving evidences of high civilisation. Elegance of taste
and perfect domestic arrangements supplied every form of
rational comfort and enjoyment. My old friend Sir John
Ross, of Arctic celebrity, was settled at Stockholm as chief
consul for Her Majesty. He introduced me to several of the
leading English merchants, from whom I received much
kind attention. Mr. Erskine invited me to spend a day or
two at his beautiful villa in the neighbourhood. It was
situated on the side of a mountain, and overlooked a lake
that reminded me very much of Loch Katrine. Fine timber
grew about, in almost inaccessible places, on the tops of
precipices, and in shelves and cliffs among the rocks. The
most important result of my visit was an introduction to
Baron Tam, the proprietor and chief director of the great
Dannemora Iron Mine.

I was at once diverted for a time from my voyage to
Copenhagen. I was most desirous of seeing with my own
eyes this celebrated mine. The baron most willingly furnished me with letters of introduction to his managers, and
I proceeded to Dannemora by way of Upsala. I was much

interested by this city, by its cathedral, containing the tomb of Gustavus Vasa, and by its many historical associations. But I was still more impressed by Old Upsala, about three miles distant. This is a place of great antiquity. It is only a little hamlet now, though at one time it must have been the centre of a large population. The old granite church was probably at one time a pagan temple. Outside, and apart from it, is a wooden bell-tower, erected in comparatively modern times. In a wooden box inside the church is a wooden painted god, a most unlikely figure to worship. And yet the Swedes in remote parts of the country carefully preserve their antique wooden gods.

The great sacrifices to Odin were made at Old Upsala. Outside the church, in a row, are three great mounds of earth, erected in commemoration of Odin, Thor, and Freia— hence our Wednesday, Thursday, and Friday. These mounds, of about 60 feet high and 232 feet in diameter, were in former times used as burying-places for the great and valiant. I went down into a cottage near the tumuli, and drank a bumper of mead to the memory of Thor from a very antique wooden vessel. I made an especial reverential obeisance to Thor, because I had a great respect for him as being the great Hammerman, and one of our craft,—the Scandinavian Vulcan.

I drove back to Upsala, and remained there for the night. It is a sleepy silent place. The only sound I heard was the voice of the watchman calling out the small hours of the morning from his station on the summit of the cathedral tower. As the place is for the most part built of wood, this precaution in the shape of a watchman who can see all points of the city is a very necessary one.

Next morning I hired a small sort of gig of a very primitive construction, with a boy for driver. His duty was to carry me to the next post-house, and there leave me to be carried forward by another similar conveyance. But

the pony No. 2 was about a mile off, occupied in drawing a
plough, so that I had to wait until the job was over.   In
about an hour or so I was again under weigh.   And so on,
da capo, until about six in the evening, when I found myself
within sight of the great mine.

The post-house where I was set down was an inn, though
without a signboard.   The landlady was a bright, cheery,
jolly woman.   She could not speak a word of English, nor I
a word of Dannemora Swedish.   I was very thirsty and
hungry, and wanted something to eat.   How was I to com-

THE ORDER FOR DINNER.

municate my wishes to the landlady?   I resorted, as I often
did, to the universal language of the pencil.   I took out my
sketch-book, and in a few minutes I made a drawing of a
table, with a dish of smoking meat upon it, a bottle and a
glass, a knife and fork, a loaf, a salt-celler, and a corkscrew.
She looked at the drawing and gave a hearty laugh.   She
nodded pleasantly, showing that she clearly understood
what I wanted.   She asked me for the sketch, and went
into the back-garden to show it to her husband, who
inspected it with great delight.   I went out and looked
about the place, which was very picturesque.   After a short
time, the landlady came to the door and beckoned me in,
and I found spread out on the table everything that I desired

——a broiled chicken, smoking hot from the gridiron, a bottle of capital home-brewed ale, and all the *et ceteras* of an excellent repast. I made use of my pencil in many other ways. I always found that a sketch was as useful as a sentence. Besides, it generally created a sympathy between me and my entertainers.

My visit to the Dannemora Mine at Osterby was one of peculiar interest. I may in the first place say that the immense collection of iron at that point has been the result of the upheaval of a vast volume of molten igneous ore, which has been injected into the rock, or deposited in masses under the crust of the earth. In some cases the quarried rock yields from 50 to 70, and even as much as 90 per cent. of iron. The Dannemora Mine is a vast quarry open to the sky. When you come near it the place looks like a deep pit, with an unfathomable bottom. Ghost-like, weird-looking pinnacles of rocks stand out from its profound depths ; but beyond these you see nothing but wreaths of smoke curling up from below. The tortuous chasm in the earth, caused by the quarries beneath, is about half a mile long, and about a thousand feet wide.

The first process of the workmen in the quarries below is devoted to breaking into small fragments the great masses of ore scattered about by the previous night's explosions. These are sent to the surface in great tubs attached to wire ropes, which are drawn up by gins worked by horses. Other miners are engaged in boring blast holes in the ore, which displays itself in great wide veins in the granite sides of the vast chasm. These blast holes are charged with gunpowder, each with a match attached. At the end of the day the greater number of the miners are drawn up in the cages or tubs, while a few are left below to light the slow-burning matches attached to about a hundred charged bore holes. The rest of the miners are drawn up, and then begins the tremendous bombardment. I watched the

progress of it from a stage projecting over the wild-looking
yawning gulph.   It was grand to hear the succession of
explosions that filled the bottom of the mine far beneath
me.   Then the volumes of smoke, through the surface of
which masses of rock were sometimes sent whirling up into
the clear blue sky, and fell back again into the pit below.
Such an infernal cannonade I have never witnessed.   In
some respects it reminded me of the crater of Vesuvius,
from which such dense clouds of steam and smoke and fire
are thrown up.   In the course of the night, the suffocating
smoke and sulphureous gases had time to pass away, and
next morning the workmen were ready to begin their
operations as before.

The wonderfully rich iron ore extracted from this great
mine is smelted in blast furnaces with wood charcoal.   The
charcoal is, of course, entirely free from sulphur.   When sent
to Sheffield the iron is placed in fire brick troughs closely
surrounded by powdered charcoal.   After a few days' exposure to a red heat, the iron is converted into splendid
steel, which has given such a reputation to that great manufacturing town.   It is also the steel from which the firm of
Stubbs and Company, of Warrington (to which I have already
referred), produce their famous P. S. files.

After the explosions had ceased at the mine, I went with
one of the managers to see the great forge.   It was a most
picturesque sight to see the forgemen at work with the tilt
hammers under the glowing light of the furnaces.   I inspected the machinery and forge works throughout, and had
thus the opportunity of seeing the whole proceeding, from
the blasting and quarrying of the ore in the mine, the
forging and rolling of the worked iron into their proper
lengths, down to the final stamp or " mark " driven in by the
blow of the tilt hammer at the end of each bar.   Having
now thoroughly examined everything connected with this
celebrated iron mine, I prepared to set out for Stockholm in

DANNEMORA IRON MINE. AFTER A DRAWING BY JAMES NASMYTH.

the same way as I had come.   To prepare the landlord for
my setting out, I again resorted to my pencil.   I made a
drawing of the little gig and pony, with the sun rising, and
the hour at which I wished to start.   He understood it in
a moment, and next morning the trap was at the door at
the specified time.

Before I left Stockholm I made a careful and elaborate
panoramic sketch of the city, as a companion to the one
I had made of Genoa from the harbour a few years before.
I made it from the summit of the King's Park, which is the
favourite pleasure-ground of the people.   I was ferried
across in a little paddle-wheel boat, worked by Dalecarlian
women in their peculiar costumes.   The King's Park, or
Djurgärd, is doubly beautiful, not only from its panoramic
view of the city, the Mälar Lake, and the arm of the Baltic,
which comes up to the Skeppsbron Quay, but also from
the magnificent oak trees with which it is studded.   These
noble trees, as foreground objects, are perfect pictures.   The
masses of rock are grand, and the drives are beautifully
kept.   No wonder that the Swedes are so proud of this
beautiful park, for it is the finest in Europe.

I left Stockholm for Gottenburg by steamer.   This is
one of the most picturesque routes in Sweden.   First, we
passed through the Mälar Lake—one of the most beautiful
pieces of water in the world.   It contains no less than
fourteen hundred islands, mostly covered with wood.   Of
course we did not see one twentieth part of the lake ; we
only steamed along its eastern shore for about twenty
miles on our way to Södertelye, where the Gotha Canal
begins.   We then reached the small Maren Lake, and afterwards an arm of the Baltic.   We passed numberless islands
and rocks and reached the Slatbacken Fiord, which we entered.
Beautiful scenery surrounds the entrance to the fiord.   In
the morning, after rising up the locks between Mariehop and
Wenneberga, and passing through Lakes Roxen and Boren,

we found ourselves at Motala, near the entrance to the
Wettern Lake.

Motala is a place of great importance in the manufacturing industry of Sweden. When I visited it the ironfoundry was in charge of my friend Mr. Caulson. I had
known him several years before in London, and had the highest opinion of his ability as a constructive engineer. He was
surrounded at Motala with everything in the way of excel-
lently arranged workshops, good machine tools, as well as
abundant employment for them. Indeed, this is the largest
ironfoundry in Sweden, where iron steamers, steam-engines,
and rolling mills are made. From its central position it
has a great future before it.

The steamer crosses the lake to Carlsborg, at the entrance to the fiord and canal that leads to Lakes Wiken and
Wenern. The latter is an immense lake—in fact, an inland
sea. During a great part of the time we were out of sight
of land. At length we reached Wenersborg, and passed
down the Charles Canal. A considerable time is required
to enable the steamer to pass from lock to lock—nine locks
in all—down to the level of the Gotha River. During
that time an opportunity was afforded us for seeing the
famous Trollhätten Falls—a very fine piece of Nature's
workmanship.

Before leaving the subject of Sweden, I feel that I must
say a word or two about the Swedish people. I admired
them exceedingly. They are tall, fair, good-looking. They
are among the most civil and obliging people that I have ever
met. I never encountered a rude word or a rude look from
them. In their homes, they are simple and natural. I
liked the pleasing softness of their voices, so sweet and
musical—" a most excellent thing in woman." There was
a natural gentleness in their deportment. All classes, even
the poorest, partook of it. Their domestic habits are excellent. They are fond of their homes ; and, above all things,

they are clean and tidy.   They strew the floors of their
ground apartments with spruce pine twigs, which form a
natural carpet as well as give out a sweet balsamic perfume.
These are swept away every morning and replaced with new.

With all their virtues the Swedes are a most self-

helping   people.
They are hard-work-
ing and honest, true and
straightforward.   In matters
of commerce they are men of
their word.   They are clear-
headed, honest - minded, and
keen in their desire for knowledge.   Their natural simple
common sense enables them to

PART OF TROLLHÄTTEN FALLS.

clear away all parasitical and traditional rubbish from their
minds, and to stand before us as men of the highest excellence.   All happiness and prosperity to dear old Sweden !

I set out from Gottenburg to Helsingborg, along the shores

of the Kattegat. From Helsingborg I crossed the Sound
by a small steamer to Elsinore, famous for its connection
with Hamlet, Prince of Denmark. The old dreary-looking
castle still stands there. From Elsinore I went to Copenhagen, and occupied myself for a few days in visiting the
wonderful museums. There I saw, in the Northern Antiquities Collection, the unwritten history of civilisation in
the stone, bronze, and iron tools which have brought the
world to what it is now. This museum is perfectly unrivalled.
I saw there the first section of kitchen-middens — that is,
the refuse of oyster shells, fish-bones, and other stuff thrown
out by the ancient inhabitants of the country after their
meals ; then the accumulations of rude stone implements,
kelts, arrow-heads, and such like ; then the articles of the
Bronze Age, with war trumpets ; and then the articles of the
early Iron Age, which also contain some remarkable golden
war horns. These are followed by the middle Iron Age, and
then by the later Iron Age. This part of the collection is
superb. But it is impossible for me to describe the wonders
of the museum.

I was greatly interested too by the collection of articles
at the Rosenburg Castle. This is the only museum at
Copenhagen which is not free ; but the price charged is very
small. It contains an extraordinary collection of royal
clothes (what would *Sartor Resartus* say ?), armour, furniture,
drinking vessels, and all manner of antiquities connected
with the Kings of Denmark.

I was especially interested by the collection of royal
drinking vessels, from the earliest, made of wood, down to
the latest, grand gold and silver flagons. What most
amused me in respect to these boozing implements was the
pegs that marked the depths down to which the stalwart
Dane was able to swig at one pull an enormous draught
of wine. In some cases the name and date of the heavy
drinker was engraved on the flagon to record his topical

feat. "Take him a peg down" was the ordinary saying, and the words have become a proverb amongst ourselves. For we unquestionably have derived a great deal of our drinking capabilities from our ancestors the Danes.

The whole of the museums at Copenhagen are excellent. Besides those I have mentioned, are the Ethnographic Museum—the best of its kind ; the Museum of Coins, the most complete I have seen; the Thorwaldsen Museum ; the Mineralogical Museum ; the Zoological Museum, and many more. The custodians are always most kind and civil ; and when they see any visitor interested in the collection, they take the greatest pleasure in going round with him and pointing out the beauty and rarity of the articles, imparting at the same time most interesting information.

Holding the memory of Tycho Brahé in the highest regard, as one of the great pioneers of astronomy, I was much interested by a contemporary portrait of him in the Town Hall; but still more so by the remains of his observatory at the top of the great Round Tower, where he carried on his careful observations by instruments of his own making and design. These, with many additions, he afterwards transported to the island of Hveen, where the remains of his castle and observatory are still to be seen. While I was mounting the Round Tower I could not but think of the footsteps of the great astronomer who has made it classic ground.

I left Copenhagen for Hamburg by coach. After passing through the island of Zealand, I was ferried across to the island of Fyen, and after that proceeded along the mainland of Sleswick and Holstein. I was much pleased with what I saw of the people of these provinces. Their farmhouses and cottages were wonderfully clean and neat. The women were all engaged in scrubbing and polishing. I believe I saw more brass, in the shape of bright door-knockers, during my journey than I had seen in all England. Even the

x

brass and iron hoops round the milk pails, by constant scrubbing, looked like gold and silver. Every window had its neat dimity curtains edged with snow-white trimming. The very flower-pots were painted red, to fetch up their brightness to the general standard. I never saw a more cheerful and happy-looking people than those whom I saw between Copenhagen and Hamburg. They seemed to me to be very like the people of England—especially in the northern and eastern parts—in their oval faces, their bright blue eyes, and their light and golden hair, as well as their active minds and bodies, which enable them to do their work with hearty cheerful energy.

I went from Hamburg to Amsterdam by steamer; and after doing a few days' business I went to take a peep at the fine collections of pictures there, as well as at the Hague. Then I proceeded to Rotterdam, and took ship for England by the Batavian steamer. I reached home safely after my prolonged tour. Everything was going on well at the Bridgewater Foundry. The seeds which I had sown in the northern countries of Europe were already springing up plentifully in orders for machine tools; and the clang of the hammer and the whirl of the lathes and planing machines were never still from morning till night.

CHAPTER XVII.

MORE ABOUT BRIDGEWATER FOUNDRY—WOOLWICH ARSENAL.

THE rapid extension of railways and steam navigation, both at home and abroad, occasioned a largely increased demand for machinery of all kinds. Our order-book was always full; and every mechanical workshop felt the impulse of expanding trade. There was an increased demand for skilled mechanical labour—a demand that was far in excess of the supply. Employers began to outbid each other, and wages rapidly rose. At the same time the disposition to steady exertion on the part of the workmen began to decline.

This state of affairs had its usual effect. It increased the demand for self-acting tools, by which the employers might increase the productiveness of their factories without having resort to the costly and untrustworthy method of meeting the demand by increasing the number of their workmen. Machine tools were found to be of much greater advantage. They displaced hand-dexterity, and muscular force. They were unfailing in their action. They could not possibly go wrong in planing and turning, because they were regulated by perfect modelling and arrangements of parts. They were always ready for work, and never required a Saint Monday.

As the Bridgewater Foundry had been so fortunate as to earn for itself a considerable reputation for mechanical contrivances, the workshops were always busy. They were crowded with machine tools in full action, and exhibited to all comers their effectiveness in the most satisfactory manner. Every facility was afforded to those who desired to see them at work ; and every machine and machine tool that was turned out became in the hands of its employers the progenitor of a numerous family.

Indeed, on many occasions I had the gratification of' seeing my mechanical notions adopted by rival or competitive machine constructors, often without acknowledgment; though, notwithstanding this point of honour, there was room enough for all. Though the parent features were easily recognisable, I esteemed such plagiarisms as a sort of left-handed compliment to their author. I also regarded them as a proof that I had hit the mark in so arranging my mechanical combinations as to cause their general adoption ; and many of them remain unaltered to this day.

The machine tools when in action did not require a skilled workman to guide or watch them. All that was necessary to superintend them was a well-selected labourer. The self-acting machine tools already possessed the requisite ability to plane, to turn, to polish, and to execute the work when firmly placed *in situ*. The work merely required to be shifted from time to time, and carefully fixed for another action of the machine.

Besides selecting clever labourers, I made an extensive use of active handy boys to superintend the smaller class of self-acting machine tools. To do this required very little exertion of muscular force, but only observant attention. In this way the tool did all the working (for the thinking had before been embodied in it), and it turned out all manner of geometrical forms with the utmost correctness. This sort of training educated the perceptive faculties of the lads, and trained their ideas to perfect truth of form, at the same time that it gave them an intimate acquaintance with the nature of the materials employed in mechanical structures. The rapidity with which they acquired the efficiency of thoroughly practical mechanics was most surprising.

As the lads grew in strength they were promoted to the higher classes of work. We gave to the foreman of each department the right to recommend to a special rise of wages any lad who showed an extra intelligent earnestness and assiduity in superintending his machine. This produced an active spirit of emulation, which not only advanced their efficiency but relieved the foreman from a source of irritation in the discharge of his duties. I have already referred to the subject in a former portion of this narrative ; but it cannot be too strongly urged upon the attention of proprietors of mechanical works. Besides making first-rate workmen, this method prevents the lads from getting into habits of workshop dishonesty, skulking, and other annoyances. My system of non-binding of apprentices was the " perfect cure," if I may so speak. All that existed between us was mutual satisfaction with each other, and that alone proved from first to last in every respect a perfect bond.

So completely was the workmen in attendance on self-acting machines relieved from the necessity of labour, that many of the employers, to keep the men from falling asleep, allowed them to attend to other machines within their powers of superintendence. This kept them fully awake. The workmen cheerfully acquiesced in this arrangement, as a relief from tedium, and especially when a shilling extra was added to their wages for each machine superintended. All went well for a time, for men as well as masters. But now came the difficulty. The system was opposed to the rules of the Trades' Union. Their committee held, that setting one man to superintend more than one machine was keeping out of employment some other man who ought to be employed. And yet, at the time that the objection was made, such persons were not to be had. The

increased demand for skilled labour had employed every
spare workman.

Nevertheless the system, in the eyes of the Union, "must
be put down." The demand was made that every machine
must have a Union man to superintend it, and that he
must be paid the full Union regulation wages. All labourers
and lads were to be discharged, and Union men employed
in their places. As the times were good, and the workshops
were full of orders, it was thought by the Union that the
time had come to put the matter to the test. The campaign
was opened by the organisation of a powerful body, entitled
"The Amalgamated Society of Mechanical Engineers." It
included every class of workmen employed in the trade—
ironfounders, turners, fitters, erectors, pattern-makers, and
such like. All were invited to make common cause against
the employers.

In order to make a conspicuous demonstration of their
power, the Council of the Union first attacked the extensive
firm of Platt Brothers, Oldham. The Council sent them a
mandate to discharge all their labourers or other "illegal
hands" from their works—all who were employed in superintending their vast assortment of machinery—and to fill
their places with "legal mechanics" at the then regulation
wages. The plan of the Union was to attack the employers
one by one—to call out the hands of one particular workshop
until the employers were subdued and obeyed the commands
of the Union; and then to attack another employer in the
same way. The sagacity of this policy very much resembled
that of the ostrich, which hides its head in a hole and thinks
it is concealed. The employers knew the drift of the policy,
and took steps to circumvent it.

A mutual defence association was formed, and a decree
was issued that, unless the demand of the Council against
Platt's factory was withdrawn by a certain day, every employer would at once close his concern. The Union, never-

theless, stuck to their guns—but only for a time. A strike
took place. The works of some of the most extensive
employers of labour were closed. Everything was paralysed
for a time ; the men went about with their hands in their
pockets, while the women and children at home were wanting food. After a few weeks the funds of the Amalgamated
Society became so reduced that the men gradually retreated
from the contest.

Meanwhile, such concerns as contrived to keep their
workmen in full employment—of whom we were one—made
use of the occasion to act on the healthy system of what
I have termed "Free trade in ability." We added, so far
as we could, to the number of intelligent labourers, advanced
them to the places which the Unionist workmen had left at
the order of their Council, and thus kept our men on full
wages until the strike was over. This was the last contest
I had with Trades' Unions. One of the results was that I
largely increased the number of self-acting machines, and
gave a still greater amount of employment to my unbound
apprentices. I placed myself in an almost impregnable
position, and showed that I could conduct my business with
full activity and increasing prosperity, and at the same time
maintain good-feeling between employed and employer.

Another important point was this,—that I always took
care to make my foremen comfortable, and consequently
loyal. A great part of a man's success in business consists
in his knowledge of character. It is not so much what he
himself does, as what he knows his heads of departments
can do. He must know them intimately, take cognisance
of the leading points of their character, pick and choose
from them, and set them to the work which they can most
satisfactorily superintend. Edward Tootal, of Manchester,
said to me long before, "Never give your men cause to
look over the hedge." He meant that I should never give
them any reason for looking for work elsewhere. It was

a wise saying, and I long remembered it. I always endeavoured to make my men and foremen as satisfied as possible
with their work, as well as with their remuneration.

I never had any cause to regret that I had struck out an
independent course in managing the Bridgewater Foundry.
The works were always busy. A cheerful sort of content-
ment and activity pervaded the entire establishment. Our
order-book continued to be filled with the most satisfactory
class of entries. The railway trucks in the yard, and the
canal barges at the wharf, presented a busy scene,—showing the influx of raw material and the output of finished
work. This happy state of affairs went on in its regular
course without any special incident worthy of being mentioned. The full and steady influx of prosperity that had
been the result of many years of interesting toil and cheerful exertion, had caused the place to assume the aspect of a
smoothly working self-acting machine.

Being blessed with a sound constitution, I was enabled
to perform all my duties with hearty active good-will. And
as I had occasional journeys to make in connection with our
affairs and interests, these formed a very interesting variety
in the ordinary course of my daily work. The intimate and
friendly intercourse which I was so fortunate as to cultivate
with the heads of the principal engineering firms of my
time, kept me well posted up in all that was new and
advanced in the way of improvements in mechanical processes. I had at the same time many pleasant opportunities
of making suggestions as to further improvements, some of
which took root and yielded results of no small importance.
These visits to my friends were always acceptable, if I
might judge from the hearty tone of welcome with which I
was generally received.

I do not know what may be the case in other classes of
businesses or professions, but as regards engineer mechanists
and metal workers generally, there is an earnest and frank

intercommunication of ideas—an interchange of thoughts and suggestions—which has always been a source of the highest pleasure to me, and which I have usually found thoroughly reciprocated. The subjects with which engineers have to deal are of a wide range, and jealousy in intercommunication is almost entirely shut out. Many of my friends were special " characters." For the most part they had made their own way in the world, like myself. I found among them a great deal of quaint humour. Their talk was quite unconventional; and yet their remarks were well worth treasuring up in the memory as things to be thought about and pondered over. Sometimes they gave the key to the comprehension of some of the grandest functions in Nature, and an insight into the operation of those invariable laws which regulate the universe. For all Nature is, as it were, a grand museum, ruled over by an ever present Almighty Master,—of whose perfect designs and works we are only as yet obtaining hasty and imperfect glimpses.

But to return to my humbler progress. From an early period of my efforts as a mechanical engineer, I had been impressed with the great advantages that would result from the employment of small high pressure steam-engines of a simple and compact construction. These, I thought, might suit the limited means and accommodation of small factories and workshops where motive power was required. The highly satisfactory results which followed the employment of steam-engines of this class, such as I supplied shortly after beginning business in Manchester, led to a constantly increasing demand for them. They were used for hoisting in and out the weighty bales of goods from the lofty Manchester warehouses. They worked the " lifts," and also the pumps of the powerful hydraulic presses used in packing the bales.

These little engines were found of service in a variety of ways. When placed in the lower parts of the building the waste steam was utilised in warming the various apartments of the house.   The steam was conveyed in iron pipes, and thus obviated the risk of fire which attended the use of stoves and open fire-grates.   I remember being much pleased with seeing a neat arrangement of a " hot-closet " heated by the waste steam conveyed from the bottom of the building. This was used for holding the dinners and teas of the minor clerks and workpeople.   Another enclosed place, heated by waste steam, was used for drying wet clothes and jackets during rainy weather.   Much attention was paid by the employers to their workpeople in these respects.   The former exhibited a great deal of kindly thoughtfulness.   But men and master were alike.   It was a source of the greatest pleasure to me, when looking round the warehouses and factories, to see the intelligent steady energy that pervaded every department, from the highest to the lowest.

I never lost sight of the importance of extending the use of my small steam-engine.   It was the most convenient method of applying steam power to individual machines. Formerly, the power to drive a small machine was derived from a very complicated arrangement of shafting and gearing brought from a distant engine.   But by my system I conveyed the power to the machine by means of a steam pipe, which enabled the engine to which it was attached to be driven either fast or slow, or to be stopped or started, just as occasion required.   It might be run while all the other machines were at rest; or, in the event of a break-down of the main engine of the factory, the small engine might still be kept going, or even assist in the repairs of the large one.

An important feature in this mode of conveying power by means of piping—in place of gearing and shifting belts and belt pulleys—was the ease with which the steam could be conveyed into intricate parts of the building.   The pipes which I used were of wrought-iron, similar to those used in

conveying gas.   They could be curved to suit any peculiarity
of the situation; and when the pipes were lapped with felt,
or enclosed in wooden troughs filled with sawdust, the loss
of heat by radiation was reduced to a minimum.   The loss
of power was certainly much less than in the friction of a
long and perhaps tortuous line of shafting.   With steam of
50 lbs. to the inch, a pipe of one-inch bore will convey
sufficient steam to give forth five horse-power at a distance
of two or three hundred feet from the boiler.[1]

I adopted the same practice in working the refined and
complex machines used in printing coloured patterns on
calico.   A great variety of colours have to be transferred
by a combination of rollers—each carrying its proper colour
—which is printed on the calico with the utmost exactness, so
as to result in the complete pattern.   My system of having
a separate engine to give motion to these colour-printing
machines was found to be of great service, and its value
was recognised by its speedy and almost universal adoption.
Every connection with the main shaft, with its gearing and
belts and pulleys—by which colour-printing had before
been accomplished—was entirely done away with, and each
machine had its own special engine.   The former practice
had led to much waste, and the printing was often confused
and badly done.   The power was conveyed from a great
central steam-engine; the printing machines were ranged
by the side of a long gallery, and by means of a " clutch "
each machine was started at once into action.   The result
of this was a considerable shock to the machine, and an
interference with the relative adjustments of the six or
eight colour rollers, which were often jerked out of their

---

[1] In the case of rambling premises, such as iron shipbuilding yards, the
conveyance of steam by well protected pipes put underground for the purpose
of driving engines to work punching and plate-shearing machines (which have
to be near at hand when the work is required), has very great practical
advantages.

exact relative adjustment. Then the machines had to be
stopped and the rollers readjusted, and sometimes many
yards of calico had been spoiled before this could be
accomplished.

These difficulties were now entirely removed. When all
was adjusted, the attendant of the print-machine had only to
open slightly the steam admission valve of his engine, and
allow it to work the machine gently at its first off-go; and
when all was seen to be acting in perfect concert, to open
the valve further and allow the machine to go at the full
speed. The same practice was adopted in slowing off the
machine, so as to allow the attendant to scrutinise the
pattern and the position of the work, or in stopping the
machine altogether. So satisfactory were the results of the
application of this mode of driving calico-printing machines,
that it was adopted for the like processes as applied to other
textile fabrics; and it is now, I believe, universally applied
at home as well as abroad.

I may also add that the waste steam, as it issued from
the engine after performing its mechanical duty there, was
utilised in a most effective manner by heating a series of
steam-tight cylinders, over which the printed cloth travelled
as it issued from the printing machine, when it was speedily
and effectively dried. In these various improvements in
calico printing I was most ably seconded by Mr. Joseph
Lese of Manchester, whose practical acquaintance with all
that related to that department of industry rendered him of
the greatest service. There was no "Invention," so to
speak, in this almost obvious application of the steam-engine
to calico printing. It required merely the faculty of observation, and the application of means to ends. The main
feature of the system, it will be observed, was in enabling
the superintendent of each machine to have perfect control over it,—to set it in motion and to regulate its speed
without the slightest jerk or shock to its intricate mechan-

ism. In this sense the arrangement was of great commercial value.

I had another opportunity of introducing my small engine system into the Government Arsenal at Woolwich. In 1847 the attention of the Board of Ordnance was directed to the inadequacy of the equipment of the workshops there. The mechanical arrangements, the machine tools, and other appliances, were found insufficient for the economical production of the apparatus of modern warfare. The Board did me the honour to call upon me to advise with them, and also with the heads of departments at the arsenal. Sir Thomas Hastings, then head of the Ordnance, requested me to accompany him at the first inspection. I made a careful survey of all the workshops, and although the machinery was very interesting as examples of the old and primitive methods of producing war material, I found that it was better fitted for a Museum of Technical Antiquity than for practical use in these days of rapid mechanical progress. Everything was certainly very far behind the arrangements which I had observed in foreign arsenals.

The immediate result of my inspection of the workshops and the processes conducted within them was, that I recommended the introduction of machine tools specially adapted to economise labour, as well as to perfect the rapid production of war material. In this I was heartily supported by the heads of the various departments. After several conferences with them, as well as with Sir Thomas Hastings, it was arranged that a large extension of the workshop space should be provided. I was so fortunate as to make a happy suggestion on this head. It was, that by a very small comparative outlay nearly double the workshop area might be provided—by covering in with light iron roofs the long wide roadway spaces that divided the parallel ranges of workshops from each other.

This plan was at once adopted. Messrs. Fox and

Henderson, the well-known railway roofing contractors, were entrusted with the order; and in a very short time the arsenal was provided with a noble set of light and airy workshops, giving ample accommodation for present requirements, as well as surplus space for many years to come. In order to supply steam power to each of these beautiful workshops, and for working the various machines placed within them, I reverted to my favourite system of small separate steam-engines. This was adopted, and the costly ranges of shafting that would otherwise have been necessary were entirely dispensed with.

A series of machine tools of the most improved modern construction, specially adapted for the various classes of work carried on in the arsenal, together with improved ranges of smiths' forge hearths, blown by an air blast supplied by fans of the best construction, and a suitable supply of small hand steam hammers, completed the arrangements; and quite a new era in the forge work of the arsenal was begun. I showed the managers and the workmen the docile powers of the steam hammer, in producing in a few minutes, by the aid of dies, many forms in wrought-iron that had heretofore occupied hours of the most skilful smiths, and that, too, in much more perfect truth and exactitude. Both masters and men were delighted with the result: and as such precise and often complex forms of wrought-iron work were frequently required by hundreds at a time for the equipment of naval gun carriages and other purposes, it was seen that the steam hammer must henceforward operate as a powerful instrument in the productions of the arsenal.

In the introduction of all these improvements I received the frank and cordial encouragement of the chief officers of the Board of Ordnance and Admiralty. My suggestions were zealously carried out by Colonel J. N. Colquhoun, then head of the chief mechanical department of the Ordnance works at Woolwich. He was one of the most clear-headed and intelligent men I have ever met with. He had in a special degree that happy power of inspiring his zeal and energy into all who worked under his superintendence, whether foremen or workmen. A wonderfully sympathetic effect is produced when the directing head of the establishment is possessed of the valuable faculty of cheerful and well-directed energy. It works like an electric thrill, and soon pervades the whole department. I may also mention General Dundas, director of the Royal Gun Foundry, and General Hardinge, head of the Royal Laboratories.[1] This latter department included all processes connected with explosives. It was superintended by Captain Boxer, an officer of the highest talent and energy, who brought everything under his control to the highest pitch of excellence. I must also add a most important person, my old and much esteemed friend John Anderson, then general director of the Machinery of the arsenal. He was an admirable mechanic, a man of clear practical good sense and judgment, and he eventually raised himself to the highest position in the public service.

The satisfactory performance of the machinery which had been supplied to the workshops of the royal dockyards and arsenals, led to further demands for similar machinery for foreign Governments. Foreign visitors were allowed freely to inspect all that had been done. Whatever may be said of the wisdom of this proceeding, it is certainly true that no mechanical improvement can long be kept secret nowadays. Everything is published and illustrated in our engineering journals. And if the foreigners had not been allowed to obtain their new machines from England, they were pro-

---

[1] The term "Laboratory" may appear an odd word to use in connection with machinery and mechanical operations. Yet its original signification was quite appropriate, inasmuch as it related to the preparation of explosive substances, such as shells, rockets, fusees, cartridges, and percussion caps, where chemistry was as much concerned as mechanism in producing the required results.

vided with facilities enough for constructing them for themselves.   At all events, one result of the improved working of the new machines at the Royal Arsenal at Woolwich, was the receipt of large orders by our firm for the supply of foreign Governments.   For instance, that of Spain employed us liberally, principally for the equipment of the royal dockyards of Ferrol and Cartagena.   These orders came to us through Messrs. Zuluatta Brothers, who conducted their proceedings with us in a prompt and business-like way for many years.   Through the same firm we obtained orders to furnish machinery for the Spanish royal dockyard at Havana.

In 1849 we received an extensive order from the Russian Government.   This was transmitted to us through the Imperial Consulate in London.   The machinery was required for the equipment of a very extensive rope factory at the naval arsenal of Nicolaiev, on the Black Sea.   This order included all the machinery requisite for the factory, from the heckling of the hemp to the twisting of the largest ropes and cables required in the Russian naval service. The design and organisation of this machinery in its minutest detail caused me to make a special study of the art of rope-making.   It was a comparatively new subject to me ; but I found it full of interest.   It was a difficulty, and therefore to be overcome.   And in this lies a great deal of the pleasure of contriving and inventing.

During the progress of the work I had the advantage of the frequent presence of an able Russian officer, Captain Putchkraskey, whose intelligent supervision was a source of much satisfaction.   We had also occasional visits from Admiral Kornileff, a man of the highest order of intelligence. He was not only able to appreciate our exertions to execute the order in first-rate style, but to enter into all the special details and contrivances of the work while in progress.   I had often occasion to meet Russian officers while at the

Bridgewater Foundry.   They were usually men of much ability, selected by the Russian Government to act as their agents abroad, in order to keep them well posted up in all that had a bearing upon their own interests.   They certainly reflected the highest credit on their Government, as proving their careful selection of the best men to advance the interests of Russia.

During the visit of the Grand Duke Constantine to England about that time, he resided for some days with the Earl of Ellesmere at Worsley Hall, about a mile and a half from Bridgewater Foundry.   We were favoured with several visits from the Grand Duke, accompanied by Baron Brunnow, Admiral Heyden, and several other Russian officials.   They came by Lord Ellesmere's beautiful barge, which drew up alongside our wharf, where the party landed and entered the works.   The Grand Duke carefully inspected the whole place, and expressed himself as greatly pleased with the complete mastery which man had obtained over obdurate materials, through the unfailing agency of mechanical substitutes for manual dexterity and muscular force.

I was invited to meet this distinguished party at Worsley Hall on more than one occasion, and was much pleased with the frank and intelligent conversation of the Grand Duke, in his reference to what he had seen in his visits to our works.   It was always a source of high pleasure to me to receive visits from Lord Ellesmere, as he was generally accompanied by men of distinction who were well able to appreciate the importance of what had been displayed before their eyes.   The visits, for instance, of Rajah Brooke, the Earl of Elgin, the Duke of Argyll, Chevalier Bunsen, and Count Flahault, stand out bright in my memory.

But to return to my rope-making machinery.   It was finished to the satisfaction of the Russian officers.   It was sent off by ship to the Black Sea, in July 1851, and fitted up at Nicolaiev shortly after.   I received a kind and press-

Y

ing invitation from Admiral Kornileff to accompany him
on the first trip of a magnificent steamer which had been
constructed in England under his supervision.   His object
was, not only that I might have a pleasant voyage in his
company, but that I might see my machinery in full action
at Nicolaiev, and also that I might make a personal survey
of the arsenal workshops at Sebastopol.   It would, no doubt,
have been a delightful trip, but it was not to be.   The
unfortunate disruption occurred between our Government and
that of Russia, which culminated in the disastrous Crimean
War.   One of the first victims was Admiral Kornileff.   He
was killed by one of our first shots while engaged in placing
some guns for the defence of the entrance to the harbour of
Sebastopol.

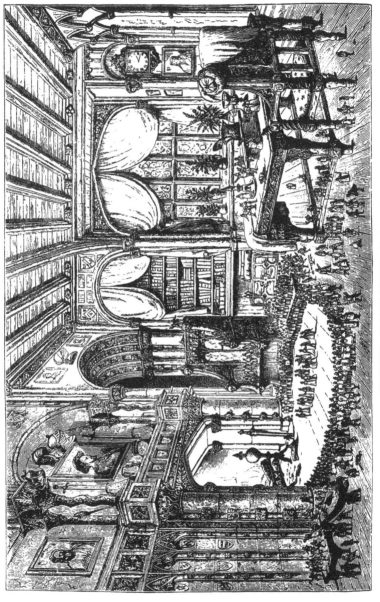

THE FAIRIES. BY JAMES NASMYTH. (FACSIMILE.)

CHAPTER XVIII.

ASTRONOMICAL PURSUITS.

LET me turn for a time from the Foundry, the whirr of the self-acting tools, and the sound of the steam hammers, to my quieter pursuits at home. There I had much tranquil enjoyment in the company of my dear wife. I had many hobbies. Drawing was as familiar to me as language. Indeed, it was often my method of speaking. It has always been the way in which I have illustrated my thoughts. In the course of my journeys at home and abroad I made many drawings of places and objects, which were always full of interest, to *me* at least; and they never ceased to bring up a store of happy thoughts.

Now and then I drew upon my fancy, and with pen and ink I conjured up " The Castle of Udolpho," " A Bit of Old England," " The Fairies are Out," and " Everybody for Ever." The last is crowded with thousands of figures and heads, so that it is almost impossible to condense the drawing into a small compass. To these I added " The Alchemist," " Old Mortality," " Robinson Crusoe," and a bit of English scenery, which I called " Gathering Sticks." I need not say with how much pleasure I executed these drawings in my evening hours. They were not " published," but I drew them with lithographic ink, and had them printed by Mr. Maclure. I afterwards made presents of the series to some of my most intimate friends.

In remembrance of the great pleasure which I had derived from the perusal of Washington Irving's fascinating works, I sent him a copy of my sketches. His answer was charming and characteristic. His letter was dated " Sunnyside," Massachusetts, where he lived. He said (17th January 1859) :—

" DEAR SIR—Accept my most sincere and hearty thanks for the exquisite fancy sketches which you have had the kindness to send me, and for the expressions of esteem and regard in the letter which accompanied them. It is indeed a heartfelt gratification to me to think that I have been able by any exercise of my pen to awaken such warm and delicate sympathies, and to call forth such testimonials of pleasure and approbation from a person of your cultivated taste and intellectual elevation. With high respect and regard, I remain, my dear sir, your truly obliged friend, WASHINGTON IRVING."

Viscount Duncan, afterwards Earl Camperdown, also acknowledged receipt of the drawings in a characteristic letter. He said :—" We are quite delighted with them, especially with ' The Fairies,' which a lady to whom I showed them very nearly stole, as she declared that it quite realised her dreams of fairyland. I am only surprised that amidst your numerous avocations you have found time to execute such detailed works of art ; and I shall have much pleasure in being reminded as I look at the drawings that the same hand and head that executed them invented the steam hammer, and many other gigantic pieces of machinery which will tend to immortalise the Anglo-Saxon race."

But my most favourite pursuit, after my daily exertions at the Foundry, was Astronomy. There were frequently clear nights when the glorious objects in the Heavens were seen in most attractive beauty and brilliancy. I cannot find words to express the thoughts which the impressive grandeur of the Stars, seen in the silence of the night, sug-

gested to me; especially when I directed my Telescope,
even at random, on any portion of the clear sky, and considered that *each* Star of the multitude it revealed to me,
was a SUN! the centre of a system! Myriads of such stars,
invisible to the unassisted eye, were rendered perfectly
distinct by the aid of the telescope. The magnificence of
the sight was vastly increased when the telescope was
directed to any portion of the Milky Way. It revealed
such countless multitudes of stars that I had only to sit
before the eyepiece, and behold the endless procession of
these glorious objects pass before me. The motion of the
earth served but to change this scene of inexpressible magnificence, which reached its climax when some such object
as the "Cluster in Hercules" came into sight. The com-
ponent stars are so crowded together there as to give the
cluster the appearance of a *gray spot*; but when examined
with a telescope of large aperture, it becomes resolved into
such myriads of stars as to defy all attempts to count them.
Nothing can convey to the mind, in so awful and impressive
a manner, the magnificence and infinite extent of Creation,
and the inconceivable power of its Creator!

I had already a slight acquaintance with Astronomy.
My father had implanted in me the first germs. He was a
great admirer of that sublimest of sciences. I had obtained
a sufficient amount of technical knowledge to construct
in 1827 a small but very effective reflecting telescope
of six inches diameter. Three years later I initiated Mr.
Maudsley into the art and mystery of making a reflecting
telescope. I then made a speculum of ten inches diameter,
and but for the unhappy circumstance of his death in
1831, it would have been mounted in his proposed observatory at Norwood. After I had settled down at Fireside,
Patricroft, I desired to possess a telescope of considerable
power in order to enjoy the tranquil pleasure of surveying
the heavens in their impressive grandeur at night.

As I had all the means and appliances for casting specula at the factory, I soon had the felicity of embodying all my former self-acquired skill in this fine art by producing a very perfect casting of a ten-inch diameter speculum. The alloy consisted of fifteen parts of pure tin and thirty-two parts of pure copper, with one part of arsenic. It was cast with perfect soundness, and was ground and polished by a machine which I contrived for the purpose. The speculum was so brilliant that when my friend William Lassell saw it, he said " it made his mouth water." It was about this time (1840) that I had the great happiness of becoming acquainted with Mr. Lassell, and profiting by his devotion to astronomical pursuits and his profound knowledge of the subject.[1] He had acquired much technical skill in the construction of reflecting telescopes, and the companionship between us was thus rendered very agreeable. There was

---

[1] Mr. Lassell was a man of superb powers. Like many others who have done so much for astronomy, he started as an amateur. He was first apprenticed to a merchant at Liverpool. He began business as a brewer. Eventually he devoted himself to astronomy and astronomical mechanics. When in his twenty-first year he began constructing reflecting telescopes for himself. He proceeded to make a Newtonian of nine inches' aperture, which he erected in an observatory at his residence near Liverpool, happily named "Starfield." With this instrument he worked diligently, and detected the sixth star in the trapezium of *Orion*. In 1844 he conceived the bold idea of constructing a reflector of two feet aperture, and twenty feet local length, to be mounted equatorially. Sir John Herschel, in mentioning Mr. Lassell's work, did me the honour of saying " that in Mr. Nasmyth he was fortunate to find a mechanist capable of executing in the highest perfection all his conceptions, and prepared by his own love of astronomy and practical acquaintance with astronomical observations, and with the construction of specula, to give them their full effect." With this fine instrument Mr. Lassell discovered the satellite of *Neptune*. He also discovered the eighth satellite of *Saturn*, of extreme minuteness, as well as two additional satellites of *Uranus*. But perhaps his best work was done at Malta with a much larger telescope, four feet in aperture, and thirty-seven feet focus, erected there in 1861. He remained at Malta for three years, and published a catalogue of 600 new nebulæ, which will be found in the *Memoirs of the Royal Astronomical Society*.

an intimate exchange of opinions on the subject, and my friendship with him continued during forty successive years. I was perhaps a little ahead of him in certain respects. I had more practical knowledge of casting, for I had begun when a boy in my bedroom at Edinburgh. In course of time I contrived many practical " dodges " (if I may use such a word), and could nimbly vault over difficulties of a special kind which had hitherto formed a barrier in the way of amateur speculum makers when fighting their way to a home-made telescope.

I may mention that I know of no mechanical pursuit in connection with science, that offers such an opportunity for practising the technical arts, as that of constructing from first to last a complete Newtonian or Gregorian Reflecting Telescope. Such an enterprise brings before the amateur a succession of the most interesting and instructive mechanical arts, and obliges the experimenter to exercise the faculty of delicate manipulation. If I were asked what course of practice was the best to instil the finest taste for refined mechanical work, I should say, set to and make for yourself from first to last a reflecting telescope with a metallic speculum. Buy nothing but the raw material, and work your way to the possession of a telescope by means of your own individual labour and skill. If you do your work with the care, intelligence, and patience that is necessary, you will find a glorious reward in the enhanced enjoyment of a night with the heavens—all the result of your own ingenuity and handiwork. It will prove a source of abundant pleasure and of infinite enjoyment for the rest of your life.

I well remember the visit I received from my dear friend Warren de la Rue in the year 1840. I was executing some work for him with respect to a new process which he had contrived for the production of white lead. I was then busy with the casting of my thirteen-inch speculum. He watched my proceedings with earnest interest and most

careful attention. He told me many years after, that it was
the sight of my special process of casting a sound speculum
that in a manner caused him to turn his thoughts to practical
astronomy, a subject in which he has exhibited such noble
devotion as well as masterly skill.   Soon after his visit I
had the honour of casting for him a thirteen-inch speculum,
which he afterwards ground and polished by a method of
his own.   He mounted it in an equatorial instrument of
such surpassing excellence as enabled him, aided by his
devotion and pure love of the subject, to record a series of
observations and results which will hand his name down to
posterity as one of the most faithful and patient of astronomical observers.

But to return to my own little work at Patricroft.   I
mounted my ten-inch home-made reflecting telescope, and
began my survey of the heavens.   Need I say with what
exquisite delight the harmony of their splendour filled me.
I began as a learner, and my learning grew with experience.
There were the prominent stars, the planets, the Milky Way
—with thousands of far-off Suns—to be seen.   My observations were at first merely general; by degrees they became
particular.   I was not satisfied with enjoying these sights
myself; I made my friends and neighbours sharers in my
pleasure; and some of them enjoyed the wonders of the
heavens as much as I did.

In my early use of the telescope I had fitted the speculum into a light square tube of deal, to which the eye-piece
was attached, so as to have all the essential parts of the
telescope combined together in the most simple and portable
form.   I had often to move it from place to place in
my small garden at the side of the Bridgewater Canal, in
order to get it clear of the trees and branches which inter-
cepted some object in the heavens which I wished to see.
How eager and enthusiastic I was in those days !   Sometimes I got out of bed in the clear small hours of the

FIRESIDE, PATRICROFT. AFTER A DRAWING BY JAMES NASMYTH.

morning, and went down to the garden in my night-shirt. I would take the telescope in my arms and plant it in some suitable spot, where I might get a peep at some special planet or star then above the horizon.

It became bruited about that a ghost was seen at Patricroft ! A barge was silently gliding along the canal near midnight, when the boatman suddenly saw a figure in white. " It moved among the trees with a coffin in its arms !" The apparition was so sudden and strange that he immediately concluded that it was a ghost. The weird sight was reported all along the canal, and also at Wolverhampton, which was the boatman's headquarters. He told the people at Patricroft on his return journey what he had seen, and great was the excitement produced. The place was haunted ; there was no doubt about it ! After all, the rumour was founded on fact, for the ghost was merely myself in my night-shirt, and the coffin was my telescope, which I was quietly shifting from one place to another in order to get a clearer sight of the heavens at midnight.

My ambition expanded. I now resolved to construct a reflecting telescope of considerably greater power than that which I possessed. I made one of twenty inches diameter, and mounted it on a very simple plan, thus removing many of the inconveniences and even personal risks that attend the use of such instruments. It had been necessary to mount steps or ladders to get at the eye-piece, especially when the objects to be observed were at a high elevation above the horizon. I now prepared to do some special work with this instrument. In 1842 I began my systematic researches upon the Moon. I carefully and minutely scrutinised the marvellous details of its surface, a pursuit which I continued for many years, and still continue with ardour until this day. My method was as follows :—

I availed myself of every favourable opportunity for carrying on the investigation. I made careful drawings

with black and white chalk on large sheets of gray-tinted
paper, of such selected portions of the Moon as embodied the
most characteristic and instructive features of her wonderful
surface.   I was thus enabled to graphically represent the
details with due fidelity as to form, as well as with regard to
the striking effect of the original in its masses of light and
shade.   I thus educated my eye for the special object by
systematic and careful observation, and at the same time
practised my hand in no less careful delineation of all that
was so distinctly presented to me by the telescope—at the
side of which my sheet of paper was handily fixed.   I
became in a manner familiar with the vast variety of those
distinct manifestations of volcanic action, which at some
inconceivably remote period had produced these wonderful
features and details of the moon's surface.   So far as could
be observed, there was an entire absence of any agency of
change, so that their formation must have remained absolutely intact since the original cosmical heat of the moon
had passed rapidly into space.   The surface, with all its
wondrous details, presents the same aspect as it did probably millions of ages ago.

   This consideration vastly enhances the deep interest with
which we look upon the moon and its volcanic details.   It is
totally without an atmosphere, or of a vapour envelope, such
as the earth possesses, and which must have contributed to
the conservation of the cosmical heat of the latter orb.   The
moon is of relatively small mass, and is consequently inferior
in heat-retaining power.   It must thus have parted with its
original stock of cosmical heat with such rapidity as to
bring about the final termination of those surface changes
which give it so peculiar an aspect.   In the case of the
earth the internal heat still continues in operation, though
in a vastly reduced degree of activity.   Again, in the case
of the moon, the total absence of water as well as atmosphere
has removed from it all those denudative activities which,

in the earth, have acted so powerfully in effecting changes of its surface as well as in the distribution of its materials. Hence the appearance of the wonderful details of the moon's surface presents us with objects of inconceivably remote antiquity.

Another striking characteristic of the moon's surface is the enormous magnitude of its volcanic crater formations.

GENERAL STRUCTURE OF LUNAR CRATERS.

In comparison with these, the greatest on the surface of the earth are reduced to insignificance. Paradoxical as the statement may at first appear, the magnitude of the remains of the primitive volcanic energy in the moon is simply due to the smallness of its mass. Though only about one-eightieth part of the size of the earth, the force of gravity on the moon's surface is only about one-sixth. And as eruptive force is quite independent, *as a force*, of the law of gravitation, and as it acted with its full energy on matter, which in the moon is little heavier than cork, it was dispersed in divergent flight from the vent of the volcanoes, free from any atmospheric resistance, and thus secured an enormously wider dispersion of the ejected scoriæ. Hence the building up of those enormous ring-formed craters which are seen in such vast numbers on the moon's surface—some

PICO. AN ISOLATED LUNAR MOUNTAIN 11,000 FEET HIGH.

of them being no less than a hundred miles in diameter, with which those of Etna and Vesuvius are the merest molehills in comparison.

I may mention, in passing, that the frequency of a central cone within these ring-shaped lunar craters supplies us with one of the most distinct and unquestionable evidences of the true nature and mode of the formation of volcanoes. They are the result of the expiring energy of the volcanic discharge, which, when near its termination, not having

sufficient energy to eject the matter far from its vent, becomes deposited around it, and thus builds up the central
cone as a sort of monument to commemorate its expiring
efforts. In this way it recalls the exact features of our own
terrestrial craters, though the latter are infinitely smaller in
comparison. When we consider how volcanoes are formed
—by the ejection and exudation of material from beneath the
solid crust—it will be seen how the lunar eminences are
formed ; that is, by the forcible projection of fluid molten
matter through cracks or vents, through which it makes its
way to the surface.

It was in reference to this very interesting subject that
I made a drawing of the great isolated volcanic mountain
Pico, about 11,000 feet high. It exhibits a very different
appearance from that of our mountain ranges, which are for
the most part the result of a tangential action. The hard
stratified crust of the earth has to adapt itself to the shrunken
diameter of the once much hotter globe. This tangential
action is illustrated in our own persons, when age causes
the body to shrink in bulk, while the skin, which does not
shrink, but has to accommodate itself to the shrunken interior,
and so forms *wrinkles*—the wrinkles of age. This theory
opens up a chapter in geology and physiology well worthy
of consideration. It may alike be seen in the old earth, in
an old apple, and in an old hand.[1]

While earnestly studying the details of the moon's surface, it was a source of great additional interest to me to
endeavour to realise in the mind's eye the possible *landscape
effect* of their marvellous elevations and depressions. Here
my artistic faculty came into operation. I endeavoured to
illustrate the landscape scenery of the Moon, in like manner
as we illustrate the landscape scenery of the Earth. The
telescope revealed to me distinctly the volcanoes, the craters,

---

[1] The shrunken hand on the other side is that of Mr. Nasmyth, photographed by himself.—ED.

the cracks, the projections, the hollows—in short, the light
and shade of the moon's surface.   One of the most prominent conditions of the awful grandeur of lunar scenery is the
brilliant light of the sun, far transcending that which we

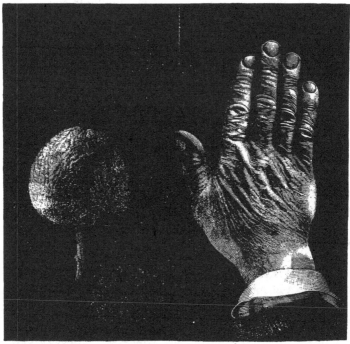

SHRUNKEN APPLE AND HAND [1]

experience upon the earth.   It is enhanced by the contrast

---

[1] These illustrations serve to illustrate one of the most potent of geological
agencies which has given the earth's surface its grandest characteristics.   I
mean the elevation of mountain ranges through the contraction of the globe
as a whole.   By the action of gravity the former larger surface crushes down,
as it were, the contracting interior ; and the superfluous matter, which belonged to a bigger globe, arranges itself by tangential displacement, and
accommodates itself to the altered or decreased size of the globe.   Hence our
mountain ranges, which though apparently enormous when seen near at hand,
are merely the *wrinkles* on the face of the earth.

with the jet-black background of the lunar heavens,—the result of the total absence of atmosphere. One portion of the moon, on which the sun is shining, is brilliantly illuminated, while all in shade is dark.

LUNAR MOUNTAINS AND EXTINCT VOLCANIC CRATERS.

SCALE OF 30000 MILES

While the disc of the sun appears a vast electric light of overpowering rayless brilliancy, every star and planet in the black vault of the lunar heavens is shining with *steady* brightness at all times; as, whether the Sun be present or absent during the long fourteen days' length of the lunar

day or night, no difference on the absolutely black aspect
of the lunar heavens will appear. That aspect will be
eternal there. No modification of the darkness of shadows
in the moon can result from the illuminative effect, as in our
case on the earth, from light reflected *into* shadows by the
blue sky of our earthly day.[1] The intensity of the contrast
between light and shade must lend another powerful aspect
to the scenery of the moon, although deprived of all those
charming effects which artists term " aerial perspective," by
which relative distances are often rendered cognisable with
such tender and exquisite beauty. But the absence of
atmosphere on the moon causes the most distant objects to
appear as near as the nearest; while the comparatively rapid
curvature of the moon, owing to its being a globe only one-
fourth the diameter of the earth, necessarily limits very
considerably the range of view.

It is the combination of all these circumstances, which
we know with absolute certainty must exist in the Moon,
that leads to the contemplation of her marvellous surface.
The subject, as revealed by the aid of powerful telescopes,
presents one of the grandest and most deeply interesting
subjects that can occupy the thoughts of man; not only
as regards the physical constitution and the peculiar structure of her surface, as that of our nearest planetary neighbour, but also as our serviceable attendant by night.

Then there are the Tides, so useful to man, preserving
the sanitary condition of the river mouths and tide-swept
shores. We must be grateful for the Moon's existence on
that account alone. She is the grand scavenger and practical
sanitary commissioner of the earth. Then what business she
transacts! She lifts hundreds of ships and barges, filled

---

[1] A small degree of illumination is, however, given to some portions of the
Moon's surface by the *Earth-shine*, when the earth is in such a position with
regard to the moon, as to reflect some light on to it, as the moon does to the
earth.

with valuable cargoes, up our tidal rivers, to the commercial cities seated upon their banks. She performs a vast amount of mechanical drudgery. She is the most effective of all Tugs, though not of steam ; and now that we understand the convertibility and conservation of force, we may be able to use her Tide-producing functions through the agency of electricity. It is even possible that the Tides may yet light our streets and houses !

Is the moon inhabited ? It seems to me that the entire absence of atmosphere and water forbids the supposition— at least of any form of life with which we are acquainted. This adverse condition, from the moon's day being equal to fourteen of our days ; the sun shining with much more brilliancy of effect in the moon than on the earth, where atmosphere and moisture act as an important agent in modifying its scorching rays ; whilst no such agency exists in the moon. The sun shines there without inter- mission for fourteen days and nights. During that time the heat must accumulate to almost the melting point of lead ; while, on the other hand, the absence of the sun for an equal period must be followed by a period of intense cold, such as we have no experience of, even in the Arctic regions. The highest authorities state that the cold during the Moon's night must reach as low as 250 degrees below the freezing point of water. These considerations, I think, conclusively prove that the existence of any form of life in the moon is utterly impossible.

The first occasion on which I exhibited my series of drawings of the Moon, together with a map six feet in diameter of its entire visible surface, was at the meeting of the British Association at Edinburgh in 1850. I always looked forward to these meetings with great pleasure, and attended them with supreme interest. My dear wife always accompanied me. It was our scientific holiday. It was also our holiday of friendship. We met many of our old

z

friends, and made many new friends. Alas, how many of
them have departed ! Herschel, Faraday, Robinson, Taylor,
Phillips, Brewster, Rosse, Fairbairn, Lassell, and a host of
minor stars, who, although perhaps wanting in the brightness or magnitude of those I have named, made good amends
by the warmth of their cheerful rays. We saw the younger
lights emerging above the horizon ; the men who still continue to shed their glory over the meetings of the Association.

How delightful was our visit to Edinburgh in 1850.
It was " my own romantic town." I remembered its salient
features so well. There was the broad mass of the Old
Town, with its endless diversity of light and shade. There
was the grand old fortress, with its towers and turrets and
black portholes. Towards evening the distant glories of the
departing sun threw forward, in dark outline, the wooded
hills of Corstorphine. The rock and Castle assumed a new
aspect every time I looked at them. The long-drawn gardens filling the valley between the Old Town and the New,
and the thickly-wooded scars of the Castle rock, were a
charm of landscape and a charm of art. Arthur's Seat, like
a lion at rest, seemed perfect witchcraft. And from the
streets in the New Town, or from Calton Hill, what singular
glances of beauty were observed in the distance—the gleaming waters of the Firth, and the blue shadows among the
hills of Fife.

I remembered it all, from the days on which I sat, as a
child, beside the lassies watching the " claes " on the Calton
Hill, and hearing the chimes of St. Giles's tinkling across the
Nor' Loch from the Old Town ; the walks, when a boy, in
the picturesque country round Edinburgh, with my father
and his scientific and artistic friends ; my days at the High
School, and then my evenings at the School of Arts ; my
castings of brass in my bedroom, and the technical training
I enjoyed in the workshop of my old schoolfellow ; my

roadway locomotive and its success; and finally, the making of my tools and machines intended for Manchester, at the foundry of my dear old friend Douglass. It all came back to me like a dream. And now, after some twenty years, I had returned to Edinburgh on a visit to the British Association. Many things had been changed—many relatives and friends had departed—and still Edinburgh remained to me as fascinating as ever.

The excursions formed our principal source of enjoyment during these scientific gatherings. The season was then at its happiest. Nature was in her most enjoyable condition, and the excursionists were usually in their holiday mood. The meeting of the British Association at Edinburgh was presided over by the Duke of Argyll. The geologists visited the remarkable displays of volcanic phenomena with which the neighbourhood of Edinburgh singularly abounds. Indeed, Edinburgh owes much of its picturesque beauty to volcanoes and earthquake upheavings. Our excursions culminated in a visit to the Bass Rock. The excursion had been carefully planned, and was successfully carried out. The day was beautiful, and the party was of the choicest. After reaching the little cove of Canty Bay, overlooked by the gigantic ruins of Tantallon Castle, we were ferried across to the Bass, through a few miles of that capricious sea, the Firth of Forth, near to where it joins the German Ocean. We were piloted by that fine old British tar, Admiral Malcolm, while the commissariat was superintended by General Pasley.

We were safely landed on that magnificent sea-girt volcanic rock—the Bass. After inspecting the ruins of what was once a castellated State prison, where the Covenanters were immured for conscience' sake, we wandered up the hill towards the summit. There we were treated to a short lecture by Professor Owen on the Solan Goose, which was illustrated by the clouds of geese flying over us. They freely exhibited their habits on land as well as in mid-air, and skimmed the dizzy crags with graceful and apparently effortless motions. The vast variety of seafowl screamed their utmost, and gave a wonderfully illustrative chorus to the lecture. It was a most impressive scene. We were high above the deep blue sea of the German Ocean, the waves of which leapt up as if they would sweep us away. into the depths below.

Another of our delightful excursions was made under the guidance of my old and dear friend Robert Chambers.[1] The object of this excursion was to visit the remarkable series of grooved and scratched rocks which had been discovered on the western edge of the cliff-like boundary of the Corstorphine Hills.[2] The glacial origin of these groovings on the rocks was then occupying the attention of geologists. It was a subject that Robert Chambers had carefully studied, both in the Lowlands, in the Highlands, in Rhineland, in Switzerland, and in Norway. He had also published his *Ancient Sea Margins* and his *Tracings of the North of Europe* in illustration of his views. He was now enabled to show us these groovings and scratchings on the rocks near Edinburgh. In order to render the records more accessible, he had the heather and mossy turf carefully

---

[1] I cannot pass over the mention of Robert Chambers's name, without adding that I was on terms of the most friendly intimacy with him from a very early period of his life to its termination in 1871. I remember when he made his first venture in business in Leith Walk. By virtue of his industry, ability, and energy, he became a prosperous man. I had the happiness of enjoying his delightful and instructive society on many occasions. We had rare cracks on all subjects, but especially respecting old places and old characters whom we had known at Edinburgh. His natural aptitude to catch up the salient and most humorous points of character, with the quaint manner in which he could describe them, gave a vast charm to his company and conversation. Added to which, the wide range and accuracy of his information, acquired by his own industry and quick-witted penetration, caused the hours spent in his society to remain among the brightest points in my memory.

[2] They had been first seen, some twenty years before, by Sir James Hall, one of the geologic lights of Edinburgh.

removed—especially from some of the most distinct evidences of glacial rock-grooving. Thus no time was lost, and
we immediately saw the unquestionable markings. Such
visits as these are a thousand times more instructive and
interesting than long papers read at scientific meetings.
They afford the best opportunity for interchange of ideas,
and directly produce an emphatic result; for one cannot
cavil about what he has seen with his own eyes and felt
with his own hands.

We returned to the city in time to be present at a most
interesting lecture by Hugh Miller on the Boulder Clay.
He illustrated it by some scratched boulders which he had
collected in the neighbourhood of Edinburgh. He brought
the subject before his audience in his own clear and admirable *viva voce* style. The Duke of Argyll was in the chair,
and a very animated discussion took place on this novel
and difficult subject. It was humorously brought to a
conclusion by the Rev. Dr. Fleming, a shrewd and learned
geologist. Like many others, he had encountered great
difficulties in arriving at definite conclusions on this mysterious subject. He concluded his remarks upon it by
describing the influence it had in preventing his sleeping at
night. He was so restless that his wife became seriously
alarmed. "What's the matter wi' ye, John? are ye ill?"
"Ou no," replied the doctor, "*it's only that confounded
Boulder Clay!*" This domestic anecdote brought down the
house, and the meeting terminated in a loud and hearty
laugh.

I, too, contributed my little quota of information to the
members of the British Association. I had brought with
me from Lancashire a considerable number of my large
graphic illustrations of the details of the Moon's surface. I
gave a *viva voce* account of my lunar researches at a crowded
meeting of the Physical Section A. The novel and interesting subject appeared to give so much satisfaction to my

audience that the Council of the Association desired me to
repeat the account at one of the special evenings, when the
members of all the various sections were generally present.
It was quite a new thing for me to appear as a public
lecturer; but I consented.   The large hall of the Assembly
Rooms in George Street was crowded with an attentive
audience.   The Duke of Argyll was in the chair.   It is a
difficult thing to give a public lecture—especially to a
scientific audience.   To see a large number of faces turned
up, waiting for the words of the lecturer, is a somewhat
appalling sight.   But the novelty of the subject and the
graphic illustrations helped me very much.   I was quite
full of the Moon.   The words came almost unsought; and
I believe the lecture went off very well, and terminated
with "great applause."   And thus the meeting of the
British Association at Edinburgh came to an end.

   This, however, was not the end of our visit to Scotland.
I was strongly urged by the Duke of Argyll to pay him a
visit at his castle at Inveraray.   I had frequently before
had the happiness of meeting the Duke and Duchess at the
Earl of Ellesmere's mansion at Worsley Hall.   He had
made us promise that if we ever came to Scotland we were
not to fail to pay him a visit.   It was accordingly arranged
at Edinburgh that we should carry out our promise, and
spend some days with him at Inveraray before our return
home.   We were most cordially welcomed at the castle, and
enjoyed our visit exceedingly.   We had the pleasure of
seeing the splendid scenery of the Western Highlands—the
mountains round the head of Loch Fyne, Loch Awe, and the
magnificent hoary-headed Ben Cruachan, requiring a base of
more than twenty miles to support him,—besides the beautiful and majestic scenery of the neighbourhood.

   But my chief interest was in the specimens of high
geological interest which the Duke showed me.   He had
discovered them in the Island of Mull, in a bed of clay

shale, under a volcanic basaltic cliff over eighty feet high, facing the Atlantic Ocean. He found in this bed many beautifully perfect impressions of forest tree leaves, chiefly of the plane-tree class. They appeared to have been enveloped in the muddy bottom of a lake, which had been sealed up by the belching forth from the bowels of the earth of molten volcanic basaltic lava, and which indeed formed the chief material of the Island of Mull. This basaltic cliff now fronts the Atlantic, and resists its waves like a rock of iron. To see all the delicate veins and stalklets, and exact forms of what had once been the green fresh foliage of a remotely primeval forest, thus brought to light again, as preserved in their clay envelope, after they had lain for ages and ages under what must have been the molten outburst of some tremendous volcanic discharge, and which now formed the rock-bound coast of Mull, filled one's mind with an idea of the inconceivable length of time that must have passed since the production of these wonderful geological phenomena.

I felt all the more special interest in these specimens, as I had many years before, on my return visit from Londonderry, availed myself of the nearness of the Giant's Causeway to make a careful examination of the marvellous volcanic columns in that neighbourhood. Having scrambled up to a great height, I found a thick band of hematitic clay underneath the upper bed of basalt, which was about sixty feet thick. In this clay I detected a rich deposit of completely charred branches of what had once been a forest tree. The bed had been burst through by the outburst of molten basalt, and converted the branches into charcoal. I dug out some of the specimens, and afterwards distributed them amongst my geological friends. The Duke was interested by my account, which so clearly confirmed his own discovery. On a subsequent occasion I revisited the Giant's Causeway in company with my dear wife. I again scrambled

up to the hematitic bed of clay under the basaltic cliff, and
dug out a sufficient quantity of the charred branches, which
I sent to the Duke, in confirmation of his theory as to the
origin of the leaf-beds at Mull.[1]

In the year following the meeting of the British Association at Edinburgh, the great Exhibition of all nations at
London took place.   The Commissioners appointed for
carrying out this noble enterprise had made special visits
to Manchester and the surrounding manufacturing districts
for the purpose of organising local committees, so that the
machinery and productions of each might be adequately
represented in the World's Great Industrial Exhibition.
The Commissioners were met with enthusiasm ; and nearly
every manufacturer was found ready to display the results
of his industry.   The local engineers and tool-makers were
put upon their mettle, and each endeavoured to do his best.
Like others, our firm contributed specimens of our special
machine tools, and a fair average specimen of the steam
hammer, with a 30 cwt. hammer-block.

I also sent one of my very simple and compact steam-
engines, in the design of which I had embodied the form of
my steam hammer—placing the crank where the anvil of
the hammer usually stands.   The simplicity and grace of

[1] I received the following reply from the Duke of Argyll, dated "Inveraray, Nov. 19, 1850":—

"MY DEAR SIR—Am I right in concluding, from the description which
you were so kind as to send to me, that the lignite bed, with its super-
incumbent basalts, lies *above* those particular columnar basalts which form
the far-famed Giant's Causeway?   I see from your sketch that basalts of
great thickness, and in some veiws beautifully columnar, do underlie the
lignite bed ; but I am not quite sure that these columnar basalts are those
precisely which are called the Causeway.   I had never heard before that the
Giant's Causeway rested on chalk, which all the basalts in your sketch do.

"I have been showing your drawing of 'Udolpho Castle' and 'The Astrologer's Tower' to the Duchess of Sutherland, who is enchanted with the
beauty of the architectural details, and wishes she had seen them before
Dunrobin was finished ; for hints might have been taken from bits of your
work.—Very truly yours,   ARGYLL."

THE ASTROLOGER'S TOWER—A DAY DREAM. BY JAMES NASMYTH. (FACSIMILE.)

this arrangement of the steam engine was much admired. Its merits were acknowledged in a way most gratifying to me, by its rapid adoption by engineers of every class, especially by marine engineers. It has been adopted for driving the shaft of screw-propelled steamships of the largest kind. The comparatively small space it occupies, its compactness, its *get-at-ability* of parts, and the action of gravity on the piston, which, working vertically, and having no undue action in causing wearing of the cylinder on one side (which was the case with horizontal engines), has now brought my steam-hammer engine into almost universal use.[1] The Commissioners, acting on the special recommendation of the jury, awarded me a medal for the construction of this form of steam-engine.[2] As it was merely a judicious arrangement of the parts, and not, in any correct sense of the term, an invention, I took out no patent for it, and left it free to work its own way into general adoption. It has since been used for high as well as low pressure steam—an arrangement which has come

---

[1] Sir John Anderson, in his Report on the machine tools, textile, and other machinery exhibited at Vienna in 1873, makes the following observations :—" Perhaps the finest pair of marine engines yet produced by France, or any country, were those exhibited by Schneider and Company, the leading firm in France. These engines were not large, but were perfect in many respects ; yet comparatively few of those who were struck with admiration seemed to know that the original of this style of construction came from the same mind as the Steam Hammer. Nasmyth's 'Infant Hercules' was the forerunner of all the steam hammer engines that have yet been made from that type, which is now being so extensively employed for working the screw propeller of steam vessels."

[2] The Council of the Exhibition thus describe the engine in the awards:— "Nasmyth, J., Patricroft, Manchester, a small portable direct-acting steam-engine. The cylinder is fixed, vertical and inverted, the crank being placed beneath it, and the piston working downwards. The sides of the frame which support the cylinder serve as guides, and the bearings of the crank-shaft and fly-wheel are firmly fixed in the bed-plate of the engine. The arrangement is compact and economical, and the workmanship practically good and durable.'

into much favour on account of the great economy of fuel which results from using it.

A Council Medal was also awarded to me for the Steam Hammer. But perhaps what pleased me most was the Prize Medal which I received for my special hobby—the drawings of the Moon's surface. I sent a collection of these, with a map, to the Exhibition. They attracted considerable attention, not only because of their novelty, but because of the accurate and artistic style of their execution. The Jurors, in making the award, gave the following description of them : " Mr. Nasmyth exhibits a well-delineated map of the Moon on a large scale, which is drawn with great accuracy, the irregularities upon the surface being shown with much force and spirit; also separate and enlarged representations of certain portions of the moon as seen through a powerful telescope : they are all good in detail, and very effective."

These drawings attracted the special notice of the Prince Consort. Shortly after the closing of the Exhibition, in October 1851, the Queen and the Prince made a visit to Manchester and Liverpool, during which time they were the guests of the Earl of Ellesmere at Worsley Hall. Finding that I lived near at hand, the Prince expressed his desire to the Earl that I should exhibit to Her Majesty some of my graphic lunar studies. On receiving a note to that effect from the Countess of Ellesmere, I sent a selection of my drawings to the Hall, and proceeded there in the evening. I had then the honour of showing them to the Queen and the Prince, and explaining them in detail. Her Majesty took a deep interest in the subject, and was most earnest in her inquiries. The Prince Consort said that the drawings opened up quite a new question to him, which he had not before had the opportunity of considering. It was as much as I could do to answer the numerous keen and incisive questions which he put to me. They were all so distinct and cogent. Their object was, of course, to draw from me the necessary explanations on this rather recondite subject. I believe, however, that notwithstanding the presence of Royalty, I was enabled to place all the most striking and important features of the Moon's surface in a clear and satisfactory manner before Her Majesty and the Prince.

I find that the Queen in her Diary alludes in the most gratifying manner to the evening's interview. In the *Life of the Prince Consort* (vol. ii. p. 398), Sir Theodore Martin thus mentions the subject:—" The evening was enlivened by the presence of Mr. Nasmyth, the inventor of the steam hammer, who had extensive works at Patricroft. He exhibited and explained the map and drawings in which he had embodied the results of his investigations of the conformations of the surface of the Moon. The Queen in her Diary dwells at considerable length on the results of Mr. Nasmyth's inquiries. The charm of his manner, in which the simplicity, modesty, and enthusiasm of genius are all strikingly combined, are warmly dwelt upon. Mr. Nasmyth belongs to a family of painters, and would have won fame for himself as an artist—for his landscapes are as true to Nature as his compositions are full of fancy and feeling—had not science and mechanical invention claimed him for their own. His drawings were submitted on this occasion, and their beauty was generally admired."[1]

---

[1] In his lecture on the "Geological Features of Edinburgh and its Neighbourhood," in the following year, Hugh Miller, speaking of the Castle Rock, observed:—" The underlying strata, though geologically and in their original position several hundred feet *higher* than those which underlie the Castle esplanade, are now, with respect to the actual level, nearly 200 feet lower. In a lecture on what may be termed the geology of the Moon, delivered in the October of last year before Her Majesty and Prince Albert by Mr. Nasmyth, he referred to certain appearances on the surface of that satellite that seemed to be the results, in some very ancient time, of the sudden falling in of portions of an unsupported crust, or a retreating nucleus of molten matter; and took occasion to suggest that some of the great slips and shifts on the surface of our own planet, with their huge downcasts, may have had a similar origin. The suggestion is at once bold and ingenious."

The next time I visited Edinburgh was in the autumn
of 1853. Lord Cockburn, an old friend, having heard that
I was sojourning in the city, sent me the following letter,
dated " Bonally, 3d September," inviting me to call a meeting
of the Faithful :—

" MY DEAR SIR—Instead of being sketching, as I thought, in
Switzerland, I was told yesterday that you was in Auld Reekie. Then
why not come out here next Thursday, or Friday, or Saturday, and
let us have a Hill day?   I suppose I need not write to summon the
Faithful, because not having been in Edinburgh except once for above
a month, I don't know where the Faithful are.   But you must know
their haunts, and it can't give you much trouble to speak to them.   I
should like to see Lauder here.   And don't forget the Gaberlunzie.[1]
Ever,   H. COCKBURN."

The meeting came off.   I collected a number of special
friends about me, and I took my wife to the meeting
of the Faithful.   There were present David Roberts,
Clarkson Stanfield, Louis and Carl Haag, Sir George
Harvey, James Ballantine, and D. O. Hill—all artists.
We made our way to Bonny Bonally, a charming residence, situated at the foot of the Pentland Hills.[2] The
day was perfect, in all respects " equal to bespoke."   With
that most genial of men, Lord Cockburn, for our guide, we
wandered far up the Pentland Hills.   After a rather toilsome walk we reached a favourite spot.   It was a semicircular hollow in the hillside, scooped out by the sheep for
shelter.   It was carpeted and cushioned with a deep bed of
wild thyme, redolent of the very essence of rural fragrance.

We sat down in a semicircle, our guide in the middle.
He said in his quaint peculiar way, " Here endeth the first

[1] James Ballantine, author of *The Gaberlunzie's Wallet*. In August 1865
Mr. Ballantine wrote to me saying : "If ever you are in Auld Reekie I
should feel proud of a call from you.   I have not forgotten the delightful
day we spent together many years ago at Bonny Bonally with the eagle-eyed
Henry Cockburn !"

[2] The house was afterwards occupied by the lamented Professor Hodgson,
the well-known Political Economist.

lesson." After gathering our breath, and settling ourselves
to enjoy our well-earned rest, we sat in silence for a time.
The gentle breeze blew past us, and we inhaled the fragrant
air. It was enough for a time to look on, for the glorious
old city was before us, with its towers, and spires, and lofty
buildings between us and the distance. On one side
Arthur's Seat, and on the other the Castle, the crown of
the city. The view extended far and wide—on to the
waters of the Forth and the blue hills of Fife. The view
is splendidly described by "Delta" :—

> "Traced like a map, the landscape lies
>   In cultured beauty, stretching wide :
> Here Pentland's green acclivities,—
>   There ocean, with its swelling tide,—
> There Arthur's Seat, and, gleaming through
> Thy southern wing, Dun Edin blue !
> While, in the Orient, Lammer's daughters,—
>   A distant giant range, are seen ;
>   North Berwick Law, with cone of green,
> And Bass amid the waters."

Then we began to crack, our host leading the way with his
humorous observations. After taking our fill of rest and
talk, we wended our way down again, with the "wimplin'
burn" by our side, fresh from the pure springs of the hill,
whispering its welcome to us.

We had earned a good appetite for dinner, which was
shortly laid before us. The bill of fare was national, and
included a haggis :—

> "Fair fa' your honest, sonsie face,
>   Great chieftain o' the puddin' race !
> Weel are ye wordy o' a grace
>   As lang's my arm !"

The haggis was admirably compounded and cooked, and was
served forth by our genial host with all appropriate accom-
paniments. But the most enjoyable was the conversation

of Lord Cockburn, who was a master of the art—quick, ready, humorous, and full of wit. At last, the day came to a close, and we wended our way towards the city.

Let me, however, before concluding, say a few words in reference to my dear departed friend David Oswald Hill. His name calls up many recollections of happy hours spent in his company. He was, in all respects, the incarnation of geniality. His lively sense of humour, combined with a romantic and poetic constitution of mind, and his fine sense of the beautiful in Nature and art, together with his kindly and genial feeling, made him, all in all, a most agreeable friend and companion. " D. O. Hill," as he was generally called, was much attached to my father. He was a very frequent visitor at our Edinburgh fireside, and was ever ready to join in our extemporised walks and jaunts, when he would overflow with his kindly sympathy and humour. He was a skilful draughtsman, and possessed a truly poetic feeling for art. His designs for pictures were always attractive, from the fine feeling exhibited in their composition and arrangement. But somehow, when he came to handle the brush, the result was not always satisfactory—a defect not uncommon with artists. Altogether, he was a delightful companion and a staunch friend, and his death made a sad blank in the artistic society of Edinburgh.

CHAPTER XIX.

MORE ABOUT ASTRONOMY.

ASTRONOMY, instead of merely being an amusement, became my chief study. It occupied many of my leisure hours. Desirous of having the advantage of a Reflecting Telescope of large aperture, I constructed one of twenty inches diameter. In order to avoid the personal risk and inconvenience of having to mount to the eye-piece by a ladder, I furnished the telescope tube with trunnions, like a cannon, with one of the trunnions hollow so as to admit of the eye-piece. Opposite to it a plain diagonal mirror was placed, to transmit the image to the eye. The whole was mounted on a turn-table, having a seat opposite to the eye-piece, as will be seen in the engraving on the other side.

The observer, when seated, could direct the telescope to any part of the heavens without moving from his seat. Although this arrangement occasioned some loss of light, that objection was more than compensated by the great convenience which it afforded for the prosecution of the special class of observations in which I was engaged; namely, that of the Sun, Moon, and Planets.

I wrote to my old friend Sir David Brewster, then living at St. Andrews, in 1849, about this improvement, and he duly congratulated me upon my devotion to astronomical science. In his letter to me he brought to mind many precious memories.

LARGE TELESCOPE, ON TRUNNION TURN-TABLE.

"I recollect," he said, "with much pleasure the many happy hours that I spent in your father's house ; and ever since I first saw you in your little workshop at Edinburgh,—then laying the foundation of your future fortunes,—I have felt a deep interest in your success, and rejoiced at your progress to wealth and reputation.

"I have perused with much pleasure the account you have sent me of your plan of shortening and moving large telescopes, and I shall state to you the opinion which I have formed of it. If you will look into the article 'Optics' in the *Edinburgh Encyclopedia* (vol. xv. p. 643), you will find an account of what has been previously done to reduce by one-half the length of reflecting telescopes. The advantage of substituting, as you propose, a convex for a plane mirror arises from two causes—that a spherical surface is more easily executed than a plane one ; and that the spherical observation of the larger speculum, if it be spherical, will be diminished by the opposite aberration of the convex one. This advantage, however, will disappear if the plane mirror of the old construction is accurately plane ; and in your case, if the large speculum is parabolic and the small one elliptical in their curvature.

"The only objection to your construction is the loss of light : first of one-fourth of the whole incident light by *obstruction*, and then one-half of the remainder by *reflection* from the convex mirror, thus reducing 100 rays of incident light to $37\frac{1}{2}$ before the pencil is thrown out of the tube by a prism or a third reflector. This loss of light, it is true, may be compensated by an additional inch or two to the margin of the large speculum ; but still it is the best part of the large speculum that is made unproductive by the eclipse of it by the convex speculum.

"With regard to the mechanical contrivance which you propose for working the instrument, I think it is singularly ingenious and beautiful, and will compensate for any imperfection in the optical arrangements which are rendered necessary for its adoption. The application of the railway turn-table is very happy, and not less so is the extraction of the image through the hollow trunnions.

"I am much obliged to you for the beautiful drawing of the apparatus for grinding and polishing specula, invented by Mr. Lassell and constructed by yourself. I shall be glad to hear of your further progress in the construction of your telescope ; and I trust that I shall have the pleasure of meeting you and Mr. Lassell at the Birmingham meeting of the British Association."

In the course of the same year (1849) I sent a model of my Trunnion turn-table for exhibition at a lecture at the Royal Institution, given by my old friend Edward Cowper,

2 A

whom I have already referred to. In the model I had
placed a neat little figure of the observer, but the head had
unfortunately been broken off during its carriage to London.
Mrs. Nasmyth had made the wearing apparel; but Edward
Cowper wrote to her, before the lecture, that he had put
"Sir Fireside Brick" all to rights in that respect. His
letter after the lecture was quite characteristic.

"The lecture," he said, "went off very well last night. All the
models performed their duty, and were duly applauded for doing so.
My new equatorial was approved of by astronomers and by instrument-
makers. The last gun I fired was a howitzer, but mounted swivel-gun
fashion ; or a sort of revolving platform, or something like a turn-table
proper—the gunner at the side of the carriage. Do you know anything of the kind ? Bang ! Invented by one Nasmyth. Bang !
The observer is sitting at ease ; the stars are brought down to you
instead of your creeping up a scaffolding after the stars. Well, the
folks came to the table after the lecture, and 'The Nasmyth Telescope'
kept banging away for a quarter of an hour, and was admired by
everybody. The loss of light was not much insisted on, but it was
said that you ran the risk of error of form in three surfaces instead of
two. I see that Sir J. South states that Lord Rosse would increase
the light of his telescope from five to seven by adopting Herschel's
plan.

"De La Rue was quite delighted. He said, 'Well, I congratulate
you on a most splendid lecture—I cannot call it anything else.' My
father, who takes very little interest in these things, said, 'Well,
Edward has made me understand more about telescopes than I ever
did in my life.' The theatre was full, gallery and all. They were
very attentive, and I never felt more comfortable in a lecture. I am
happy to say that, having administered a dose of cement to Mrs.
Nasmyth's friend, Sir Fireside Brick of Green Lanes, he is now in a
convalescent state. The lecture is to be repeated in another fortnight.
With many thanks for your kind assistance, yours very sincerely,
                                                        "EDWARD COWPER."

In the course of my astronomical inquiries I had occasion to consider the causes of the sun's light. I observed
the remarkable phenomena of the variable and sometimes
transitory brightness of the stars. In connection with
geology, there was the evidence of an arctic or glacial

climate in regions where such cannot now naturally exist;
thus giving evidence of the existence of a condition of
climate, for the explanation of which we look in vain for
any at present known cause. I wrote a paper on the
subject, which I sent to the Astronomical Society. It was
read in May 1851. In that paper I wrote as follows:—

"A course of observations on the solar spots, and on the remarkable features which from time to time appear on the sun's surface,
which I have examined with considerable assiduity for several years,
had in the first place led me to entertain the following conclusion:
namely, that whatever be the nature of solar light, its main source
appears to result from an action induced on the *exterior surface* of the
solar sphere,—a conclusion in which I doubt not all who have attentively pursued observations on the structure of the sun's surface will
agree.

"Impressed with the correctness of this conclusion, I was led to
consider whether we might not reasonably consider the true source of
the latent element of light to reside, *not in the solar orb,* but in space
itself; and that the grand function and duty of the sun was to act as
an agent for bringing forth into vivid existence its due portion of the
illuminating or luciferous element, which element I suppose to be
diffused throughout the boundless regions of space, and which in that
case must be perfectly exhaustless.

"Assuming, therefore, that the sun's light is the result of some
peculiar action by which it brings forth into *visible* existence the element of light, which I conceive to be latent in, and diffused throughout space, we have but to imagine the existence of a very probable
condition, namely, the *unequal* diffusion of this light-yielding element,
to catch a glimpse of a reason why our sun may, in common with his
solar brotherhood, in some portions of his vast stellar orbit, have
passed, and may yet have to pass, through regions of space, in which
the light-yielding element may either abound or be deficient, and so
cause him to beam forth with increased splendour, or fade in brilliancy, just in proportion to the richness or poverty of this supposed
light-yielding element as may occur in those regions of space through
which our sun, in common with every stellar orb, has passed, is now
passing, or is destined to pass, in following up their mighty orbits.

"Once admit that this light-yielding element resides in *space,* and
that it is *not* equally diffused, we may then catch a glimpse of the
cause of the variable and transitory brightness of stars, and more
especially of those which have been known to beam forth with such

extraordinary splendour, and have again so mysteriously faded away ; many instances of which abound in historical record.

" Finally, in reference to such a state of change having come over our sun, as indicated by the existence of a glacial period, as is now placed beyond doubt by geological research, it appears to me no very wild stretch of analogy to suppose that in such former periods of the earth's history our sun may have passed through portions of his stellar orbit in which the light-yielding element was deficient, and in which case his brilliancy would have suffered the while, and an arctic climate in consequence spread from the poles towards the equator, and thus leave the record of such a condition in glacial handwriting on the everlasting walls of our mountain ravines, of which there is such abundant and unquestionable evidence. As before said, it is the existence of such facts as we have in stars of transitory brightness, and the above-named evidence of an arctic climate existing in what are now genial climates, that renders some adequate cause to be looked for. I have accordingly hazarded the preceding remarks as suggestive of a cause, in the hope that the subject may receive that attention which its deep interest entitles it to obtain.

" This view of the source of light, as respects the existence of the luciferous element throughout space, accords with the Mosaic account of creation, in so far as that light is described as having been created in the first instance *before the sun* was called forth.[1]

---

[1] Dr. Siemens read a paper before the Royal Society in March 1882, on " A New Theory of the Sun." His views in some respects coincide with mine. Interstellar space, according to Dr. Siemens, is filled with attenuated matter, consisting of highly rarefied gaseous bodies—including hydrogen, oxygen, nitrogen, carbon, and aqueous vapour ; that these gaseous compounds are capable of being dissociated by radiant solar energy while in a state of extreme attenuation ; and that the vapours so dissociated are drawn towards the sun in consequence of solar rotation, are flashed into flame in the photosphere, and rendered back into space in the condition of products of combustion. With respect to the influence of the sun's light on Geology, Dr. Siemens says: " The effect of this continuous outpour of solar materials could not be without very important influences as regards the geological conditions of our earth. Geologists have long acknowledged the difficulty of accounting for the amount of carbonic acid that must have been in our atmosphere at one time or another in order to form with lime those enormous beds of dolomite and limestone of which the crust of our earth is in great measure composed. It has been calculated that if this carbonic acid had been at one and the same time in our atmosphere it would have caused an elastic pressure fifty times that of our present atmosphere ; and if we add the carbonic acid that must have been absorbed in vegetation in order to form our coal-beds we should

Soon after my paper was read, Lord Murray of Henderland, an old friend, then a Judge on the Scottish Bench, wrote to me as follows:—" I shall be much obliged to you for a copy, if you have a spare one, of your printed note on Light. It is expressed with great clearness and brevity. If you wish to have a quotation for it, you may have recourse to the blind Milton, who has expressed your views in his address to Light :—

> " ' Hail, holy Light ! offspring of heaven first-born !
> Or of the Eternal co-eternal beam
> May I express thee unblamed ? since God is light,
> And never but in unapproachèd light
> Dwelt from eternity—dwelt then in thee,
> Bright effluence of bright essence increate ! ' "

About the same time Sir Thomas Mitchell, Surveyor-General of Australia, communicated his notions on the subject. " My dear Sir," he wrote, " Your kind and valuable communications are as welcome to me as the sun's light, and I now thank you most gratefully for the last, with its two enclosures. These, and especially your views as to the source of light, afford me new scope for satisfactory thinking—a sort of treasure one can always carry about, and, unlike other treasures, is most valuable in the solitude of a desert. The beauty of your theory as to the nature of the source of light is, that it rather supports all preconceived notions respecting the soul, heaven, and an immortal state."

I still continued the study of astronomy. The sun, moon, and planets yielded to me an inexhaustible source of delight. I gazed at them with increasing wonder and awe. Among the glorious objects which the telescope reveals, the most impressive is that of the starry heavens in a clear

---

probably have to double that pressure. Animal life, of which we had abundant traces in these ' measures,' could not have existed under such conditions, and we are almost forced to the conclusion that the carbonic acid must have been derived from *an external source*."

dark night.   When I directed my 20-inch reflecting telescope almost at random to any part of the firmament, especially to any portion of the Milky Way, the sight of
myriads of stars brought into view within the field of the
eye-piece was overpoweringly sublime.

When it is considered that every one of these stars
which so bewilderingly crowd the field of vision is, according to rational probability, and, I might even say, absolute
certainty, a Sun as vast in magnitude as that which gives
light to our globe, and yet situated so inconceivably deep in
the abyss of space as to appear minute points of light even
to the most powerful telescope, it will be seen what a sublime aspect of nature appears before us.   Turn the telescope
to any part of the heavens, it is the same.

Let us suppose ourselves perched upon the farthest star
which we are enabled to see by the aid of the most powerful telescope.   There, too, we should see countless myriads
of Suns, rolling along in their appointed orbits, and thus on
and on throughout eternity.   What an idea of the limitless
extent of Creative Power—filling up eternal space with
His Almighty Presence !   The human mind feels its perfect
impotency in endeavouring to grasp such a subject.

I also turned my attention to the microscope.   In
1851 I examined, by the aid of this instrument, the infu-
soria in the Bridgewater Canal.   I found twenty-seven of
them, of the most varied form, colour, and movements.
This was almost as remarkable a revelation as the mighty
phenomena of the heavens.   I found these living things
moving about in the minutest drop of water.   The sight of
the wonderful range of creative power—from the myriads
of suns revealed by the telescope, to the myriads of moving
organisms revealed by the microscope—filled me with
unutterably devout wonder and awe.

Moreover, it seemed to me to confer a glory even upon
the instruments of human skill, which elevated man to the

Unseen and the Divine. When we examine the most minute organisms, we find clear evidence in their voluntary powers of motion that these creatures possess *a will*, and that such Will must be conveyed by a nervous system of an infinitesimally minute description. When we follow out such a train of thought, and contrast the myriads of suns and planets at one extreme, with the myriads of minute organised atoms at the other, we cannot but feel inexpressible wonder at the transcendent range of Creative Power.

Shortly after, I sent to the Royal Astronomical Society a paper on another equally wonderful subject, "The Rotatory Movements of the Celestial Bodies."[1] As the paper is not very long, and as I endeavoured to illustrate my ideas in a familiar manner, I may here give it entire :—

"What first set me thinking on this subject was the endeavour to get at the reason of why water in a basin acquires a rotatory motion when a portion of it is allowed to escape through a hole in the bottom. Every well-trained philosophical judgment is accustomed to observe illustrations of the most sublime phenomena of creation in the most minute and familiar operations of the Creator's laws, one of the most characteristic features of which consists in the absolute and wonderful integrity maintained in their action whatsoever be the range as to magnitude or distance of the objects on which they operate.

"For instance, the minute particles of dew which whiten the grass-blade in early morn are moulded into spheres by the identical law which gives to the mighty sun its globular form !

"Let us pass from the rotation of water in a basin to the consideration of the particles of a nebulous mass just summoned into existence by the fiat of the Creator—the law of gravitation coexisting.

"The first moment of the existence of such a nebulous mass would be inaugurated by the election of a centre of gravity, and, instantly after, every particle throughout the entire mass of such nebulæ would tend to and converge towards that centre of gravity.

"Now let us consider what would be the result of this. It appears to me that the inevitable consequence of the convergence of the particles towards the centre of gravity of such a nebulous mass would

---

[1] "Suggestions respecting the Origin of the Rotatory Movements of the Celestial Bodies and the Spiral Forms of the Nebulæ, as seen in Lord Rosse's Telescope."

not only result in the formation of a nucleus, but by reason of the
physical impossibility that all the converging particles should arrive
at the focus of convergence in directions perfectly radial and diametrically opposite to each other, however slight the degree of deviation
from the absolute diametrically opposite direction in which the converging particles coalesce at the focus of attraction, a twisting action
would result, and Rotation ensue, which, once engendered, be its intensity ever so slight, from that instant forward the nucleus would continue to revolve, and all the particles which its attraction would
subsequently cause to coalesce with it, would do so in directions
tangential to its surface, and not diametrically towards its centre.

"In due course of time the entire of the remaining nebulous mass
would become affected with rotation from the more rapidly moving
centre, and would assume what appears to me to be their inherent
normal condition, namely, spirality, as the prevailing character of their
structure ; and as that is *actually* the aspect which may be said to
characterise the majority of those marvellous nebulæ, as revealed to
us by Lord Rosse's magnificent telescope, I am strongly impressed with
the conviction that such reasons as I have assigned have been the cause
of their spiral aspect and arrangement.

"And by following up the same train of reasoning, it appears to
me that we may catch a glimpse of the primeval cause of the rotation
of every body throughout the regions of space, whether they be nebulæ,
stars, double stars, or planetary systems.

"The primary cause of rotation which I have endeavoured to
describe in the preceding remarks is essentially cosmical, and is the
direct and immediate offspring of the action of gravitation on matter
in a diffused, nebulous, and, as such, highly mobile condition.

"It will be obvious that in the case of a nebulous mass, whose
matter is unequally distributed, that in such a case several sub-centres
of gravity would be elected, that is to say, each patch of nebulous
matter would have its own centre of gravity ; but these in their turn
subordinate to that of the common centre of gravity of the whole
system, about which all such outlaying parts would revolve. Each of
the portions above alluded to would either be attracted by the superior
mass, and pass in towards it as a *wisp* of nebulous matter, or else
establish perfect individual and distinct rotation within itself, and
finally revolve about the great common centre of gravity of the
whole.

"Bearing this in mind, and referring to some of the figures of the
marvellous spiral nebulæ which Lord Rosse's telescope has revealed to
us, I shall now bring these suggestions to a conclusion. I have avoided
expanding them to the extent I feel the subject to be worthy and

capable of ; but I trust such as I have offered will be sufficient to convey a pretty clear idea of my views on this sublime subject, which I trust may receive the careful consideration its nature entitles it to. Let any one carefully reflect on the reason why water assumes a rotatory motion when a portion of it is permitted to escape from an aperture in the bottom of the circular vessel containing it ; if they will do so in the right spirit, I am fain to think they will arrive at the same conclusion as the contemplation of this familiar phenomenon has brought me to.

"BRIDGEWATER FOUNDRY, *June* 7, 1855."

I was present at a meeting of the Geological Society at Manchester in 1853, in the discussions of which I took part. I was very much impressed by an address of the Rev. Dr. Vaughan (then Principal of the Independent College at Manchester), which is as interesting now as it was then. After referring to the influence which geological changes had produced upon the condition of nations, and the moral results which oceans, mountains, islands, and continents have had upon the social history of man, he went on to say : " Is not this island of ours indebted to these great causes ? Oh, that blessed geological accident that broke up a strait between Calais and Dover ! It looks but a little thing ; it was a matter to take place ; but how mighty the moral results upon the condition and history of this country, and, through this country's influence, upon humanity ! Bridge over the space between,[1] and you have directly the huge continental barrack-yard system all over England. And once get into the condition of a great continental military power, and you get the arbitrary power ; you cramp down the people, and you unfit them from being what they ought to be—FREE ! And all the good influences together at work in this country could not have secured us against this, but for that blessed separation between this Isle and the Continent."

---

[1] Tunnels were not thought of at that time.

In 1853 I was appointed a member of the Small Arms
Committee for the purpose of remodelling and, in fact, re-
establishing the Small Arms Factory at Enfield.   The wonderful success of the needle gun in the war between Prussia
and Denmark in 1848 occasioned some alarm amongst our
military authorities as to the state of affairs at home.   The
Duke of Wellington to the last proclaimed the sufficiency of
" Brown Bess " as a weapon of offence and defence; but
matters could no longer be deferred.   The United States
Government, though possessing only a very small standing
army, had established at Springfield a small arms factory,
where, by the use of machine tools specially designed to
execute with the most unerring precision all the details of
muskets and rifles, they were enabled to dispense with mere
manual dexterity, and to produce arms to any amount.   It
was finally determined to improve the musketry and rifle
systems of the English army.   The Government resolved to
introduce the American system, by which Arms might be
produced much more perfectly, and at a great diminution of
cost.   It was under such circumstances that the Small Arms
Committee was appointed.

Colonel Colt had brought to England some striking
examples of the admirable machine tools used at Springfield,
and he established a manufactory at Pimlico for the production of his well-known revolvers.   The committee resolved
to make a personal visit to the United States' Factory at
Springfield.   My own business engagements at home prevented my accompanying the members who were selected;
but as my friend John Anderson (now Sir John), acted as
their guide, the committee had in him the most able and
effective helper.   He directed their attention to the most
important and available details of that admirable establishment.   The United States Government acted most liberally,
in allowing the committee to obtain every information on
the subject; and the heads of the various departments, who

were intelligent and zealous, rendered them every attention and civility.

The members of the mission returned home enthusiastically delighted with the results of their inquiry. The committee immediately proceeded with the entire remodelling of the Small Arms Factory at Enfield. The workshops were equipped with a complete series of special machine tools, chiefly obtained from the Springfield factory. The United States Government also permitted several of their best and most experienced workmen and superintendents to take service under the English Government.

Such was the origin of the Enfield rifle. The weapon came as near to absolute perfection as possible. It was perfect in action, durable, and excellent in every respect. Even in its conversion to the breech-loader it is still one of the best weapons   It is impossible to give too much praise to Sir John Anderson and Colonel Dixon for the untiring and intelligent zeal with which they carried out the plans, as well as for the numerous improvements which they introduced. These have rendered the Enfield Small Arms Factory one of the most perfect and best regulated establishments in the kingdom.

CHAPTER XX.

I HAD been for some time contemplating the possibility of retiring altogether from business. I had got enough of the world's goods, and was willing to make way for younger men. But I found it difficult to break loose from old associations. Like the retired tallow-chandler, I might wish to go back "on melting days." I had some correspondence with my old friend David Roberts, Royal Academician, on the subject. He wrote to me on the 2d June 1853, and said :—

"I rejoice to learn, from the healthy tone that breathes throughout your epistle, that you are as happy as every one who knows you wishes you to be, and as prosperous as you deserve. Knowing, also, as I do, your feeling for art and all that tends to raise and dignify man, I most sincerely congratulate you on the prospect of your being able to retire, in the full vigour of manhood, to follow out that sublime pursuit, in comparison with which the painter's art is but a faint glimmering. 'The Landscape of other worlds' you alone have sketched for us, and enlightened us on that with which the ancient world but gazed upon and worshipped in the symbol of Astarte, Isis, and Diana. We are matter-of-fact now, and have outlived childhood. What say you of a photograph of those wonderful drawings ?　　It may come to that." [1]

But I had something else yet to do in my special vocation.　In 1854 I took out a patent for puddling iron by means of steam.　Many of my readers may not know that cast iron is converted into malleable iron by the process called puddling.　The iron, while in a molten state, is violently stirred and agitated by a stiff iron rod, having its end bent like a hoe or flattened hook, by which every portion of the molten metal is exposed to the oxygen of the air, and the supercharge of carbon which the cast iron contains is thus "burnt out."　When this is effectually done the iron becomes malleable and weldable.

This state of the iron is indicated by a general loss of fluidity, accompanied by a tendency to gather together in globular masses.　The puddler, by his dexterous use of the end of the rabbling bar, puts the masses together, and, in fact, welds the new-born particles of malleable iron into puddle balls of about three-quarters of a hundredweight each.　These are successively removed from the pool of the puddling furnace, and subjected to the energetic blows of the steam hammer, which drives out all the scoriae lurking within the spongy puddle balls, and thus welds them into compact masses of malleable iron.　When reheated to a welding heat, they are rolled out into flat bars or round rods, in a variety of sizes, so as to be suitable for the consumer.

The manual and physical labour of the puddler is tedious, fatiguing, and unhealthy.　The process of puddling occupies about an hour's violent labour, and only robust young men can stand the fatigue and violent heat.　I had frequent opportunities of observing the labour and unhealthiness of the process, as well as the great loss of time

---

[1] It did indeed "come to that," for I shortly after learned the art of photography, chiefly for this special purpose.

required to bring it to a conclusion.   It occurred to me that
much of this could be avoided by employing some other
means for getting rid of the superfluous carbon, and bringing the molten cast iron into a malleable condition.

The method that occurred to me was the substitution of
a small steam pipe in the place of the puddler's rabbling
bar.   By having the end of this steam pipe bent downwards,
so as to reach the bottom of the pool, and then to discharge
a current of steam *beneath the surface of the molten cast iron*,
I thought that I should by this simple means supply a
most effective carbon-oxidating agent, at the same time that
I produced a powerful agitating action within the pool.
Thus the steam would be decomposed and supply oxygen to
the carbon of the cast iron, while the mechanical action of
the rush of steam upwards would cause so violent a com-
motion throughout the pool of melted iron as to exceed the
utmost efforts of the labour of the puddler.   All the gases
would pass up the chimney of the puddling furnace, and the
puddler would not be subject to their influence.   Such was
the method specified in my patent of 1854.[1]

My friend, Thomas Lever Rushton, proprietor of the
Bolton Ironworks, was so much impressed with the soundness of the principle, as well as with the great simplicity of
carrying the invention into practical effect, that he urged
me to secure the patent, and he soon after gave me the
opportunity of trying the process at his works.   The results
were most encouraging.   There was a great saving of labour
and time compared with the old puddling process ; and the
malleable iron produced was found to be of the highest order
as regarded strength, toughness, and purity.   My process
was soon after adopted by several iron manufacturers with
equally favourable results.   Such, however, was the energy
of the steam, that unless the workmen were most careful to

---

[1] Specification of James Nasmyth—Employment of steam in the process of
puddling iron.   May 4, 1854 ; No. 1001.

regulate its force and the duration of its action, the waste of iron by undue oxidation was such as in a great measure to neutralise its commercial gain as regarded the superior value of the malleable iron thus produced.

Before I had time or opportunity to remove this commercial difficulty, Mr. Bessemer had secured his patent of the 17th of October, 1855. By this patent he employed a blast of air to do the same work as I had proposed to accomplish by means of a blast of steam, forced up beneath the surface of the molten cast iron. He added some other improvements, with that happy fertility of invention which has always characterised him. The results were so magnificently successful as to totally eclipse my process, and to cast it comparatively into the shade. At the same time I may say that I was in a measure *the pioneer* of his invention, that I initiated a new system, and led up to one of the most important improvements in the manufacture of iron and steel that has ever been given to the world.

Mr. Bessemer brought the subject of his invention before the meeting of the British Association at Cheltenham in the autumn of 1856. There he read his paper " On the Manufacture of Iron into Steel without Fuel." I was present on the occasion, and listened to his statement with mingled feelings of regret and enthusiasm—of regret, because I had been so clearly anticipated and excelled in my performances ; and of enthusiasm—because I could not but admire and honour the genius who had given so great an invention to the mechanical world. I immediately took the opportunity of giving my assent to the principles which he had propounded. My words were not reported at the time, nor was Mr. Bessemer's paper printed by the Association, perhaps because it was thought of so little importance.[1] But

---

[1] On the morning of the day on which the paper was to be read, Mr. Bessemer was sitting at breakfast at his hotel, when an ironmaster (to whom we was unknown) said, laughing, to a friend within his hearing, " Do you

on applying to Mr., now Sir Henry Bessemer, he was so
kind as to give me the following as his recollection of the
words which I used on the occasion.

"I shall ever feel grateful," says Sir Henry, "for the
noble way in which you spoke at the meeting at Cheltenham of my invention. If I remember rightly, you held up
a piece of my malleable iron, saying words to this effect:
'Here is a true British nugget! Here is a new process
that promises to put an end to all puddling; and I may
mention that at this moment there are puddling furnaces in
successful operation where my patent hollow steam Rabbler
is at work, producing iron of superior quality by the introduction of jets of steam in the puddling process. I do not,
however, lay any claim to this invention of Mr. Bessemer;
but I may fairly be entitled to say that I have advanced
along the road on which he has travelled so many miles,
and has effected such unexpected results that I do not
hesitate to say that I may go home from this meeting and
tear up my patent, for my process of puddling is assuredly
superseded.'"

After giving an account of the true origin of his process,
in which he met with failures as well as successes, and at
last recognised the decarburation of pig iron by atmospheric
air, Sir Henry proceeds to say:—

"I prepared to try another experiment, in a crucible having no
hole in the bottom, but which was provided with an iron pipe put
through a hole in the cover, and passing down nearly to the bottom
of the crucible. The small lumps and grains of iron were packed
around it, so as nearly to fill the crucible. A blast of air was to be

know that there is somebody come down from London to read us a paper on
making steel from cast iron without fuel? Did you ever hear of such non-
sense?" The title of the paper was perhaps a misnomer, but the correctness
of the principles on which the pig iron was converted into malleable iron,
as explained by the inventor, was generally recognised, and there seemed
every reason to anticipate that the process would before long come into
general use.

forced down the pipe so as to rise up among the pieces of granular iron and partially decarburise them. The pipe could then be withdrawn, and the fire urged until the metal with its coat of oxyde was fused, and cast steel thereby produced.

"While the blowing apparatus for this experiment was being fitted up, I was taken with one of those short but painful illnesses to which I was subject at that time. I was confined to my bed, and it was then that my mind, dwelling for hours together on the experiment about to be made, suggested that instead of trying to decarburise the granulated metal by forcing the air down the vertical pipe among the pieces of iron, the air would act much more energetically and more rapidly if I first melted the iron in the crucible, and *forced the air down the pipe below the surface of the fluid metal*, and thus burn out the carbon and silicum which it contained.

"This appeared so feasible, and in every way so great an improvement, that the experiment on the granular pieces was at once abandoned, and, as soon as I was well enough, I proceeded to try the experiment of forcing the air under the fluid metal. The result was marvellous. Complete decarburation was effected in half an hour. The heat produced was immense, but, unfortunately, more than half the metal was blown out of the pot. This led to the use of pots with large hollow perforated covers, which effectually prevented the loss of metal. These experiments continued from January to October 1855. I have by me on the mantelpiece at this moment, a small piece of rolled bar iron which was rolled at Woolwich arsenal, and exhibited a year later at Cheltenham.

"I then applied for a patent, but before preparing my provisional specification (dated October 17, 1855), I searched for other patents to ascertain whether anything of the sort had been done before. I then found your patent for puddling with the steam rabble, and also Martin's patent for the use of steam in gutters while molten iron was being conveyed from the blast furnace to a finery, there to be refined in the ordinary way prior to puddling.

"I then tried steam in my cast steel process, alone, and also mixed with air. I found that it cooled the metal very much, and of itself could not be used, as it always produced solidification. I was nevertheless advised to claim the use of steam as well as air in my particular process (lest it might be used against me), at the same time disclaiming its employment for any purpose except in the production of fluid malleable iron or steel. And I have no doubt it is to this fact that I referred when speaking to you on the occasion you mention. I have deemed it best that the exact truth—so far as a short history can give it—should be given at once to you, who are so true and candid. Had it not been for you and Martin, I should probably never have proposed the use of steam in my process, but the use of air came by degrees, just in the way I have described."

It was thoroughly consistent with Mr. Bessemer's kindly feelings towards me, that, after our meeting at Cheltenham, he made me an offer of one-third share of the value of his patent. This would have been another fortune to me. But I had already made money enough. I was just then taking down my sign-board and leaving business. I did not need to plunge into any such tempting enterprise, and I therefore thankfully declined the offer.

Many long years of pleasant toil and exertion had done their work. A full momentum of prosperity had been given to my engineering business at Patricroft. My share in the financial results accumulated with accelerated rapidity, to an amount far beyond my most sanguine hopes. But finding, from long continued and incessant mental efforts, that my nervous system was beginning to become shaken, especially in regard to an affection of the eyes, which in some respects damaged my sight, I thought the time had arrived for me to retire from commercial life.

Some of my friends advised me to " slack off," and not to retire entirely from Bridgewater Foundry. But to do so was not in my nature. I could not be indifferent to any concern in which I was engaged. I must give my mind and heart to it as before. I could not give half to leisure, and half to business. I therefore concluded that a final decision was necessary. Fortunately I possessed an abundant and various stock of hobbies. I held all these in reserve to fall back upon. They would furnish me with an almost inexhaustible source of healthy employment. They might give me occupation for mind and body as long as I lived. I bethought me of the lines of Burns :—

> " Wi' steady aim some Fortune chase ;
> Keen hope does ev'ry sinew brace ;

> Thro' fair, thro' foul, they urge the race,
>   And seize the prey :
> Then cannie, in some cozie place,
>   They close the day." [1]

It was no doubt a great sorrow for me and my dear wife to leave the Home in which we had been so happy and prosperous for so many years. It was a cozy little cottage at Patricroft. We had named it "Fireside." It was small, but suitable for our requirements. We never needed to enlarge it, for we had no children to accommodate. It was within five minutes' walk of the Foundry, and I was scarcely ever out of reach of the Fireside, where we were both so happy. It had been sanctified by our united love for thirteen years. It was surrounded by a nice garden, planted with trees and shrubs. Though close to the Bridgewater Canal, and a busy manufacturing population was not far off, the cottage was perfectly quiet. It was in this garden, when I was arranging the telescope at night, that I had been detected by the passing boatman as "The Patricroft Ghost."

When we were about to leave Patricroft, the Countess of Ellesmere, who, as well as the Earl, had always been our attached friends, wrote to my wife as follows :—
"I can well understand Mr. Nasmyth's satisfaction at the emancipation he looks forward to in December next. But I hope you do not expect us to share it! for what is so much natural pleasure to you is a sad loss and privation to us. I really don't know how we shall get on at Worsley without you. You have nevertheless my most sincere and hearty good wishes that the change may be as grateful to you both as anything in this world can be."

Yet we had to tear ourselves away from this abode of peace and happiness. I had given notice to my partner [2] that

---

[1] "Letter to James Smith," 18th verse.

[2] The "Partner" here referred to, was my excellent friend Henry Garnett, Esq., of Wyre Side, near Lancaster. He had been my sleeping partner or

it was my intention to retire from business at the end of
1856. The necessary arrangements were accordingly made
for carrying on the business after my retirement. All was
pleasantly and satisfactorily settled several months before I
finally left; and the character and prosperity of the Bridgewater Foundry have been continued to the present day.

But where was I to turn to for a settled home? Many
years before I had seen a charming picture by my brother
Patrick of "A Cottage in Kent." It took such a hold of
my memory and imagination that I never ceased to
entertain the longing and ambition to possess such a cottage
as a cozy place of refuge for the rest of my life. Accordingly, about six months before my final retirement, I
accompanied my wife in a visit to the south. In the first
place we made a careful selection from the advertisements in
the *Times* of "desirable residences" in Kent. One in particular appeared very tempting. We set out to view it.
It seemed to embody all the conditions that we had pictured
in our imagination as necessary to fulfil the idea of our
"Cottage in Kent." It had been the property of F. R. Lee,
the Royal Academician. With a few alterations and
additions it would entirely answer our purpose. So we
bought the property.

I may mention that when I retired from business, and
took out of it the fortune that I had accumulated during
my twenty-two years of assiduous attention and labour,
I invested the bulk of it in Three per cent Consols. The
rate of interest was not high, but it was nevertheless secure.
High interest, as every one knows, means riskful security.
I desired to have no anxiety about the source of my income, such as might hinder my enjoying the rest of my
days in the *active leisure* which I desired. I had for some
time before my retirement been investing in consols, which

"Co." for nearly twenty years, and the most perfect harmony always existed
between us.

my dear wife termed "the true antibilious stock," and I
have ever since had good reason to be satisfied with that
safe and tranquillising investment.   All who value the
health-conserving influence of the absence of financial worry
will agree with me that this antibilious stock is about the
best.

The "Cottage in Kent" was beautiful, especially in its
rural surroundings.   The view from it was charming, and

HAMMERFIELD, PENSHURST.

embodied all the attractive elements of happy-looking
English scenery.   The noble old forest trees of Penshurst
Park were close alongside, and the grand old historic mansion of Penshurst Place was within a quarter of a mile's
distance from our house.   There were many other beautiful
parks and country residences in our neighbourhood; the railway station, which was within thirty-five minutes' pleasant
walk, enabling us to be within reach of London, with its
innumerable attractions, in little more than an hour and

a quarter.   Six acres of garden-ground at first surrounded
our cottage, but these were afterwards expanded to sixteen ;
and the whole was made beautiful by the planting of trees
and shrubs over the grounds.   In all this my wife and myself took the greatest delight.

From my hereditary regard for hammers—two broken
hammer-shafts being the crest of our family for hundreds of
years—I named the place " Hammerfield ; " and so it
remains to this day.   The improvements and additions to
the house and the grounds were considerable.   A greenhouse was built, 120 feet long by 32 feet wide.   Roomy
apartments were added to the house.   The trees and
shrubs planted about the grounds were carefully selected.
The conifera class were my special favourites.   I arranged
them so that their natural variety of tints should form the
most pleasing contrasts.   In this respect I introduced the
beech-tree with the happiest effect.   It is bright green in
spring, and in the autumn it retains its beautiful ruddy-
tinted leaves until the end of winter, when they are again
replaced by the new growth.

The warm tint of the beech contrasts beautifully with
the bright green of the conifera, especially of the *Lawson-
iania* and the *Douglassi*—the latter being one of the finest
accessions to our list of conifers.   It is graceful in form, and
perfectly hardy.   I also interspersed with these several
birch-trees, whose slender and graceful habit of growth
forms so fine a contrast to the dense foliage of the conifers.
To thus paint, as it were, with trees, is a high source of
pleasure in gardening.   Among my various enjoyments this
has been about the greatest.

During the time that the alterations and enlargements
were in progress we rented a house for six months at
Sydenham, close to the beautiful grounds of the Crystal
Palace.   This was a most happy episode in our lives, for,
besides the great attractions of the place, both inside and

out, there were the admirable orchestral daily concerts, at which we were constant attendants. We had the pleasure of listening to the noble compositions of the great masters of music, the perfectly trained band being led by Herr Manns, who throws so much of his fine natural taste and enthusiastic spirit into the productions as to give them every possible charm.

From a very early period of my life I have derived the highest enjoyment from listening to music, especially to *melody*, which is to me the most pleasing form of composition. When I have the opportunity of listening to such kind of music, it yields me enjoyment that transcends all others. It suggests ideas, and brings vividly before the mind's eye scenes that move the imagination. This is, to me, the highest order of excellence in musical composition.

I used long ago, and still continue, to whistle a bit, especially when engaged in some pleasant occupation. I can draw from my mental repository a vast number of airs and certain bits of compositions that I had once heard. I possess that important qualification for a musician—" a good ear;" and I always worked most successfully at a mechanical drawing when I was engaged in whistling some favourite air. The dual occupation of the brain had always the best results in the quick development of the con- structive faculty. And even in circumstances where whistling is not allowed I can *think* airs, and enjoy them almost as much as when they are distinctly audible. This power of the brain, I am fain to believe, indicates the natural existence of the true musical faculty. But I had been so busy during the course of my life that I had never any opportunity of learning the practical use of any musical instrument. And here I must leave this interesting subject.

So soon as I was in due possession of my house, I had speedily transported thither all my art treasures—my tele- scopes, my home stock of tools, the instruments of my own

construction, made from the very beginning of my career as
a mechanic, and associated with the most interesting and
active parts of my life.   I lovingly treasured them, and gave
them an honoured place in the workshop which I added to
my residence.   There they are now, and I often spend a
busy and delightful hour in handling my tools.   It is
curious how the mere sight of such objects brings back to
the memory bygone incidents and recollections.   Friends
long dead seem to start up while looking at them.   You
almost feel as if you could converse with the departed.   I
do not know of anything so touchingly powerful in vividly
bringing back the treasured incidents and memories of
one's life as the sight of such humble objects.   Every one
has, no doubt, a treasured store of such material records
of a well - remembered portion of his past life.   These
strike, as it were, the keynote to thoughts that bring back
in vivid form the most cherished remembrances of our
lives.

On many occasions I have seen at sale rooms long
treasured hoards of such objects thrown together in a heap
as mere rubbish.   And yet these had been to some the
sources of many pleasant thoughts and recollections.   But
the last final break-up has come, and the personal belong-
ings of some departed kind heart are scattered far and wide.
These touching relics of a long life, which had almost become part of himself, are " knocked down " to the highest
bidder.   It is indeed a sad sight to witness the uncared-
for dispersion of such objects—objects that had been
lovingly stored up as the most valued of personal treasures.
I could have wished that, as was the practice in remote
antiquity, such touching relics were buried with the dead,
as their most fitting repository.   Then they might have
left some record, instead of being desecrated by the harpies
who wait at sales for such " job lots."

Behold us, then, settled down at Hammerfield for life.

We had plenty to do. My workshop was fully equipped. My hobbies were there, and I could work them to my heart's content. The walls of our various rooms were soon hung with pictures, and other works of art, suggestive of many pleasant associations of former days. Our library bookcase was crowded with old friends, in the shape of books that had been read and re-read many times, until they had almost become part of ourselves. Old Lancashire friends made their way to us when " up in town," and expressed themselves delighted with our pleasant house and its beautiful surroundings.

The continuous planting of the shrubs and trees gave us great pleasure. Those already planted had grown luxuriantly, fed by the fertile soil and the pure air. Indeed, in course of time they required the judicious use of the axe in order to allow the fittest to survive and grow at their own free will. Trees contrive to manage their own affairs without the necessity of much labour or interference. The " survival of the fittest " prevails here as elsewhere. It is always a pleasure to watch them. There are many ordinary old-fashioned roadside flowering plants which I esteem for their vigorous beauty, and I enjoy seeing them assume the careless grace of Nature.

The greenhouse is also a source of pleasure, especially to my dear wife. It is full of flowers of all kinds, of which she is devotedly fond. They supply her with subjects for her brush or her needle. She both paints them and works them by her needle in beautiful forms and groups. This is one of her many favourite hobbies. All this is suitable to our fireside employments, and makes the days and the evenings pass pleasantly away.

CHAPTER XXI.

ACTIVE LEISURE.

WHEN James Watt retired from business towards the close of his useful and admirable life, he spoke to his friends of occupying himself with "ingenious trifles," and of turning "some of his idle thoughts" upon the invention of an arithmetical machine and a machine for copying sculpture. These and other useful works occupied his attention for many years.

It was the same with myself. I had good health (which Watt had not) and abundant energy. When I retired from business I was only forty-eight years old, which may be considered the prime of life. But I had plenty of hobbies, perhaps the chief of which was Astronomy. No sooner had I settled at Hammerfield than I had my telescopes brought out and mounted. The fine clear skies with which we were favoured, furnished me with abundant opportunities for the use of my instruments. I began again my investigations on the Sun and the Moon, and made some original discoveries, of which more anon.

Early in the year 1858 I received a pressing invitation from the Council of the Edinburgh Philosophical Society to give a lecture before their members on the Structure of the Lunar Surface. As the subject was a favourite one with me, and as I had continued my investigations and increased my store of drawings since I had last appeared before an Edinburgh audience, I cheerfully complied with their request. I accordingly gave my lecture before a crowded meeting in the Queen Street Lecture Hall.

The audience appeared to be so earnestly interested by the subject that I offered to appear before them on two successive evenings and give any *viva voce* explanations about the drawings which those present might desire. This deviation from the formality of a regular lecture was attended with the happiest results. Edinburgh always supplies a highly-intelligent audience, and the cleverest and brightest were ready with their questions. I was thus enabled to elucidate the lecture and to expand many of the most interesting points connected with the moon's surface, such as might formerly have appeared obscure. These questioning lectures gave the highest satisfaction. They satisfied myself as well as the audience, who went away filled with the most graphic information I could give them on the subject.

But not the least interesting part of my visit to Edinburgh on this occasion was the renewed intercourse which I enjoyed with many of my old friends. Among these were my venerable friend Professor Pillans, Charles Maclaren (editor of the *Scotsman*), and Robert Chambers. We had a long "dander"[1] together through the Old Town, our talk being in broad Scotch. Pillans was one of the fine old Edinburgh Liberals, who stuck to his principles through good report and through evil. In his position as Rector of the High School, he had given rare evidence of his excellence as a classical scholar. He was afterwards promoted to be a Professor in the University. He had as his pupils some of the most excellent men of my time. Amongst his intimate friends were Sydney Smith, Brougham, Jeffrey, Cockburn—men who gave so special a character to the Edinburgh society of that time.

---

[1] *Dander*—to saunter, to roam, to go from place to place.

We had a delightful stroll through some of the most remarkable parts of the Old Town, with Robert Chambers as our guide. We next mounted Arthur's Seat to observe some of the manifestations of volcanic action, which had given such a remarkable structure to the mountain. On this subject, Charles Maclaren was one of the best living expounders. He was an admirable geologist, and had closely observed the features of volcanic action round his native city. Robert Chambers then took us to see the glacial grooved rocks on another part of the mountain. On this subject he was a master. It was a vast treat to me to see those distinct evidences of actions so remotely separated in point of geological time—in respect to which even a million of years is a humble approximate unit.[1]

What a fine subject for a picture the group would have made! with the great volcanic summit of the mountain behind, the noble romantic city in the near distance, and the animated intelligent countenances of the demonstrators, with the venerable Pillans eagerly listening—for the Professor was then in his eighty-eighth year. I had the happiness of receiving a visit from him at Hammerfield in the following year. He was still hale and active; and although I was comparatively a boy to him, he was as bright and clear-headed as he had been forty years before.

In the course of the same year I accompanied my wife and my sister Charlotte on a visit to the Continent. It was

---

[1] It is to our ever-dropping climate, with its hundred and fifty-two days of annual rain, that we owe our vegetable mould with its rich and beauteous mantle of sward and foliage. And next, stripping from off the landscape its sands and gravels, we see its underlying boulder-clays, dingy and gray, and here presenting their vast ice-borne stones, and there its iceberg pavements. And these clays in turn stripped away, the bare rocks appear, various in colour and uneven in surface, but everywhere grooved and polished, from the sea level and beneath it, to the height of more than a thousand feet, by evidently the same agent that careered along the pavements and transported the great stones."—HUGH MILLER'S *Geological Features of Edinburgh and its Neighbourhood.*

their first sojourn in foreign parts. I was able, in some respects, to act as their guide. Our visit to Paris was most agreeable. During the three weeks we were there, we visited the Louvre, the Luxembourg, Versailles, and the parts round about. We made many visits to the Hôtel Cluny, and inspected its most interesting contents, as well as the Roman baths and that part of the building devoted to Roman antiquities. We were especially delighted with the apartments of the Archbishop of Paris, now hung with fine old tapestry and provided with authentic specimens of mediæval furniture. The quaint old cabinets were beautiful studies; and many artists were at work painting them in oil. Everything was in harmony. When the sun shone in through the windows in long beams of coloured light, illuminating portions of the antique furniture, the pictures were perfect. We were much interested also by the chapel in which Mary Queen of Scots was married to the Dauphin. It is still in complete preservation. The Gothic details of the chapel are quite a study; and the whole of these and the contents of this interesting Museum form a school of art of the best kind.

From Paris we paid a visit to Chartres, one of the most magnificent cathedrals in France. Its dimensions are vast, its proportions are elegant, and its painted glass is unequalled. Nothing can be more beautiful than its three rose-windows. But I am not writing a guide-book, and I must forbear. After a few days more at Paris we proceeded south, and visited Lyons, Avignon, and Nismes, on our way to Marseilles. I have already described Nismes in my previous visit to France. I revisited the Roman amphitheatre, the Maison Quarré, that perfect Roman temple, which, standing as it does in an open square, is seen to full advantage. We also went to see the magnificent Roman aqueduct at Pont du Gard. The sight of the noble structure well repays a visit. It consists of three tiers of arches. Its magnitude, the skilful fitting of its enormous blocks, makes a powerful impression on the mind. It has stood there, in that solitary wooded valley, for upwards of sixteen centuries; and it is still as well fitted for conveying its aqueduct of water as ever. I have seen nothing to compare with it, even at Rome. It throws all our architectural buildings into the shade. On our way back from Marseilles to Paris we visited Grenoble and its surrounding beautiful Alpine scenery. Then to Chambery, and afterwards to Chamounix, where we obtained a splendid view of Mont Blanc. We returned home by way of Geneva and Paris, vastly delighted with our most enjoyable journey.

I return to another of my hobbies. I had an earnest desire to acquire the art and mystery of practical photography. I bought the necessary apparatus, together with the chemicals; and before long I became an expert in the use of the positive and negative collodion process, including the printing from negatives, in all the details of that wonderful and delightful art. To any one who has some artistic taste, photography, both in its interesting processes and glorious results, becomes a most attractive and almost engrossing pursuit. It is a delightful means of educating the eye for artistic feeling, as well as of educating the hands in delicate manipulation. I know of nothing equal to photography as a means of advancing one's knowledge in these respects. I had long meditated a work "On the Moon," and it was for this purpose more especially that I was earnest in endeavouring to acquire the necessary practical skill. I was soon enabled to obtain photographic copies of the elaborate models of parts of the moon's surface, which I had long before prepared. These copies were hailed by the highest authorities in this special department of astronomical research as the best examples of the moon's surface which had yet been produced.

In reference to this subject, as well as to my researches into the structure of the sun's surface, I had the inestimable

happiness of securing the friendship of that noble philosopher, Sir John Herschel. His visits to me, and my visits
to him, have left in my memory the most cherished and
happy recollections. Of all the scientific men I have had
the happiness of meeting, Sir John stands supremely at the
head of the list. He combined profound knowledge with
perfect humility. He was simple, earnest, and companionable. He was entirely free from assumptions of superiority,
and, still learning, would listen attentively to the humblest
student. He was ready to counsel and instruct, as well as
to receive information. He would sit down in my workshop, and see me go through the various technical processes
of casting, grinding, and polishing specula for reflecting telescopes. That was a pleasure to him, and a vast treat to me.

I had been busily occupied for some time in making
careful investigations into the dark spots upon the Sun's
surface. These spots are of extraordinary dimensions, sometimes more than 100,000 miles in diameter. Our world
might be dropped into them. I observed that the spots
were sometimes bridged over by a streak of light, formed of
willow-leaf shaped objects. They were apparently possessed
of voluntary motion, and moved from one side to the other.
These flakes were evidently the immediate sources of the
solar light and heat. I wrote a paper on the subject, which
I sent to the Literary and Philosophical Society of Manchester.[1] The results of my observations were of so novel

---

[1] *Memoirs of the Literary and Philosophical Society of Manchester*, 3d
series, vol. i. p. 407. My first discovery of the "Willow-leaf" objects on
the Sun's surface was made in June 1860. I afterwards obtained several
glimpses of them from time to time. But the occasions are very rare
when the bright sun can be seen in a tranquil atmosphere free from vibra-
tions, and when the delicate objects on its surface can be *clearly defined*.
It was not until the 5th of June 1864 that I obtained the finest sight of
the Sun's spots and the Willow-leaf objects ; it was then that I made a
careful drawing of them, from which the annexed faithful engraving has been
produced. Indeed I never had a better sight of this extraordinary aspect of
the Sun than on that day.

a character that astronomers for some time hesitated to accept them as facts. Yet Sir John Herschel, the chief of astronomers, declared them to be "a most wonderful discovery."

I received a letter from Sir John, dated Collingwood, 21st of May, 1861, in which he said:

"I am very much obliged to you for your note, and by the sight of your drawings, which Mr. Maclaren was so kind as to bring over here the other day. I suppose there can be no doubt as to the reality of the willow-leaved flakes, and in that case they certainly are the most marvellous phenomena that have yet turned up—I had almost said in all Nature—certainly in all Astronomy.

"What can they be? Are they huge phosphorised fishes? If so, what monsters! Or are they crystals? a kind of igneous snow-flakes? floating in a fluid of their own, or very nearly their own, specific gravity? Some kind of solidity or coherence they must have, or they would not retain their shape in the violent movements of the atmosphere which the change of the spots indicate.

"I observe that in the bridges all their axes have an approximate parallelism, and that in the penumbra they are dispersed, radiating from the inside and the outside of the spot, giving rise to that striated appearance which is familiar to all observers of the spots.

"I am very glad that you have pitched your tent in this part of the world, and I only wish it were a little nearer. You will anyhow have the advantage at Penshurst of a much clearer atmosphere than in the north; but here, nearer the coast, I think we are still better off.

"Mr. Maclaren holds out the prospect of our meeting you at Pachley at no distant period, and I hope you will find your way ere long to Collingwood. I have no instruments or astronomical apparatus to show you, but a remarkably pretty country, which is beginning to put on (rather late) its gala dress of spring."

Sir John afterwards requested my permission to insert in his *Outlines of Astronomy*, of which a new edition was about to appear, a representation of "the willow-leaved structure of the Sun's surface,"—which had been published in the Manchester transactions,—to which I gladly gave my assent. Sir John thus expresses himself on the subject: —"The curious appearance of the 'pores' of the Sun's surface has lately received a most singular and unex-

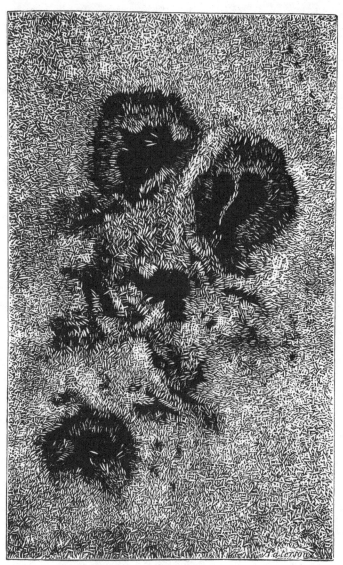

SUN SPOTS AS SEEN BY JAMES NASMYTH, 5TH JUNE 1864.

pected interpretation from the remarkable discovery of Mr. J. Nasmyth, who, from a series of observations made with a reflecting telescope of his own construction under very high magnifying powers, and under exceptional circumstances of tranquillity and definition, has come to the conclusion that these pores are the polygonal interstices between certain luminous objects of an exceedingly definite shape and general uniformity of size, whose form (at least as seen in projection in the central portions of the disc) is that of the oblong leaves of a willow tree. These cover the whole disc of the Sun (except in the space occupied by spots) in countless millions, and lie crossing each other in every imaginable direction. . . . This most astonishing revelation has been confirmed to a certain considerable extent, and with some modifications as to the form of the objects, their exact uniformity of size and resemblance of figure, by Messrs. De la Rue, Pritchard, and Stone in England, and M. Secchi in Rome."

On the 25th of February 1864, I received a communication from Mr. W. J. Stone, first assistant at the Royal Observatory, Greenwich.

"The Astronomer-Royal," he says, "has placed in my hands your letter of February 20. Your discovery of the ' willow leaves ' on the Solar photosphere having been brought forward at one of the late meetings of the Royal Astronomical Society, my attention was attracted to the subject. At my request, the Astronomer-Royal ordered of Mr. J. Simms a reflecting eye-piece for our great equatorial. The eye-piece was completed about the end of January last, and at the first good opportunity I turned the telescope on the Sun.

"I may state that my impression was, and it appears to have been the impression of several of the assistants here, that the willow leaves stand out *dark* against the luminous photosphere. On looking at the Sun, I was at once struck with the apparent resolvability of its mottled appearance. The whole disc of the Sun, so far as I examined it, appeared to be covered over with relatively bright rice-like particles, and the mottled appearance seemed to be produced by the interlacing of these particles.

"I could not observe any particular arrangement of the particles.

2 c

but they appeared to be more numerous in some parts than in others. I have used the word 'rice-like' merely to convey a rough impression of their form.   I have seen them on two occasions since, but not so well as on the first day, when the definition was exceedingly good.

"On the first day that I saw them I called Mr. Dunkin's attention to them.   He appears to have seen them.   He says, however, that he should not have noticed them if his attention had not been called to them."

The Astronomer-Royal, in his report to the Admiralty on my discovery, said :

"An examination of the Sun's surface with the South-East Equatorial, under favourable circumstances, has convinced me of the accuracy of the description, which compares it with interlacing willow leaves or rice grains."

In March 1864 I received a letter from my friend De la Rue, dated from his observatory at Cranford, Middlesex, in which he said : " I like good honest doubting.   Before I had seen with my own eyes your willow leaves, I doubted their real existence, but I did not doubt your having seen what you had drawn.   But when I actually saw them for the first time, I could not restrain the exclamation, ' Why, here are Nasmyth's willow leaves !'   It requires a very fine state of the atmosphere to permit of their being seen, as I have seen them on three or four occasions, when their substantial reality can no longer be doubted." [1]

Sir John Herschel confirmed this information in a letter which I received from him in the following May.   He

---

[1] Let me give another letter from my friend, dated the Observatory, Cranford, Middlesex, October 26, 1864.   He said :—"I am quite pleased to learn that you like the large photograph.   The first given to any friend was destined for and sent to you.   No one has so great a claim on the fruit of my labours ; for you inoculated me with the love of star-gazing, and gave me invaluable aid and advice in figuring specula.   I daresay you may remember the first occasion on which I saw a reflecting telescope, which was then being tried on the sun in a pattern loft at Patricroft.   You may also recall the *volumes* you wrote in answer to my troublesome questions.—Yours very sincerely,   WARREN DE LA RUE."

said "that Mr. De la Rue and a foreign gentleman, Hugo
Müller, had been very successful in seeing and delineating
the 'willow leaves.' They are represented by Mr. M. as
packed together on the edge of a spot, and appear rather like
a bunch of bristles or thorns. In other respects the individual forms agree very well with your delineations."
Another observer had discovered a marvellous resemblance
between the solar spots and the hollows left by the breaking
and subsidence of bubbles, which rise when oil-varnish, which
has moisture in it, is boiled, and the streaky channels
are left by the retiring liquid. "I cannot help," adds
Sir John, "fancying a bare possibility of some upward
outbreak, followed by a retreat of some gaseous matter, or
some dilated portion of the general atmosphere struggling
upwards, and at the same time expanding outwards. I can
conceive of an up-surge of some highly-compressed matter,
which, relieved of pressure, will dilate laterally and upwards
to an enormous extent (as Poullett Scrope supposes of his
lavas full of compressed gases and steam), producing the
spots, and, in that case, the furrows might equally well arise
in the origination as in the closing in of a spot."

I had the honour and happiness of receiving a visit from
Sir John Herschel at my house at Hammerfield in the summer
of 1864. He was accompanied by his daughter. They spent
several days with us. The weather was most enjoyable. I
had much conversation with Sir John as to the Sun spots
and willow leaf shaped objects on the Sun's surface, as well
as about my drawings of the Moon. I exhibited to him my
apparatus for obtaining sound castings of specula for reflecting telescopes. I compounded the alloy, melted it, and
cast a 10-inch speculum on my peculiar common-sense
system. I introduced the molten alloy, chilled it in a
metal mould, by which every chance of flaws and imperfec-
tions is obviated. I also showed him the action and results
of my machine, by which I obtained the most exquisite

polish and figure for the speculum. Sir John was in the
highest degree cognisant of the importance of these details,
as contributing to the final excellent result. It was therefore with great pleasure that I could exhibit these practical
details before so competent a judge.

We had a great set-to one day in blowing iridescent
soap bubbles from a mixture of soap and glycerine. Some
of the bubbles were of about fifteen inches diameter. By
carefully covering them with a bell glass, we kept them for
about thirty-six hours, while they went through their
changes of brilliant colour, ending in deep blue. I contrived this method of preserving them by placing a dish of
water below, within the covering bell glass, by means of
which the dampness of the air prevented evaporation of the
bubble. This dodge of mine vastly delighted Sir John, as
it allowed him to watch the exquisite series of iridescent
tints at his tranquil leisure.

I had also the pleasure of showing him my experiment
of cracking a glass globe filled with water and hermetically
sealed. The water was then slightly expanded, on which
the glass cracked. This was my method of explaining the
nature of the action which, at some previous period of the
cosmical history of the Moon, had produced those bright radiating lines that diverge from the lunar volcanic craters. Sir
John expressed his delight at witnessing my practical illustration of this hitherto unexplained subject, and he considered
it quite conclusive. I also produced my enlarged drawings of
the Moon's surface, which I had made at the side of my
telescope. These greatly pleased him, and he earnestly urged
me to publish them, accompanied with a descriptive account
of the conclusions I had arrived at. I then determined to
proceed with the preparations which I had already made for
my long contemplated work.

Among the many things that I showed Sir John while
at Hammerfield, was a piece of white calico on which I had

got printed *one million spots.* This was for the purpose of exhibiting one million in visible form. In astronomical subjects a million is a sort of unit, and it occurred to me to

FROM A PHOTOGRAPH OF THE MOON, EXHIBITING THE BRIGHT RADIAL LINES.

show what a million really is. Sir John was delighted and astonished at the sight. He went carefully over the out-stretched piece with his rule, measured its length and breadth, and verified its correctness.[1] I also exhibited to him a

---

[1] At a recent meeting of the Metropolitan Railway Company I ex-hibited one million of letters, in order to show the number of passengers (thirty-seven millions) that had been conveyed during the previous twelve months. This number was so vast that my method only helped the meeting

diagram, which I had distributed amongst the geologists at the meeting of the British Association at Ipswich in 1851, showing a portion of the earth's curve, to the scale of one-tenth

GLASS GLOBE CRACKED BY INTERNAL PRESSURE, IN ILLUSTRATION OF THE CAUSE OF THE BRIGHT RADIAL LINES SEEN ON THE MOON.

of an inch to a mile. I set out the height of Mont Blanc, Etna, and also the depth of the deepest mine, as showing the almost incredible minimum of knowledge we possess about even the merest surface of the globe. This diagram

to understand what had been done in the way of conveyance. Mr. Macdonald, of the *Times*, supplied me with one million type impressions, contained in sixty average columns of the *Times* newspaper.

was hailed by many as of much value, as conveying a correct idea of the relative magnitude of geological phenomena in comparison with that of the earth itself.

On this subject Sir Thomas Mitchell, Surveyor-General of Australia wrote to me at the time : " I will not obtrude upon you any crude notions of my own, but merely say that you could not have sent the ' Geological Standard Scale ' to one who better deserved it, if the claim in such favour is, as I suppose, to be estimated by the amount of the time of one whole life, applied to the survey of great mountain ranges, and coasts, rivers, etc. By this long practice of mine, you may know how appreciable this satisfactory standard scale is to your humble servant."

In the winter of 1865 I visited Italy. While at Rome, in April, I had the pleasure of meeting Otto W. von Struve, the celebrated Russian astronomer. He invited me to accompany him on a visit to Father Secchi at his fine observatory of the Collegio Romano. I accepted the invitation with pleasure. We duly reached the Observatory, when Struve introduced me to the Father. Secchi gave me a most cordial and unlooked for welcome. " This," he said, " is a most extraordinary interview ; as I am at this moment making a representation of your willow-leaf shaped constituents of the Solar surface !" He then pointed to a large black board, which he had daubed over with glue, and was sprinkling over (when we came in) with rice grains. " That," said he, " is what I feel to be a most excellent representation of your discovery *as I see it*, verified by the aid of my telescope." It appeared to Father Secchi so singular a circumstance that I should come upon him in this sudden manner, while he was for the first time engaged in representing what I had (on .the spur of the moment when first seeing them) described as willow-leaf shaped objects. I thought that his representation of them, by scattering rice grains over his glue-covered black board, was apt and

admirable ; and so did Otto Struve.   This chance meeting
with these two admirable astronomers was one of the little
bits of romance in my life.

I returned to England shortly after.   Among our visitors
at Hammerfield was Lord Lyndhurst.   He was in his nine-
tieth year when he paid a visit to Tunbridge Wells.
Charles Greville, Secretary to the Privy Council, wrote
to me, saying that his Lordship complained much of the
want of society, and asked me to call upon him.   I did so,
and found him cheerful and happy.   I afterwards sent
him a present of some of my drawings.   He answered :
" A thousand thanks for the charming etchings.   I am
especially interested in Robinson Crusoe.   He looks very
comfortable, but I can't see his bed, which troubles me.
The election (' Everybody for ever !') is wonderful.   I should
not like to be there.   I hope we shall go to you again one
of these days, and have another peep into that wonderful
telescope."

To return to Sir John Herschel.   We returned his visit
at his house at Collingwood, near Hawkhurst.   I found him
in the garden, down upon his knees, collecting crocus bulbs
for next year's planting.   Like myself, he loved gardening,
and was never tired of it.   I mention this as an instance of
his simple zeal in entering practically into all that interested
him.   At home he was the happy father and lover of his
family.   One of his favourite pastimes, when surrounded by
his children in the evening, was telling them stories.   He
was most happy and entertaining in this tranquil occupation.   His masterly intellect could grasp the world and all
its visible contents, and yet descend to entertain his children
with extemporised tales.   He possessed information of the
most varied kind, which he communicated with perfect
simplicity and artlessness.   His profound astronomical
knowledge was combined with a rich store of mechanical and manipulative faculty, which enabled him to take a

keen interest in all the technical arts which so materially aid in the progress of science. I shall never forget the happy days that he spent with me in my workshop. His visits have left in my mind the most cherished recollections. Our friendly intercourse continued unbroken to the day of his death.

The following is the last letter I received from him :—

"COLLINGWOOD, *March* 10, 1871.

" MY DEAR SIR—A great many thanks for the opportunity of see-ing your most exquisite photographs from models of lunar mountains. I hope you will publish them. They will create quite an electric sensation. Would not one or two specimens of the apparently non-volcanic mountain ranges, bordering on the great plains, add to the interest ? Excuse my writing more, as I pen this lying on my back in bed, to which a fierce attack of bronchitis condemns me. With best regards to Mrs. Nasmyth, believe me yours very truly,

"J. F. W. HERSCHEL."

Scientific knowledge seems to travel slowly. It was not until the year 1875, more than fourteen years after my discovery of the willow-leaved bridges over the Sun's spots that I understood they had been accepted in America. I learned this from my dear friend William Lassell. His letter was as follows :—" I see the Americans are appre-ciating your solar observations. A communication I have lately received from the Alleghany Observatory remarks ' that he (Mr. Nasmyth) appears to have been the first to distinctly call attention to the singular individuality of the minute components of the photosphere ; and this seems in fairness to entitle him to the credit of an important dis-covery, with which his name should remain associated.' "

I proceeded to do that which Sir John Herschel had so earnestly recommended, that is, to write out my observations on the Moon. It was a very serious matter, for I had never written a book before. It occupied me many years ; though I had the kind assistance of my friend James Car-penter, then of the Royal Observatory, Greenwich. The

volcanoes and craters, and general landscape scenery of the
Moon, had to be photographed and engraved, and this caused
great labour.

At length the book entitled *The Moon, considered as a
Planet, a World, and a Satellite,* appeared in November 1874.
It was received with much favour and passed into a second
edition.   A courteous and kind review of the book appeared
in the *Edinburgh;* and the notices in other periodicals were
equally favourable.   I dedicated the volume to the Duke of
Argyll, because I had been so long associated with him in geological affairs, and also because of the deep friendship which
I entertained for his Grace.   I presented the volume to him
as well as to many other of my astronomical friends.   I
might quote their answers at great length, from the Astronomer-Royal downwards.   But I will quote two—one from
a Royal Academician and another from a Cardinal.   The
first was from Philip H. Calderon.   He said :—

" Let me thank you many times for your kind letter, and for your
glorious book.   It arrived at twelve to-day, and there has been no
painting since.   Once having taken it up, attracted by the illustrations,
I could not put it down again.   I forgot everything ; and, indeed, I
have been up in the Moon.   As soon as these few words of thanks are
given, I am going up into the Moon again.   What a comfort it is to
read a scientific work which is quite clear, and what a gift it is to write
thus !

" The photographs took my breath away.   I could not understand
how you did them, and your explanation of how you *built* the models
from your drawings only changed the wonder into admiration.   Only
an artist could have said what you say about the education of the eye
and of the hand.   You may well understand how it went home to me.
Ever gratefully yours,   PHILIP H. CALDERON."

I now proceed to the Cardinal.   I was present at one of
the receptions of the President of the Royal Society at Bur-
lington House, when I was introduced to Cardinal Manning
as " The Steam Hammer !"   After a cordial reception he
suddenly said, "But are you not also the Man in the Moon ?"
" Yes, your Eminence ! I have written a book about the

Moon, and I shall be glad if you will accept a copy of it?"
" By all means," he said, " and I thank you for the offer
very much." I accordingly sent the copy, and received the
following answer :—

" My Dear Mr. Nasmyth—When I asked you to send me your
book on the Moon, I had no idea of its bulk and value, and I feel
ashamed of my importunity, yet more than half delighted at my
sturdy begging.

" I thank you for it very sincerely. My life is one of endless work,
leaving me few moments for reading. But such books as yours refresh
me like a clover field.

" I hope I may have an opportunity of renewing our conversation.
Believe me always truly yours,   Henry, Cardinal Manning."

I may also mention that I received a charming letter
from Miss Herschel, the daughter of the late Astronomer.

" Is it possible," she said, " that this beautiful book is destined by you
as a gift to my most unworthy self ? I do not know, indeed, *how* sufficiently to thank you, or even to express my delight in being possessed
of so exquisite and valuable a work, made so valuable, too, by the most
kind inscription on the first page ! I fear I shall be very very far from
understanding the theories developed in the book, though we have
been endeavouring to gather some faint notion of them from the
reviews we have seen ; but it will be of the greatest interest for us to
try and follow them under your guidance, and with the help of these
perfectly enchanting photographs, which, I think, one could never be
tired of looking at.

" How well I remember the original photographs, and the oil
painting which you sent for dear papa's inspection, and which he did
so enjoy ! and also the experiment with the glass globe, in which he
was so interested, at your own house. We cannot but think how he
would have appreciated your researches, and what pleasure this lovely
book would have given him. Indeed, I shall treasure it especially as
a remembrance of that visit, which is so completely connected in my
thoughts with *him*, as well as with your cordial kindness, as a precious
souvenir, of which let me once more offer you my heartfelt thanks. I
remain, my dear sir, yours very truly and gratefully,
    " Isabella Herschel."

I cannot refrain from adding the communication I
received from my dear old friend William Lassell. " I do

not know," he said, " how sufficiently to thank you for your most kind letter, and the superb present which almost immediately followed it.   My pleasure was greatly enhanced by the consideration of how far this splendid work must add to your fame and gratify the scientific world.   The illustrations are magnificent, and I am persuaded that no book has ever been published before which gives so faithful, accurate, and comprehensive a picture of the surface of the Moon. The work must have cost you much time, thought, and labour, and I doubt not you will now receive a gratifying, if not an adequate reward."

After reading the book Mr. Lassell again wrote to me. " I am indebted to your beautiful book," he said, " for a deeper interest in the Moon than I ever felt before. . . . I see many of your pictures have been taken when the Moon was waning, which tells me of many a shivering exposure you must have had in the early mornings. . . . I was sorry to find from your letter that you had a severe cold, which made you very unwell.   I hope you have ere this perfectly recovered.   I suppose maladies of this kind must be expected to take rather severe hold of us now, as we are both past the meridian of life.   I am, however, very thankful for the measure of health I enjoy, and the pleasure mechanical pursuits give me.   I fully sympathise with you in the contempt (shall I say ?) which you feel for the taste of so many people who find their chief pleasure in ' killing something,' and how often their pleasures are fatal !   Two distinguished men killed only the other day in hunting.   For my part I would rather take to the bicycle and do my seventeen miles within the hour."

He proceeds : " I have no doubt your windmill is very nicely contrived, and has afforded you much pleasure in constructing it.   The only drawback to it is, that in this variable climate it is apt to strike work, and in the midst of a job of polishing I fear no increase of wages would induce it to complete its task ! If water were plentiful, you might make it pump up a quantity when the wind served, to be used as a motive power when you chose."

This reference alludes to a windmill which I erected on the top of my workshop, to drive the apparatus below. It was the mirror of a reflecting telescope which was in progress. The windmill went on night and day, and polished the speculum while I slept. In the small hours of the morning I keeked through the corner of the window blinds and saw it hard at work. I prefer, however, a small steam-engine, which works much more regularly.

It is time to come to an end of my Recollections. I have endeavoured to give a brief *résumé* of my life and labours. I hope they may prove interesting as well as useful to others. Thanks to a good constitution and a frame invigorated by work, I continue to lead, with my dear wife, a happy life. I still take a deep interest in mechanics, in astronomy, and in art. It is a pleasure to me to run up to London and enjoy the collections at the National Gallery, South Kensington, and the Royal Academy. The Crystal Palace continues to attract a share of my attention, though, since the fire, it has been greatly altered. I miss, too, many of the dear accustomed faces of the old friends we used to meet there. Still we visit it, and leave to memory the filling up of what is gone. All things change, and we with them.

The following *Dial of Life* gives a brief summary of my career. It shows the brevity of life, and indicates the tale that is soon told. The first part of the semi-circle includes the passage from infancy to boyhood and manhood. While that period lasts, time seems to pass very slowly. We long to be men, and doing men's work. What I have called *The Tableland of Life* is then reached. Ordinary observation shows that between thirty and fifty the full strength of body and mind is reached ; and at that period we energise our faculties to the utmost.

Those who are blessed with good health and a sound constitution may prolong the period of energy to sixty or or even seventy ; but Nature's laws must be obeyed, and the period of decline begins, and usually goes on rapidly. Then comes Old Age ; and as we descend the semi-circle towards eighty, we find that the remnant of life becomes

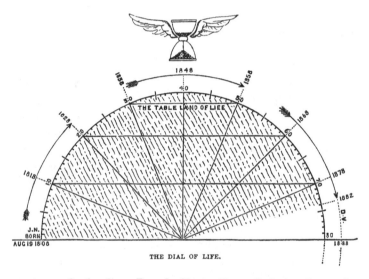

THE DIAL OF LIFE.

vague and cloudy. By shading off, as I have done, the portion of the area of the diagram according to the individual age, every one may see how much of life is consumed, and what is left—*D.V.*

Here is my brief record :—

AGE.	YEAR.	
—	1808.	BORN 19TH AUGUST.
9.	1817.	WENT TO THE HIGH SCHOOL, EDINBURGH.
13.	1821.	ATTENDED THE SCHOOL OF ARTS.
21.	1829.	WENT TO LONDON, TO MAUDSLEY'S.

AGE.	YEARS.	
23.	1831.	RETURNED TO EDINBURGH, TO MAKE MY ENGINEER'S TOOLS.
26.	1834.	WENT TO MANCHESTER, TO BEGIN BUSINESS.
28.	1836.	REMOVED TO PATRICROFT, AND BUILT THE BRIDGE-WATER FOUNDRY.
31.	1839.	INVENTED THE STEAM HAMMER.
32.	1840.	MARRIAGE.
34.	1842.	FIRST VISIT TO FRANCE AND ITALY.
35.	1843.	VISIT TO ST. PETERSBURG, STOCKHOLM, DANNEMORA.
37.	1845.	APPLICATION OF THE STEAM HAMMER PILE-DRIVER.
48.	1856.	RETIRED FROM BUSINESS, TO ENJOY THE REST OF MY LIFE IN THE ACTIVE PURSUIT OF ALL MY MOST FAVOURITE OCCUPATIONS.

I have not in this list referred to my investigations in connection with astronomy. All this will be found referred to in the text. It only remains for me to say that I append a *résumé* of my inventions, contrivances, and workshop "dodges," to give the reader a summary idea of the Active Life of a working mechanic. And with this I end my tale.

CHRONOLOGICAL LIST OF

MECHANICAL INVENTIONS AND TECHNICAL CONTRIVANCES.

By James Nasymth.

1825. *A Mode of applying Steam Power for the Traction of Canal Barges, without injury to the Canal Banks.*

A Canal having been formed to connect Edinburgh with the Forth and Clyde Canal, and so to give a direct water-way communication between Edinburgh and Glasgow, I heard much talk about the desirableness of substituting Steam for Horse power as the means of moving the boats and barges along the canal.   But, as the action of paddle wheels had been found destructive to the canal banks, no scheme of that nature could be entertained.   Although a tyro in such matters, I made an attempt to solve the problem, and accordingly prepared drawings, with a description of my design, for employing Steam power as the tractive agency for trains of canal barges, in such a manner as to obviate all risk of injury to the banks.

The scheme consisted in laying a chain along the bottom of the canal, and of passing any part of its length between three grooved and notched pulleys or rollers, made to revolve with suitable velocity by means of a small steam-engine placed in a tug-boat, to the stern of which a train of barges was attached.

The steam-engine could thus warp its way along the chain, taking it up between the rollers of the bow of the tug-boat, and dropping it into the water at the stern, so as to leave the chain at the service of the next following tug-boat with its attached train of barges. By this simple mode of employing the power of a steam-engine for canal boat traction, all risk of injury to the banks would be avoided, as the *chain* and not the *water* of the canal was the fulcrum or resistance which the steam-engine on the tug-boat operated upon in thus warping its way along the chain; and thus effectually, without slip or other waste of power, dragging along the train of barges attached to the stern of the steam-tug. I had arranged for two separate chains, so as to allow trains of barges to be conveyed along the canal in opposite directions, without interfering with each other.

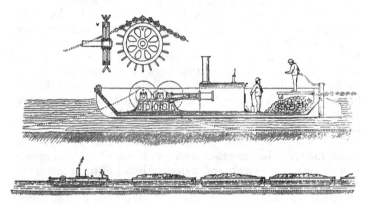

I submitted a complete set of drawings, and a full description of my design in all its details, to the directors of the Canal Company; and I received a complimentary acknowledgment of them in writing. But such was the prejudice that existed, in consequence of the injury to the canal banks resulting from the use of paddle wheels, that it extended to the use of steam power in any form, as a substitute for ordinary horse traction; and although I had taken every care to point out the essential difference of my system (as above indicated) by which all such objections were obviated, my design was at length courteously declined, and the old system of horse traction continued.

In 1845, I had the pleasure to see this simple mode of moving vessels along a definite course in most successful action at the ferry across the Hamoaze at Devonport, in which my system of

2 D

employing the power of a steam-engine on board the ferry boat, to warp its way along a submerged chain lying along the bottom of the channel from side to side of the ferry, was most ably carried out by my late excellent friend, James Rendell, Esq., C.E., and is still, I believe, in daily action, giving every satisfaction.

1826. *An Instrument for Measuring the Total and Comparative Expansion of all Solid Bodies.*

My kind friend and patron, Professor Leslie, being engaged in some investigations, in which it was essential to know the exact comparative total expansion in bulk of metals and other solid bodies, under the same number of degrees of heat, mentioned the subject in the course of conversation. The instrument at that time in use was defective in principle as well as in construction, and the results of its application were untrustworthy. As the Professor had done me the honour to request me to assist him in his experiments, I had the happiness to suggest an arrangement of apparatus, which I thought might obviate the sources of error ; and, with his approval, I proceeded to put it in operation.

My contrivance consisted of an arrangement by means of which the metal bar or other solid substance, whose total expansion under a given number of degrees of heat had to be measured, was in a manner itself converted into a thermometer. Absolutely equal bulks of each solid were placed inside a metal tube or vessel, and surrounded with an exact equal quantity of water at one and the same normal temperature. A cap or cover, having a suitable length of thermometer tube attached to it, was then screwed down, and the water of the index tube was adjusted to the zero point of the scale attached to it, the whole being at say 50° of heat, as the normal temperature in each case. The apparatus was then heated up to say 200° by immersion in water at that temperature. The expansion of the enclosed bar of metal or other solid substance under experiment caused the water to rise above the zero, and it was accordingly so indicated on the scale attached to the cap tube. In this way we had a thermometer whose bulb was for the time being filled with the solid under investigation,—the water surrounding it simply acting as the means by which the expansion of each solid under trial was rendered visible, and its amount capable of being ascertained and recorded with the utmost exactness, as the expansion of the water was in every case the same, and also that of the instrument itself which was " a constant quantity."

In this way we obtained the correct *relative* amount of expansion in bulk of all the solid substances experimented upon. That each bar of metal or other solid substance was of absolutely *equal bulk*, was readily ascertained by finding that each, when *weighed in water*, *lost* the exact same weight. The figure of this simple instrument will be found in the text (p. 120). My friend, Sir David Brewster, was so much pleased with the instrument that he published a drawing and description of it in the *Edinburgh Philosophical Journal*, of which he was then editor.

1827. *A method of increasing the Effectiveness of Steam by super-heating it on its Passage from the Boiler to the Engine.*

One of the earliest mechanical contrivances which I made was for preventing water, in a liquid form, from passing along with the steam from the boiler to the cylinder of the steam-engine. The first steam-engine I made was employed in grinding oil colours for my father's use in his paintings. When I set this engine to work for the first time I was annoyed by slight jerks which now and then disturbed the otherwise smooth and regular action of the machine. After careful examination I found that these jerks were caused by the small quantities of water that were occasionally carried along with the current of the steam, and deposited in the cylinder, where it accumulated above and below the piston, and thus produced the jerks.

In order to remove the cause of these irregularities, I placed a considerable portion of the length of the pipe which conveyed the steam from the boiler to the engine *within* the highly heated side flue of the boiler, so that any portion of water in the liquid form which might chance to pass along with the steam, might, ere it reached the cylinder, traverse this highly-heated steam pipe, and, in doing so, be converted into perfectly dry steam, and in that condition enter the cylinder. On carrying this simple arrangement into practice, I found the result to be in every way satisfactory. The active little steam-engine thenceforward performed its work in the most smooth and regular manner.

So far as I am aware, this early effort of mine at mechanical contrivance was the first introduction of what has since been termed "super-heated steam"—a system now extensively employed, and yielding important results, especially in the case of marine steam-engines. Without such means of supplying dry steam to the engines, the latter are specially liable to "break-downs," resulting from water, in the liquid form, passing into the cylinders along with the steam.

1828. *A Method of "chucking" delicate Metal-work, in order that it may be turned with perfect truth.*

In fixing portions of work in the turning-lathe one of the most important points to attend it is, that while they are held with sufficient firmness in order to be turned to the required form, they should be free from any strain which might in any way distort them. In strong and ponderous objects this can be easily accomplished by due care on the part of an intelligent workman. It is in operating by the lathe on delicate and flexible objects that the utmost care is requisite in the process of chucking, as they are easily strained out of shape by fastening them by screws and bolts, or suchlike ordinary means. This is especially the case with disc-like objects. As I had on several occasions to operate in the lathe with this class of work I contrived a method of chucking or holding them firm while receiving the required turning process, which has in all cases proved most handy and satisfactory.

This method consisted of tinning three, or, if need be, more parts of the work, and laying them down on a tinned face-plate or chuck, which had been heated so as just to cause the solder to flow. As soon as the solder is cooled and set, the chuck with its attached work may then be put in the lathe, and the work proceeded with until it be completed. By again heating the chuck, by laying upon it a piece of red-hot iron, the work, however delicate, can be simply lifted off, and will be found perfectly free from all distortion.

I have been the more particular in naming the use of three points of attachment to the chuck or face-plate, as that number is naturally free from any risk of distortion. I have on so many occasions found the great value of this simple yet most secure mode of fixing delicate work in the lathe, that I feel sure that any one able to appreciate its practical value will be highly pleased with the results of its employment.

The same means can, in many cases, be employed in fixing delicate work in the planing-machine. All that is requisite is, to have a clean-planed wrought iron or brass fixing-plate, to which the work in hand can be attached at a few suitable parts with soft solder, as in the case of the turning lathe above described.

1828. *A Method of casting Specula for Reflecting Telescopes, so as to ensure perfect Freeness from Defects, at the same time enhancing the Brilliancy of the Alloy.*

My father possessed a very excellent acromatic spy-glass of 2 inches diameter. The object-glass was made by the celebrated Ramsden. When I was about fifteen I used it to gaze at the moon, planets, and sun-spots. Although this instrument revealed to me the general characteristic details of these grand objects, my father gave me a wonderful account of what he had seen of the moon's surface by means of a powerful reflecting telescope of 12 inches diameter, made by Short—that justly celebrated pioneer of telescope-making. It had been erected in a temporary observatory on the Calton Hill, Edinburgh. These descriptions of my father's so fired me with the desire to obtain a sight of the glorious objects in the heavens through a more powerful instrument than the spy-glass, that I determined to try and make a reflecting telescope which I hoped might in some degree satisfy my ardent desires.

I accordingly searched for the requisite practical instruction in the pages of the *Encyclopædia Britannica*, and in other books that professed to give the necessary technical information on the subject. I found, however, that the information given in books—at least in the books to which I had access—was meagre and unsatisfactory. Nevertheless I set to work with all earnestness, and began by compounding the requisite alloy for casting a speculum of 8 inches diameter. This alloy consisted of 32 parts of copper, 15 parts of grain tin, and 1 part of white arsenic. These ingredients, when melted together, yielded a compound metal which possessed a high degree of brilliancy. Having made a wooden pattern for my intended 8-inch diameter speculum, and moulded it in sand, I cast this my first reflecting telescope speculum according to the best *book* instructions. I allowed my casting to cool in the mould in the slowest possible manner; for such is the excessive brittleness of this alloy (though composed of two of the toughest of metals) that in any sudden change of temperature, or want of due delicacy in handling it, it is very apt to give way, and a fracture more or less serious is sure to result. Even glass, brittle though it be, is strong in comparison with speculum metal of the above proportions, though, as I have said, it yields the most brilliant composition.

Notwithstanding the observance of all due care in respect of the annealing of the casting by slow cooling, and the utmost care and delicate handling of it in the process of grinding the surface into the requisite curve and smoothness suitable to receive the final polish,—I was on more than one occasion inexpressibly mortified by the sudden disruption and breaking up of my speculum. Thus many hours of anxious care and labour proved of no avail. I had to begin again and proceed *da capo*. I observed, however, that the surplus alloy that was left in the crucible, after I had cast my speculum, when again melted and poured out into a metal ingot mould, yielded a cake that, brittle though it might be, was yet strong in comparison with that of the speculum cast in the sand mould; and that it was also, judging from the fragments chipped from it, possessed of even a higher degree of brilliancy.

The happy thought occurred to me of substituting an open *metal* mould for the closed *sand* one. I soon had the metal mould ready for casting. It consisted of a base plate of cast-iron, on the surface of which I placed a ring or hoop of iron turned to fully the diameter of the intended speculum, so as to anticipate the contraction of the alloy. The result of the very first trial of this simple metal mould was most satisfactory. It yielded me a very perfect casting; and it passed successively through the ordeal of the first rough grinding, and eventually through the processes of polishing, until in the end it exhibited a brilliancy that far exceeded that of the sand mould castings.

The only remaining difficulty that I had to surmount was the risk of defects in the surface of the speculum. These sometimes result from the first *splash* of the melted metal as it is poured into the ring mould. The globules sometimes get oxidised before they became incorporated with the main body of the inflowing molten alloy; and dingy spots in the otherwise brilliant alloy were thus produced. I soon mastered this, the only remaining source of defect, by a very simple arrangement. In place of pouring the melted alloy direct into the ring mould, I attached to the side of it what I termed a "pouring pocket;" which communicated with an opening at the lower edge of the ring, and by a self-acting arrangement by which the mould plate was slightly tilted up, the influx of the molten alloy advanced in one unbroken tide. As soon as the entire surface of the mould plate was covered by the alloy, its weight overcame that of my up-tilting counterpoise, and allowed the entire apparatus to resume its exact level. The resulting speculum was, by these simple arrangements, absolutely perfect in soundness. It was a perfect casting, in all respects worthy of the care and labour which I invested in its future grinding and polishing, and enabled it to perform its glorious duties as the grand essential part of a noble reflecting telescope!

A. Chill plate of cast iron turned to the curve of the speculum. B. Turned hoop of wrought iron with opening at O. C. Pouring pocket. D. Counterpoise, by which the chill plate is tilted up. The largest figure in the engraving is the annealing tub of cast iron filled with sawdust, where the speculum is placed to cool as slowly as possible.

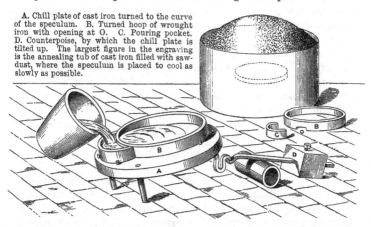

The *rationale* of the strength of speculæ cast in this metal-mould system, as compared with the treacherous brittleness of those cast in sand moulds, arises simply from the consolidation of the molten metal pool taking place first at the lower surface, next the metal base of the mould—the yet fluid alloy above satisfying the contractile requirements of that immediately beneath it; and so on in succession, until the last to consolidate is the top or upper stratum. Thus all risk of contractile tension, which is so dangerously eminent and inherent in the case of sand-mould castings, made of so exceedingly brittle an alloy as that of speculum metal, is entirely avoided. By the employment of these simple and effective improvements in the art of casting the specula for reflecting telescopes, and also by the contrivance and employment of mechanical means for grinding and polishing them, I at length completed my first 8-inch diameter speculum, and mounted it according to the Newtonian plan.   I was most amply rewarded for all the anxious labour I had gone through in preparing it, by the glorious views it yielded me of the wonderful objects in the heavens at night.   My enjoyment was in no small degree enhanced by the pleasure it gave to my father, and to many intimate friends. Amongst these was Sir David Brewster, who took a most lively and special interest in all my labours on this subject.

In later years I resumed my telescope-making enjoyments, as a delightful and congenial relaxation from the ordinary run of my business occupations.   I constructed several reflecting-telescopes, of sizes from 10-inch to 20-inch diameter specula.   I had also the pleasure of assisting other astronomical friends, by casting and grinding specula for them.   Among these I may mention my late dear friend William Lassell, and my excellent friend Warren de la Rue, both of whom have indelibly recorded their names in the annals of astronomical science.   I know of no subject connected with the pursuit of science which so abounds with exciting and delightful interest as that of constructing reflecting telescopes.   It brings into play every principle of constructive art, with the inexpressibly glorious reward of a more intimate acquaintance with the sublime wonders of the heavens.

I communicated in full detail all my improvements in the art of casting, grinding, and polishing the specula of reflecting-telescopes to the Literary and Philosophical Society of Manchester, illustrating my paper with many drawings.   But as my paper was of considerable length, and as the illustrations would prove costly to engrave, it was not published in the Society's Transactions. They are still, however, kept in the library for reference by those who take a special interest in the subject.

1829. *A Mode of transmitting Rotary Motion by means of a Flexible Shaft, formed of a Coiled Spiral Wire or Rod of Steel.*

While assisting Mr. Maudsley in the execution of a special piece of machinery, in which it became necessary to have some holes drilled in rather inaccessible portions of the work in hand, and where the employment of the ordinary drill was impossible, it occurred to me that a flexible shaft, formed of a closely-coiled spiral of steel wire, might enable us to transmit the requisite rotary motion to a drill attached to the end of this spiral shaft. Mr. Maudsley was much pleased with the notion, and I speedily put it in action by a close coiled spiral wire of about two feet in length. This was found to transmit the requisite rotary motion to the drill at the end of the spiral with perfect and faithful efficiency. The difficulty was got over, to Mr. Maudsley's great satisfaction.

So far as I am aware, such a mode of transmitting rotary motion was new and original. The device was useful, and proved of essential service in other important applications. By a suitably close coiled spiral steel wire I have conveyed rotary motion quite round an obstacle, such as is indicated in the annexed figure. It has acted with perfect faithfulness from the winch handle at A to the drill at B. Any ingenious mechanic will be able to appreciate the value of such a flexible shaft in many applications.

Four years ago I saw the same arrangement in action at a dentist's operating-room, when a drill was worked in the mouth of a patient to enable a decayed tooth to be stopped. It was said to be the last thing out in "Yankee notions." It was merely a replica of my flexible drill of 1829.

1829. *A Mode of cutting Square or Hexagonal Collared Nuts or Bolt-Heads by means of a Revolving File or Cutter.*

This method is referred to, and drawings given, in the text, pp. 145-6.

1829. *An Investigation into the Origin and Mode of writing the Cuneiform Character.*

This will be found described in the next and final chapter.

1836. *A Machine for cutting the Key-Grooves in Metal Wheels and Belt Pulleys, of* ANY *Diameter.*

The fastening of wheels and belt pulleys to shafts, so as to enable them to transmit rotary motion, is one of the most frequently-recurring processes in the construction of machinery. This is best effected by driving a slightly tapered iron or steel wedge, or "key" as it is technically termed, into a corresponding recess, or flat part of the shaft, so that the wheel and shaft thus become in effect one solid structure.

The old mode of cutting such key-grooves in the eyes of wheels was accomplished by the laborious and costly process of chipping and filing. Maudsley's mortising machine, which he contrived for the Block machinery, although intended originally to operate upon wood, contained all the essential principles and details required for acting on metals. Mr. Richard Roberts, by some excellent modifications, enabled it to mortise or cut out

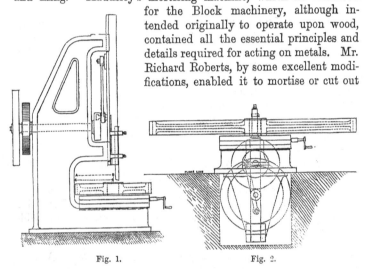

Fig. 1.   Fig. 2.

the key-grooves in metal wheels, and this method soon came into general use. This machine consisted of a vertical slide bar, to the lower end of which was attached the steel mortising tool, which received its requisite up and down motion from an adjustable crank, through a suitable arrangement of the gearing. The wheel to be operated upon was fixed to a slide-table, and gradually advanced, so as to cause the mortising tool to take successive cuts through the depth of the eye of the wheel, until the mortise or key-groove had attained its required depth.

The only drawback to this admirable machine was that its service was limited in respect to admitting wheels whose half diameter did not exceed the distance from the back of the jaw of the machine to the face of the mortise tool; so that to give to this machine the requisite rigidity and strength to resist the strain on the jaw, due to the mortising of the key-grooves, in wheels of say 6 feet diameter, a more massive and cumbrous framework was required, which was most costly in space as well as in money.

In order to obviate this inconvenience, I designed an arrangement of a key-groove mortising machine. It was capable of operating upon wheels of *any* diameter, having no limit to its capacity in that respect. It was, at the same time, possessed, in respect of the principle on which it was arranged, of the power of taking a much deeper cut, there being an entire absence of any source of springing or elasticity in its structure. This not only enabled the machine to perform its work with more rapidity, but also with more precision. Besides, it occupied much less space in the workshop, and did not cost above one-third of the machines formerly in use. It gave the highest satisfaction to those who availed themselves of its effective services.

A comparison of Fig. 1—which represents the general arrangement of the machine in use previous to the introduction of mine— with that of Fig. 2, may serve to convey some idea of their relative sizes. Fig. 1 shows a limit to the admission of wheels exceeding 6 feet diameter, Fig. 2 shows an unlimited capability in that respect.

1836. *An Instrument for finding and marking the Centres of Cylindrical Rods or Bolts about to be turned on the Lathe.*

One of the most numerous details in the structure of all classes of machines is the bolts which serve to hold the various parts together. As it is most important that each bolt fits perfectly the hole it belongs to, it is requisite that each bolt should, by the process of turning, be made perfectly cylindrical. In preparing such bolts, as they come from the forge, in order to undergo the process of turning, they have to be " centred ;" that is, each end has to receive a hollow conical indent, which must *agree with the axis of the bolt*. To find this in the usual mode, by trial and frequent error, is a most tedious process, and consumes much valuable time of the workman as well as his lathe.

In order to obviate the necessity for this costly process, I devised the simple instrument, a drawing of which is annexed. The use of this enabled any boy to find and mark with absolute exactness and rapidity the centres of each end of bolts, or suchlike objects. All that was required was to place the body of the bolt in the V-shaped supports, and to gently cause it to revolve, pressing it longitudinally against the steel-pointed marker, which scratched a neat small circle in the true centre or axis of the bolt.   This

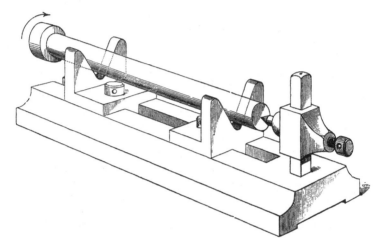

small circle had its centre easily marked by the indent of a punch, and the work was then ready for the lathe.   This humble but really important process was accomplished with ease, rapidity, and great economy.

1836. *Improvement in Steam-Engine Pistons, and in Water and Air-Pump Buckets, so as to lessen Friction and dispense with Packing.*

The desire to make the pistons of steam-engines of air-pump buckets of condensing engines perfectly steam and water tight has led to the contrivance of many complex and costly constructions for the purpose of packing them.   When we take a common-sense view of the subject, we find that in most cases the loss resulting from the extra friction neutralises the expected saving.   This is especially the case with the air-pump bucket of a condensing steam-engine, as it is in reality much more a *water* than an *air* pump.   But when it is constructed with a deep well-fitted bucket, entirely without packing, the loss sustained by such an insignificant amount of leakage as may occur from the want of packing is more than compensated by the saving of power resulting from the total absence of friction.

The first condensing steam-engine, to which I applied an air-pump bucket, *entirely without packing*, was the forty-horse-power engine, which I constructed for the Bridgewater Foundry. It answered its purpose so well that, after twenty years' constant working, the air-pump cover was taken off, out of curiosity, to examine the bucket, when it was found in perfect order. This system, in which I dispensed with the packing for air-pump buckets of condensing steam-engines, I have also applied to the pistons of the steam cylinders, especially those of high-pressure engines of the smaller vertical construction, the stroke of which is generally short and rapid. Provided the cylinder is bored true, and the piston is carefully fitted, and of a considerable depth in proportion to its diameter, such pistons will be found to perform perfectly all their functions, and with a total absence of friction as a direct result of the absence of packing. By the aid of our improved machine tools, cylinders can now be bored with such perfect accuracy, and the pistons be fitted to them with such absolute exactness, that the small quantity of water which the steam always deposits on the upper side of the piston, not only serves as a *frictionless* packing, but also serves as a lubricant of the most appropriate kind. I have applied the same kind of piston to ordinary water-pumps, with similar excellent results.

1836. *An instantaneous Mode of producing graceful Curves, suitable for designing Vases and other graceful objects in Pottery and Glass.*

The mode referred to consists in giving a rapid " switch " motion to a pencil upon a piece of paper, or a cardboard, or a smooth metal plate ; and then cutting out the curve so produced, and employing it as a pattern or "template," to enable copies to be traced from it. When placed at equal distances, and at equal angles on each side of a central line, so as to secure perfect symmetry of form according to the nature of the required design, the beauty of these " instantaneous " curves, as I term them, arises from the entire absence of any *sudden* variation in their course. This is due to the momentum of the hand when " switching " the

Fig. A.

pencil at a high velocity over the paper. By such simple means
was the beautiful curve produced, which is given above.   It

was produced " in a twinkling," if I may use the term to express the rapidity with which it was " switched." The chief source of the gracefulness of these curves consists in the almost imperceptible manner in which they pass in their course from one degree of curvature into another. I have had the pleasure of showing this simple mode of producing graceful curves to several potters, who have turned the idea to good account. The above illustrative figures have all been drawn from " templates " whose curves were " switched " in the manner of Fig. A.

1836. *A Machine for planing the smaller or detail parts of Machinery, whether Flat or Cylindrical.*

Although the introduction of the planing machine into the workshops of mechanical engineers yielded results of the highest importance in perfecting and economising the production of machinery generally, yet, as the employment of these valuable machine tools was chiefly intended to assist in the execution of the larger parts of machine manufacture, a very considerable proportion of the detail parts still continued to be executed by hand labour, in which the chisel and the file were the chief instruments employed. The results were consequently very unsatisfactory, both as regards inaccuracy and costliness.

With the desire of rendering the valuable services of the Planing Machine applicable to the smallest detail parts of machine manufacture, I designed a simple and compact modification of it, such as should enable any attentive lad to execute all the detail parts of machines in so unerring and perfect a manner as not only to rival the hand work of the most skilful mechanic, but also at such a reduced cost as to place the most active hand workman far into the background. The contrivance I refer to is usually known as " Nasmyth's Steam Arm."

None but those who have had ample opportunities of watching the process of executing the detail parts of machines, can form a correct idea of the great amount of time that is practically wasted and unproductive, even when highly-skilled and careful workmen are employed. They have so frequently to stop working, in order to examine the work in hand, to use the straight edge, the square, or the calipers, to ascertain whether they are " working correctly." During that interval, the work is making no progress ; and the loss of time on this account is not less than one-sixth of the working hours, and sometimes much more ; though all this lost time is fully paid for in wages.

But by the employment of such a machine as I describe, even when placed under the superintendence of well-selected intelligent lads, in whom the faculty of good sight and nicety of handling is naturally in a high state of perfection, any deficiency in their physical strength is amply compensated by these self-acting machines.

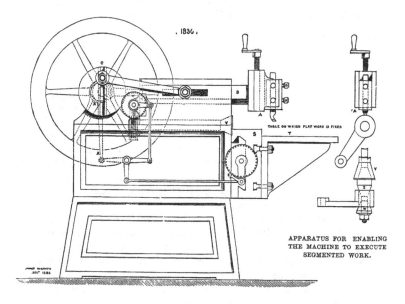

. 1836 .

TABLE ON WHICH FLAT WORK IS FIXED

APPARATUS FOR ENABLING THE MACHINE TO EXECUTE SEGMENTED WORK.

The factory engine supplies the labour or the element of Force, while the machines perform their work with practical perfection. The details of machinery are thus turned out with geometrical accuracy, and are in the highest sense *fitted* to perform their intended purposes.

1837. *Solar Ray Origin of the form of the Egyptian Pyramids, Obelisks, etc.*

This will be found described summarily in the next and final chapter.

1837. *Method of reversing the action of Slide Lathes.*

In the employment of Slide Turning Lathes, it is of great advantage to be able *to reverse the motion of the Slide* so as enable the turning tool to cut towards the Head of the Lathe or away from it, and also to be able to arrest the motion of the Slide altogether, while all the other functions of the lathe are continued in action. All these objects are attained by the simple contrivance represented in the annexed illustration. It consists of a lever E, moving on a stud-pin S, attached to the back of the head stock of the lathe T. This lever carries two wheels of equal diameter marked B and C. These wheels can pitch into a corresponding wheel A, fixed on the back end of the lay spindle. When the handle of the lever E is depressed (as seen in the drawing) the wheel B is in gear with wheel A, while C is in gear with the slide-screw wheel D, and so moves the slide 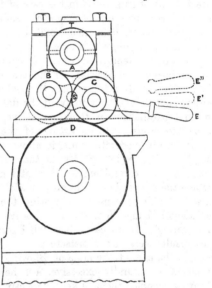 (say *from* the Head Stock of the Lathe). On the other hand, when the lever E is elevated in position E″, wheel B is taken out of gear with A, while C is put in gear with A, and B is put in gear with D ; and thus the Slide is caused to move *towards* the Head Stock of the lathe. Again, where it is desired to arrest the motion of the Slide altogether, or for a time, as occasion may require, the lever handle is put into the intermediate position E′, which entirely severs the communication between A and D, and so arrests the motion of the slide. This simple contrivance effectually served all its purposes, and was adopted by many machine tool-makers and engineers.

2 E

1838. *Self-adjusting Bearings for the Shafts of Machinery.*

A frequent cause of undue friction and heating of rapidly-rotating machinery, arises from some inaccuracy or want of due parallelism between the rotating shaft or spindle and its bearing. This is occasioned in most cases by some accidental change in the level of the supports of the bearings. Many of the bearings are situated in dark places, and cannot be seen. There are others that are difficult of access—as in the case of bearings of screw-propeller shafts. Serious mischief may result before the heating of the bearing proclaims its dangerous condition. In some cases the timber work is set on fire, which may result in serious destruction.

In order to remove the cause of such serious mischief, I designed an arrangement of bearing, which enabled it, and the shaft working in it, to mutually accommodate themselves to each other under all circumstances, and thus to avoid the danger of a want of due and mutual parallelism in their respective axis. This arrangement consisted in giving to the *exterior* of the bearing a *spherical* form, so as, within moderate limits, to allow it to accommodate itself to any such changes in regard to mutual parallelism, as above referred to. In other cases, I employed what I may call *Rocking centres*, on which the Pedestal or "Plumber Block" rested; and thus supplied a self-adjusting means for obviating the evils resulting from any accidental change in the proper relative position of the shaft and its bearing. In all cases in which I introduced this arrangement, the results were most satisfactory.

In the case of the arms of Blowing Fans, in which the rate of rotation is naturally excessive, a spherical resting-place for the bearings enabled them to keep perfectly cool at the highest speed. This was also the case in the driving apparatus for machine tools, which is generally fixed at a considerable height above the machine. These spherical or self-adjusting bearings were found of great service. The apparatus, being generally out of convenient reach, is apt to get out of order unless duly attended to. But, whether or not, the saving of friction is in itself a reason for the adoption of such bearings. This may appear a technical matter of detail; but its great practical value must be my excuse for mentioning it.

1838. *Invention of Safety Foundry Ladle.*

The safety foundry ladle is described in the text, p. 209.

1838. *Invention of the Steam Ram.*

My invention was made at this early date, long before the attack by the steam-ram *Merrimac* upon the *Cumberland*, and other ships, in Hampton Roads, United States. I brought my plans and drawings under the notice of the Admiralty in 1845 ; but nothing was done for many years. Much had been accomplished in rendering our ships shot-proof by the application of iron plates ; but it appeared to me that not one of them 'could exist above water after receiving on its side a single blow from an iron-plated steam ram of 2000 tons. I said, in a letter to the *Times,* " As the grand object of naval warfare is the destruction by the most speedy mode of the ships of the enemy, why should we continue to attempt to attain this object by making small holes in the hull of the enemy when, by one single masterly *crashing* blow from a steam ram, we can crush in the side of any armour-plated ship, and let the water rush in through a hole, 'not perhaps as wide as a church door or as deep as a well, but it will do' ; and be certain to send her below water in a few minutes."

I published my description of the steam ram and its appar· atus in the *Times* of January 1853, and again addressed the Editor on the subject in April 1862. General Sir John Burgoyne took up the subject, and addressed me in the note at the foot of this page.[1] In June 1870, I received a letter from Sir E. J.

---

[1] The following is the letter of General Sir John Burgoyne :—

WAR OFFICE, PALL MALL,
LONDON, *8th April* 1862.

" General Sir John Burgoyne presents his compliments to Mr. Nasmyth, and was much pleased to find, by Mr. Nasymth's letter in the *Times* of this day, certain impressions that he has held for some time confirmed by so good an authority.

" A difficulty seems to be anticipated by many that a steamer used as a ram with high velocity, if impelled upon a heavy ship, would, by the revulsion of the sudden shock, be liable to have much of her gear thrown entirely out of order, parts displaced, and perhaps the boilers burst. Some judgment, however, may be formed on this point by a knowledge of whether such circumstances have occurred on ships suddenly grounding ; and even so, it may be a question whether so great a velocity is necessary.

" An accident occurred some twenty years ago, within Sir John Burgoyne's immediate cognisance, that has led him particularly to consider the great power of a ship acting as a ram. A somewhat heavy steamer went, by acci-

Reed, containing the following extracts :—" I was aware previously that plans had been proposed for constructing unarmoured steam rams, but I was not acquainted with the fact that you had put forward so well-matured a scheme at so early a date ; and it has given me much pleasure to find that such is the case. It has been a cause both of pleasure and surprise to me to find that so long ago you incorporated into a design almost all the features which we now regard as essential to ramming efficiency—twin screws and moderate dimensions for handiness, numerous water-tight divisions for safety, and special strengthenings at the bow. Facts such as these deserve to be put on record. . . . Meanwhile accept my congratulations on the great skill and foresight which your ram-design displays."

Collisions at sea unhappily afford ample evidence of the fatal efficiency of the ramming principle. Even iron-clad ships have not been able to withstand the destructive effect. The *Vanguard* and the *Kurfürst* now lie at the bottom of the sea in consequence of an accidental "end-on" ram from a heavy ship going at a moderate velocity. High speed in a Steam Ram is only desirable when the attempt is made to overtake an enemy's ship ; but not necessary for doing its destructive work. A crash on the thick plates of the strongest Iron-clad, from a Ram of 2000 tons at the speed of four miles an hour, would drive them inwards with the most fatal results.

1839. *Invention of the Steam Hammer, in its general principles
and details.*

Described in text, p. 245.

1839. *Invention of the Floating Mortar, or Torpedo Ram.*

For particulars and details, see Report of Torpedo Committee.

---

dent or mismanagement, end on to a very substantial wharf wall in Kings-town Harbour, Dublin Bay. Though the force of the blow was greatly checked through the measures taken for that purpose, and indeed so much so that the vessel itself suffered no very material injury, yet several of the massive granite stones of the facing were driven some inches in, showing the enormous force used upon them.

" Superior speed will be very essential to the successful action of the ram ; but by the above circumstance we may assume that even a moderate speed would enable great effects to be produced, at least on any comparatively weak point of even ironclad ships, such as the rudder."

1839. *A Double-faced Wedge-shaped Sluice-Valve for Main Street Water-pipes.*

The late Mr. Wicksteed, engineer of the East London Water Company, having stated to me the inconvenience which had been experienced from the defects in respect of water-tightness, as well as the difficulty of opening and closing the valves of the main water-pipes in the streets, I turned my attention to the

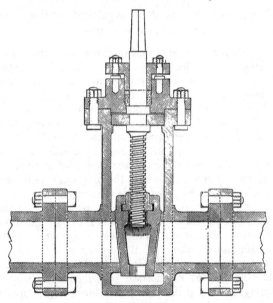

subject. The result was my contrivance of a double-faced wedge-shaped sluice-valve, which combined the desirable property of perfect water-tightness with ease of opening and closing the valve.

This was effected by a screw which raised the valve from its bearings at the first partial turn of the screw, after which there was no farther resistance or friction, except the trifling friction of the screw in its nut on the upper part of the sluice-valve. When screwed down again, it closed simultaneously the end of the entrance pipe and that of the exit pipe attached to the valve case in the most effective manner.

Mr. Wicksteed was so much pleased with the simplicity and efficiency of this valve, that he had it applied to all the main pipes of his Company. When its advantages became known, I received many orders from other water companies, and the valves have since come into general use. The prefixed figure will convey a clear idea of the construction. The wedge form of the double-faced valve is conspicuous as the characteristic feature of the arrangement.

> 1839. *A Hydraulic Matrass Press, capable of exerting a pressure of Twenty thousand tons.*

Being under the impression that there are many processes in the manufacturing arts, in which a perfectly controllable compressing power of vast potency might be serviceable, I many years ago prepared a design of an apparatus of a very simple and easily executed kind, which would supply such a desideratum. It was possessed of a range of compressing or *squeezing* power, which far surpassed anything of the kind that had been invented. As above said, it was perfectly controllable ; so as either to yield the most gentle pressure, or to possess the power of compressing to upwards of twenty thousand tons ; the only limit to its strength being in the materials employed in its construction.

The principle of this enormously powerful compressing machine is similar to that of the Hydraulic Press ; the difference consisting principally in the substitution of what I term a Hydraulic Matrass in place of the cylinder and ram of the ordinary hydraulic press. The Hydraulic Matrass consists of a water-tight vessel or flat bag formed of $\frac{1}{3}$-inch thick iron or steel plates securely riveted together ; its dimensions being 15 feet square by 3 feet deep, and having semicircular sides, which form enables the upper flat part of the Matrass to rise say to the extent of 6 inches, without any injury to the riveted joints, as such a rise or alteration of the normal form of the semicircular sides would be perfectly harmless, and not exceed their capability of returning to their normal curve when the 6-inch rise was no longer necessary, and the elevating pressure removed.

The action of this gigantic press is as follows. The Matrass A A having been filled with water, an additional quantity is supplied by a force pump, capable of forcing in water with a pressure of one ton to the square inch ; thus acting on an avail-

able surface of at least 144 square feet surface—namely, that of
the upper flat surface of the Matrass. It will be forced up by
no less a pressure than twenty thousand tons, and transfer that
enormous pressure to any article that is placed between the rising
table of the press and the upper table. When any object less
thick than the normal space is required to receive the pressure,
the spare space must be filled with a suitable set of iron flat
blocks, so as to subject the article to be pressed to the requisite
power.

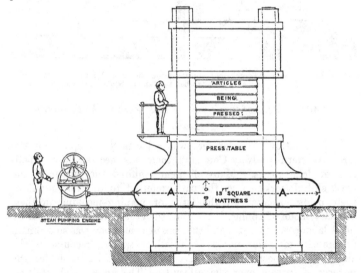

As before stated, there may be many processes in the manufacturing arts, in which such an enormous pressure may be useful ;
and this can be accomplished with perfect ease and certainty. I
trust that this account of the principles and construction of such a
machine may suggest some employment worthy of its powers. In
the general use of the Matrass press, it would be best to supply
the pressure water from an accumulator, which should be kept
constantly full by the action of suitable pumps worked by a small
steam-engine. The great press would require the high-pressure
water only now and then ; so that it would not be necessary to
wait for the small pump to supply the pressure water when the
Matrass was required to be in action.

1840. *A Tapping Square, or instrument by which Perfect Verticality of the Tapping of Screwed Holes is insured.*

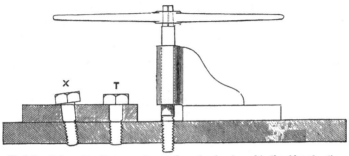

The letter X shows how Screws are frequently made when tapped in the old mode ; the letter T as they are always made when the Tapping Square is employed.

1840. *A Mode of turning Segmental Work in the Ordinary Lathe.*

In executing an order for twenty locomotive engines for the Great Western Railway Company, there was necessarily a repetition of detail parts. Many of them required the labour of the most skilful workmen, as the parts referred to did not admit of their being executed by the lathe or planing-machine in their ordinary mode of application. But the cost of their execution by hand labour was so great. and the risk of inaccuracy was so common (where extreme accuracy was essential), that I had recourse to the aid of special mechanical contrivances and machine tools for the purpose of getting over the difficulty. The annexed illustration has reference to only one class of objects in which I effected great saving in the production, as well as great accuracy in the work. It refers to a contrivance for producing by the turning-lathe the eighty bands of the eccentrics for these twenty engines. Being of a segmental form, but with a projection at each extremity, which rendered their production and finish impossible by the ordinary lathe, I bethought me of applying what is termed the *mangle motion* to the rim of a face plate of the lay, with so many pins in it as to give the required course of segmental motion for the turning tool to operate upon, between the projections C C in the illustration. I availed myself of the limited to-and-fro horizontal motion of the shaft of the mangle motion wheel, as it, at each end of the row of

pegs in the face plate (when it passes from the exterior to the
interior range of them) in giving the feed motion to the tool in the
slide rest, "turned" the segmental exterior of the eccentric hoops.
This it did perfectly, as the change of position of the small shaft
occurred at the exact time when the cut was at its termination,—
that being the correct moment to give the tool "the feed," or
advance for the taking of the next cut. The saving, in respect to
time, was 10 to 1 in comparison with the same amount of work done
by hand labour; while the "truth" or correctness of the work done

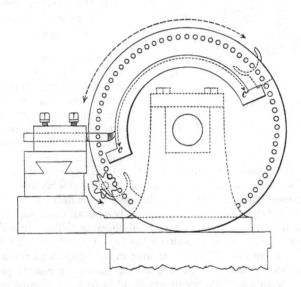

by this handy little application of the turning-lathe was absolutely
perfect. I have been the more particular in my allusion to this
contrivance, as it is applicable to any lathe, and can perform work
which no lathe without it can accomplish. The unceasing industry
of such machines is no small addition to their attractions, in respect
to the production of unquestionably accurate work.

1843. *Invention of the Steam Hammer Pile-driver.*

Described in text, p. 274.

1843. *A Universal Flexible Joint for Steam and Water-pipes.*

The chief novelty in this swivel joint is the manner in which the packing of the joints is completely inclosed, and so rendering them perfectly and permanently water-tight.

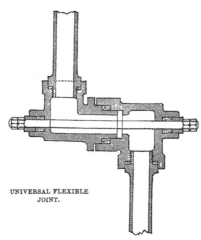

UNIVERSAL FLEXIBLE JOINT.

1844. *An Improvement in Blowing Fans and their Bearings.*

The principle on which Blowing Fans act, and to which they owe their efficiency, consists in their communicating Centrifugal action to the air within them. In order to obtain the maximum force of blast, with the minimum expenditure of power, it is requisite so to form the outside rim of the Fan-case as that each compartment formed by the space between the ends of the blades of the Fan shall in its course of rotation possess an equal *facility of exit* for the passage of the air it is discharging. Thus, in a Fan with six blades, the space between the top of the blades and the case of the Fan should increase in area in the progressive ratios of 1-2-3-4-5-6. If a Fan be constructed on this common-sense principle, we shall secure the maximum of blast from the minimum of driving power. And not only so ; but the humming sound,—so disagreeable an accompaniment to the action of the Fans (being caused by the successive sudden escape of the air from each compartment as it comes opposite the space where it can discharge its confined block of air),—will be avoided. When the outer case of a Fan is formed on the expanding or spiral principle, as above described, all these important advantages will attend its use. As the inward current of air rushes in at the circular openings on each side of the Fan-case, and would thus oppose each other if there was a free communication between them, this is effectually obviated by forming the rotating portion of the fan by a disc of iron plate, which prevents the opposite in-rushing currents from interfering with each other, and at the same time supplies a most substantial means of fastening the blades, as they are conveniently riveted to this central disc. On the whole, this arrangement of machinery supplies a most effective "Noiseless Blowing Fan."

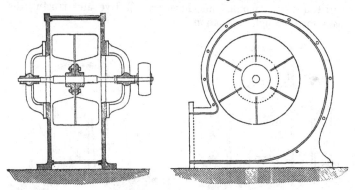

1845. *A Direct Action "Suction" Fan for the Ventilation of Coal-Mines.*

The frequency of disastrous colliery explosions induced me to give my attention to an improved method for ventilating coal-mines. The practice then was to employ a furnace, placed at the bottom of the upcast shaft of the coal-pit, to produce the necessary ventilation. This practice was highly riskful. It was dangerous as well as ineffective. It was also liable to total destruction when an explosion occurred, and the means of ventilation were thus lost when it was most urgently required.

The ventilation of mines by a current of air forced by a Fan into the workings, had been proposed by a German named George Agricola, as far back as 1621. The arrangement is found figured in his work entitled *De Re Metalica*, p. 162. But in all cases in which this system of *forcing* air through the workings and passages of a mine has been tried, it has invariably been found unsuccessful as a means of ventilation.

As all rotative Blowing Fans draw in the air at their centres, and expel it at their circumference, it occurred to me that if we were to make a communication between the upcast shaft of the mine and the centre or *suctional* part of the Fan closing the top of

the upcast shaft, a Fan so arranged would draw out the foul air
from the mine, and allow the fresh air to descend by the down-
cast shaft, and so traverse the workings.   And as a Suction Fan
so placed would be on the surface of the ground, and quite out of
the way of any risk of injury—being open to view and inspection
at all times—we should thus have an effective and trustworthy
means for thorough ventilation.

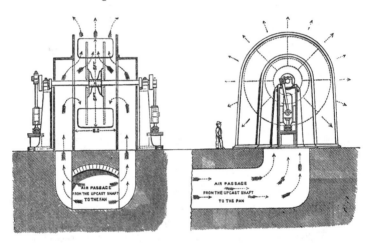

Having communicated the design for my Direct Action Suction
Fan for coal-pit ventilation to the Earl Fitzwilliam, through his
agent Mr. Hartop, in 1850, his lordship was so much pleased
with it that I received an order for one of 14 feet diameter,
for the purpose of ventilating one of his largest coal-pits.   I
arranged the steam-engine which gave motion to the large Fan,
so as to be a part of it; and by placing the crank of the engine
on the end of the Fan-shaft, the engine transferred its power to it
in the most simple and direct manner.   The high satisfaction
which this Ventilating Fan gave to the Earl, and to all connected
with his coal-mines, led to my receiving orders for several of them.

I took out no patent for the invention, but sent drawings and
descriptions to all whom I knew to be interested in coal-mine
ventilation.   I read a paper on the subject, and exhibited the
necessary drawings, at the meeting of the British Association at
Ipswich in 1851.   These were afterwards published in the

Mining Journal. The consequence is that many of my Suction Ventilating Fans are now in successful action at home and abroad.

1845. *An Improvement in the Links of Chain Cables.*

1845. *An Improved Method of Welding Iron.*

One of the most important processes in connection with the production of the details of machinery, and other purposes in which malleable iron is employed, is that termed *welding,*— namely, when more or less complex forms are, so to speak, "built up" by the union of suitable portions of malleable iron united and incorporated with each other in the process of welding. This consists in heating the parts which we desire to unite to a *white heat* in a smith's forge fire, or in an air furnace, by means of which that peculiar adhesive "wax-like" capability of sticking together is induced,—so that when the several parts are forcibly pressed into close contact by blows of a hammer, their union is rendered perfect.

But as the intense degree of heat which is requisite to induce this adhesive quality is accompanied by the production of a molten oxide of iron that clings tenaciously to the white-hot surfaces of the iron, the union will not be complete unless every particle of the adhesing molten scoriæ is thoroughly discharged and driven out from between the surfaces we desire to unite by welding. If by any want of due care on the part of the smith, the surfaces be *concave* or have hollows in them, the scoriæ will be sure to lurk in the recesses, and result in a defective welding of a most treacherous nature. Though the *exterior* may display no evidence of the existence of this fertile cause of failure, yet some undue or unexpected strain will rend and disclose the shut-up scoriæ, and probably end in some fatal break-down.

The annexed figures will perhaps serve to render my remarks on this truly important subject more clear to the reader. Fig. 1 represents an imperfectly prepared surface of two pieces of malleable iron about to be welded. The result of their concavity of form is that the scoriæ are almost certain to be shut up in the hollow part,—as the pieces will unite first at *the edges* and thus include the scoriæ, which no amount of subsequent hammering will ever dislodge. They will remain lurking between, as seen in Fig. 2. Happily, the means of obviating all such treacherous risks are as simple as they are thoroughly effective. All that has to be done to render their occurrence next to impossible is to give to the surfaces we desire to unite by welding a *convex* form as repre-

Fig. 1.

sented in Fig. 3 ; the result of which is that we thus provide an open door for the scoriæ to escape from between the surfaces,—as these unite first in the centre, as due to the convex form, and then the union proceeds outwards, until every particle of scoriæ is expelled, and the union is perfectly com-

Fig. 2.

pleted under the blows of the hammer or other compressing agency. Fig. 4 represents the final and perfect completion of the welding, which is effected by this common-

Fig. 3.

sense and simple means,—that is, by giving the surfaces a *convex* form instead of a *con-cave* one.

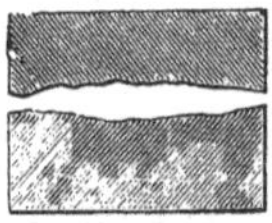

Fig. 4.

When I was called by the Lords of the Admiralty in 1846 to serve on a Committee, the object of which was to investigate the causes of failure in the wrought-iron smith work of the navy, many sad instances came

before us of accidents which had been caused by defective welding, especially in the vitally important articles of Anchors and Chain Cables. In the case of the occasional failure of chain cables, the cause was generally assigned to defective material ; but circumstances led me to the conclusion that it was a question of workmanship or maltreatment of what I knew to be of excellent material. I therefore instituted a series of experiments which yielded conclusive evidence upon the subject ; and which proved that *defective welding* was the main and chief cause of failure. In order to prove this, several apparently excellent cables were, by the aid of "the proving machine," pulled to pieces, link by link, and a careful record was kept of the nature of the fracture. The result was, that out of every 100 links pulled asunder 80 cases clearly exhibited defective welding ; while only 20 were broken through the clear sound metal. This yielded a very important lesson to those specially concerned.

1847. *A Spherical-seated Direct-weighted Safety Valve.*

Having been on several occasions called to investigate the causes of steam boiler explosions, my attention was naturally directed to the condition of the Safety Valve. I found the construction of them in many cases to be defective in principle as well as in mechanical details; resulting chiefly from the employment of a *conical* form in the valve, which necessitated the use of a guide spindle to enable it to keep in correct relative position to its corresponding conical seat, as seen at A in Fig. 1. As this guide

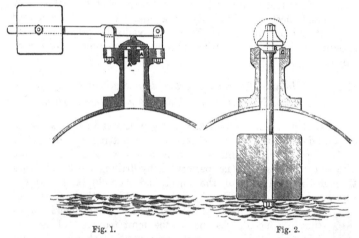

Fig. 1.   Fig. 2.

spindle is always liable to be clogged with the muddy deposit from the boiling water, which yields a very adhesive encrustation, the result is a very riskful tendency to impede the free action of the Safety Valve, and thereby prevent its serving its purpose.

With a view to remove all such causes of uncertainty in the action of this vitally important part of a steam boiler I designed a Safety Valve, having a *spherical* valve and corresponding seat, as seen in B C, Fig. 2. This form of Safety Valve had the important property of fitting to its bearing-seat in all positions, requiring no other guide than its own spherical seat to effect that essential purpose. And as the weight required to keep the valve closed until the exact desired maximum pressure of steam has been attained, is directly attached to the under side of the valve by the rod, the weight, by being inside the boiler, is placed out of reach from any attempt to tamper with it.

The entire arrangement of this Safety Valve is quite simple. It is free from all Lever Joints and other parts which might become clogged; and as there is always a slight pendulous motion in the weight by the action of the water in the boiler, the spherical surfaces of the valve and its seat are thus ever kept in perfect order. As soon as the desired pressure of steam has been reached, and the gravity of the weight overcome, the valve rises from its seat, and gives perfectly free egress to any farther accumulation of steam.   It is really quite a treat, in its way, to observe this truly simple and effective Safety Valve in action.   After I had contrived and introduced this Safety Valve, its valuable properties were speedily acknowledged, and its employment has now become very general.

1847. *A Machine for cutting out Cottar Slots and Key Groove Recesses in Parts of Machinery by a Traversing Drill.*

One of the most tedious and costly processes in the execution of the detail parts of machinery is the cutting out of Cottar Slots in piston rods, connecting rods, and key recesses in shafts. This operation used to be performed by drilling a row of holes through the solid body of the object, and then chipping away the intermediate metal between the holes, and filing the rude slot, so produced, into its required form.   The whole operation, as thus conducted, was one of the most tedious and irksome jobs that an engineer workman could be set to, and could only be performed by those possessed of the highest skill. What with broken chisels and files, and the tedious nature of the work, it was a most severe task to the very best men, not to speak of the heavy cost in wages.

In order to obviate all these disadvantages, I contrived an arrangement of a drilling machine, with a specially formed drill, which at once reduced the process to one of the easiest conducted in an engineer's workshop.   The "special" form of the Drill consisted in the removal of the centre portion of its flat cutting face by making it with a notch O.   This enabled it to cut sideways,†, as well as downwards, and thus to cut a slit or oblong hole.   No labour, as such, was required; but only the intelligent superintendence of a lad to place the work in the machine, and remove it for the next piece in its turn. The machine did the labour, and by its self-action did the work in the most perfect manner.

I may further mention that the arrangement of the machine consisted in causing the object to traverse to and fro in a straight line, of any required length, under the action of the drill. The traversing action was obtained by the employment of an adjust- able crank, which gave the requisite motion to a slide table, on which the work was fastened. The "feed" downwards of the drill was effected by the crank at the moment of its reversing the slide, as the drill reached the end of the traverse; and, as there is a slight pause of the traverse at each end of it, the "feed" for the next cutting taking place at that time, the drill has the opportunity given to perfect its cut ere it commences the next cutting traverse in succession. This action continues in regular course until the drill makes its way right through the piece of work under its action; or can be arrested at any required depth according to the requirements of the work. Soap and water as a lubricator continues to drop into the recess of the slot, and is always in its right place to assist the cutting of the drill.

As before said, the entire function of this most effective machine tool is self-acting. It only required an intelligent lad or labourer to attend to it; and as there was ample time to spare, the superintendence of two of these machines was quite within his ability. The rates of the productive powers of this machine, as compared with the former employment of hand labour, was at least ten to one; to say nothing of the superior quality of the work executed.

Such were the manifold advantages of this machine, that its merits soon became known and appreciated; and although I had taken out no patent for it, we always had an abundance of orders, as it was its own best advertisement.

1848. *A Steam Hammer Form of Steam-Engine.*

This engine is of great simplicity and get-at-ability of parts. It is specially adapted for screw-propelled steamships, and many other purposes. It is now in very general use. The outline is given on the next page.

2 F

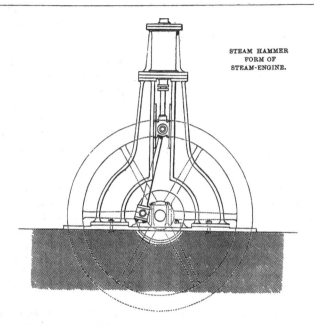

STEAM HAMMER
FORM OF
STEAM-ENGINE.

1848. *An Improved Mode of Punching large Holes in Plate Iron by slightly skewing the Face of the Punch in the Punching Machine.*

1848. *Application of Hydraulic Power to the Punching of Large Holes in Iron Bars, and Plates of Great Thickness.*

Dr. Faraday having applied to me to furnish him, for one of his lectures at the Royal Institution, with some striking example of the Power of Machinery in overcoming the resistance to penetration in the case of some such material as cold malleable iron, it occurred to me to apply the tranquil but vast power of a hydraulic press to punch out a large hole in a thick cake of malleable iron. Knowing that my excellent friend John Hick had in his works at Bolton one of the most powerful hydraulic presses then existing, contrived and constructed by his ingenious father, the late Benjamin Hick, I proceeded to Bolton, and explained Dr. Faraday's requirement, when, with his usual liberal zeal, Mr. Hick at once placed the use of his great hydraulic press at my service.

Having had a suitable cake of steam-hammered malleable iron given to me for the purpose in question, by my valued friend Thomas Lever Rushton of the Bolton Ironworks, we soon had the cake of iron placed in the great press. It was 5 inches thick, 18 inches long, and 15 inches wide. Placing a cylindrical coupling box of cast-iron on the table of the press, and then placing the thick cake of iron on it, and a short cylindrical mass of iron (somewhat of the size and form of a Stilton Cheese) on the iron cake,—the coupling box acting as the Bolster of the extemporised punching machine,—the press was then set to work. We soon saw the Stilton Cheese-like punch begin to sink slowly and quietly through the 5-inch thick cake of iron, as if it had been stiff clay. The only sound heard was when the punched-out mass dropped into the recess of the coupling below. Such a demonstration of tranquil but almost resistless power of a hydraulic press had never, so far as we were aware, been seen before. The punched cake of iron, together with the punched-out disc, were then packed off to Faraday ; and great was his delight at having his request so promptly complied with. Great also was the wonder of his audience when the punched plate was placed upon the lecture table.

This feat of Benjamin Hick's great hydraulic press set me a-thinking. I conceived the idea that the application of hydraulic press power might serve many similar purposes in dealing with ultra thick plates or bar iron,—such as the punching out of holes, and cutting thick bars and plates into definite shapes, as might be required. I suggested the subject to my friend Charles Fox, head of the firm of Fox, Henderson, and Co. He had taken a large contract for a chain bridge, the links of which were to be of thick flat iron bars, with the ends broadened out for the link-pins to pass through. He had described to me the trouble and cost they had occasioned him in drilling the holes, and in cropping the rude-shaped ends of the bars into the required form. I advised him to try the use of the hydraulic press as a punching-machine, and also as a cutting-machine to dress the ends of the great links. He did so in due time, and found the suggestion of great service and value to him in this, and in other cases of a similar kind. The saving of cost was very great, and the work was much more perfect than under the former system.

1848. *An Alternately-pegged " Shive " or Pulley for Rope Band Power Transmission.*

1848. *A Turn-table "Trunion Vision" Reflecting Telescope.*

This is so arranged that the observer can direct the Telescope and view an object in any part of the heavens without moving from his seat, which is attached to the turn-table. For explanations, see text, p. 351.

1850. *A Double or Ambidexter Self-acting Turning-Lathe, with "Dead Cutters," specially adapted for turning Bolts and suchlike detail Parts of Machinery.*

This is a very valuable tool. It requires only one attendant. It is especially useful as regards efficiency and economy. It will be sufficiently understood by mechanical engineers from the annexed drawings.

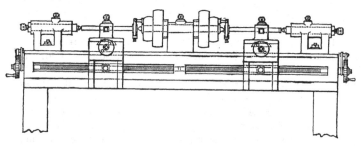

1852. *A Solid-bar "Link-Valve Motion," especially valuable for the larger class of Marine Steam-Engines.*

1854. *Steam Puddling Patent.*

This was the "pioneer" of the Bessemer process. See Bessemer correspondence, p. 365.

1854. *A reversible Rolling Mill without Fly-wheel.*

This Rolling Mill consists of two combined steam-engines, acting on cranks at right angles, the reversing of the rolls being effected by the link motion. The requisite rolling power is obtained by suitable wheel and pinion gear, so as to be entirely independent of the momentum of a fly-wheel, which is entirely dispensed with.

This invention was first brought into use by Mr. Ramsbotham at the Crewe works of the London and North Western Railway. It soon came into general use, especially for rolling long and heavy bars and plates. It enables the workmen to "see-saw" these ponderous objects, and pass them to and fro through the rolls with the utmost ease,—to the great saving of heat, time, and labour.

1854. *Drilling Tunnels through Hard Rock.*

Besides these contrivances and methods of accomplishing mechanical objects, I have on several occasions read papers, prepared drawings, and given suggestions, out of which have come so-called "inventions" made by others. For instance, at the meeting of the British Association in Liverpool in 1854, I read a paper and exhibited drawings before the Mechanical Science Section, on my method of drilling tunnels through hard rock. The paper and drawings excited considerable interest among the railway engineers who were present. I afterwards met Mr. George Newmann, C.E., who consulted me on the same subject. Several years after (21st April 1863) I received the following letter from him :—

"DEAR SIR—Some few years ago, I had the pleasure of spending an evening in your company at my relative's (Mr. C. Withington) house at Pendleton. As I was then Engineer to the Victor Emmanuel Railway, and had made a survey of the Mont Cenis for the purpose of the Tunnel, I consulted you as to the application of the machinery for that work. You suggested the driving of drills in a manner similar to a piston-rod, with other details. On my return to Savoy, I communicated these ideas to Mr. Bartlett, the contractor's agent, and I recommended him to get a small trial machine made. This he had done in a few months, and then he claimed the whole idea as his own. The system has since been carried out (see *Times*, 4th April 1863) by compressed air instead of steam. I call your attention to this, as you may contradict, if you think proper, the assertion in the article above mentioned, that the idea originated with Bartlett."

I did *not*, however, contradict the assertion. I am glad that my description and drawings proved in any way useful towards the completion of that magnificent work, the seven-mile tunnel under Mont Cenis.

1862. *Chilled Cast-Iron Shot.*

In like manner, I proposed the use of Chilled Cast-Iron Shot at a meeting of the Mechanical Science Section of the British

Association, held at Cambridge in October 1862. Up to that
time hardened steel shot had been used to penetrate thick iron
plates, but the cost was excessive, about £30 a ton. I proposed
that Chilled Cast-Iron should be substituted ; it was more simple
and inexpensive. Considerable discussion took place on the
subject ; and Sir William Fairbairn, who was President of the
Section, said that "he would have experiments made, and he
hoped that before the next meeting of the Association, the matter
would be proved experimentally." A brief report of the discussion is given in the *Times* of the 7th October, and in the
Athenæum of the 18th October, 1862. Before, however, the
matter could be put to the test of experiment, Major Palliser had
taken out his Patent for the invention of Chilled Cast-Iron Shot,
in May 1863, for which he was afterwards handsomely rewarded.

I do not wish to "grasp" at any man's inventions, but it is
right to claim my own, and to state the facts. The discussion
above mentioned took place upon a paper read by J. Aston, Esq.,
Q.C., who thus refers to the subject in his letter to me, dated the
7th January 1867 :—

"I perfectly remember the discussion which took place at the meeting of
the British Association at Cambridge in 1862, upon the material proper to be
used as projectiles. The discussion arose after a paper had been read by me
in the Mechanical Section upon "Rifled guns and projectiles adapted for
attacking armour plates." The paper was, I think, printed by the Association in their Report for 1862. You spoke, I believe, at some length on the
occasion ; and I recollect that you surprised and much interested all who
were present, by strenuously urging the use of Chilled Cast-iron for shot and
shell, intended for penetrating armour plates.

"Having embraced all opportunities, and I had many at that time, of
ascertaining all that was done in the way of improving rifled projectiles, I
entertained a very strong opinion that experiments had shown that ordinary
cast-iron was, as compared with steel, of very little value for shot and shell
to be used against iron plates. For that reason, I remember I took an opportunity, after the termination of the discussion, in which you held your own
against all comers in favour of chilled cast-iron, of questioning you closely on
the subject, and you gave me, I admitted, good reason for the opinion you
expressed. You also urged me to cause a trial to be made of chilled cast iron
for shell, such as I had shown to the section, and which (in hardened steel
shot) had been fired by Mr. Whitworth through thick iron plates. This I had
not an opportunity of doing. Term began soon after, and Temple occupations then took up all my time.

"There can be no doubt whatever that any one who may claim to have
been before you in teaching the public the use of Chilled Cast-Iron for pro-

jectiles intended to penetrate iron plates, must give proof of having so done prior to your vigorous advocacy of that material at the Cambridge Meeting in 1862.—Yours very sincerely,   J. ASTON."

In another letter, Mr. Aston says—" It is quite right of you to assert your claim to that which in fact belongs to you." I did not, however, assert my claim ; and, with these observations and extracts, I leave the matter, stating again the fact that my public communication of the invention was made in October 1862 ; and that the patent for the invention was taken out by Major Palliser in May 1863.

I have only mentioned the more prominent of my inventions and contrivances. Had I described them fully I should have required another volume. I have the satisfaction to know that many of them have greatly advanced the progress of the mechanical arts, though they may not be acknowledged as mine. I patented very few of my inventions. The others I sowed broadcast over the world of practical mechanics. My reward is in the knowledge that these " children of my brain " are doing, and will continue to do, good service in time present and in time to come.

In mechanical structures and contrivances, I have always endeavoured to attain the desired purpose by the employment of the *Fewest Parts*, casting aside every detail not absolutely necessary, and guarding carefully against the intrusion of mere *traditional* forms and arrangements. The latter are apt to insinuate themselves, and to interfere with that simplicity and directness of action which is in all cases so desirable a quality in mechanical structures. PLAIN COMMON SENSE should be apparent in the general design, as in the form and arrangement of the details ; and a general character of *severe utility* pervade the whole, accompanied with as much attention to gracefulness of form as is consistent with the nature and purpose of the structure.

SUN-RAY ORIGIN OF THE PYRAMIDS, AND THE CUNEIFORM CHARACTER.

BEFORE I take my leave of the public, I wish to put on record my speculations as to the origin of two subjects of remote antiquity, viz.: the Sun-ray origin of the Pyramids, and the origin of the Arrow-head or Cuneiform Character.

First, with respect to the Sun-ray origin of the Egyptian Pyramids.

In pursuing a very favourite subject of inquiry, namely, *the origin of forms,* no portion of it appears to me to be invested with so deep an interest as that of the Worship of the Sun,—one of the most primitive and sacred foundations of adorative religion,—affecting, as it has done, architectural structures and numerous habits and customs which have come down to us from remote antiquity, and which owe their origin to its influence.

On many occasions, while beholding the sublime effects of the Sun's Rays streaming down on the earth through openings in the clouds near the horizon, I have been forcibly impressed with the analogy they appear to suggest as to the form of the Pyramid, while the single vertical ray suggests that of the Obelisk.

In following up this subject, I was fortunate enough to find what appears to me a strong confirmation of my views, namely, that the Pyramid, as such, *was a sacred form.* I met with many examples of this in the Egyptian Collection at the Louvre at Paris ; especially in small pyramids, which were probably the objects of household worship. In one case I found a small pyramid, on the upper part of which appeared the disc of the Sun, with pyramidal rays descending from it on to figures in the Egyptian attitude of adoration. This consists in the hands held up before the eyes—an attitude expressive of the brightness of the object adored. It is associated with

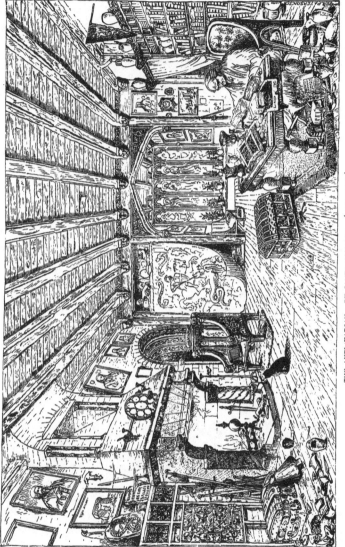

THE ANTIQUARIAN. BY JAMES NASMYTH. (FACSIMILE.)

the brightness of the Sun, and it still survives in the *Salaam*, which expresses profound reverence and respect among Eastern nations. It also survives in the disc of the Sun, which has for ages been placed like a halo behind the heads of sacred and exalted personages, as may be seen in eastern and early paintings, as well as in church windows at the present day.

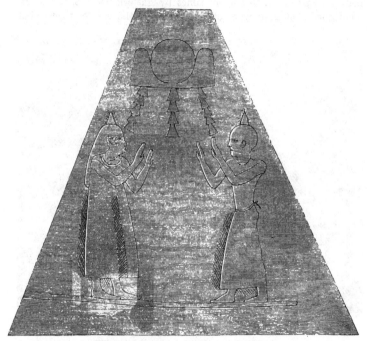

Fig. 1.—RUBBING FROM A HOUSEHOLD PYRAMID.

This is also intimately connected with lighted lamps and candles, which latter may often be met with in Continental churches, as well as in English Ritualist Churches at the present day. In Romish Continental churches they are stuck on to *pyramidal* stands, and placed before pictures and images of sacred personages. All such lighted lamps or candles are *survivals* of that most ancient form of worship,—that of THE SUN !

The accompanying illustrations will serve in some degree to

confirm the correctness of my views as to this very interesting subject. Fig. 1 is from a "rubbing" of one of the many small or "Household" pyramids in the Louvre Collection at Paris; while Fig. 2 is an attempt to illustrate in a graphic manner the derivation of the *form* of the Pyramid and Obelisk from the Sun's Rays.

In connection with the worship of the Sun and other heavenly bodies, as practised in ancient times by Eastern nations, it may be mentioned that their want of knowledge of the vast distances that

JAMES NASMYTH
1882   Fig. 2.—SUN-RAY ORIGIN OF THE PYRAMID.

separate them from the earth led them to the belief that these bodies were so near as to exert a direct influence upon man and his affairs. Hence the origin of Astrology, with all its accompanying mystifications; this was practised under the impression that the Sun, Moon, and planets, were near to the earth. The summits of mountains and "High Places" became "sacred," and were for this reason resorted to for the performance of the most important religious ceremonies.

As the "High places" could not be transported to the Temples, the cone-bearing trees, which were naturally associated with these elevated places, in a manner partook of their sacred character, and the fruit of the trees became in like manner sacred. Hence the Fir Cone became a portable emblem of their sacredness; and, accordingly, in the Assyrian Worship, so clearly represented to us in the Assyrian Sculptures in our Museums, we find the Fir Cone being presented by the priests towards the head of their kings as a high function of Beatification. So sacred was the Fir Cone, as the fruit of the sacred tree, that the priest who presents it has a reticule-shaped bag in which, no doubt, the sacred emblem was reverently deposited when not in use for the performance of these high religious ceremonies.

The same emblem "survived" in the Greek worship. I annex a tracing from a wood-engraving in *Fellows's Researches in Asia Minor*, 1852 (p. 175), showing the Fir Cone as the finial to the staff of office of the Wine-god Bacchus. To this day it is employed to stir the juice of the grape previous to fermentation, and so sanctifying it by contact with the fruit of the Sacred Tree. This is still practised by the Greeks in Asia Minor and in Greece, though introduced in times of remote antiquity. The Fir Cone communicates to most of the Greek wines that peculiar turpentine or resinous flavour which is found in them. Although the sanctification motive has departed, the resinous flavour is all that survives of a once most sacred ceremony, as having so close a relation to the worship of the Sun and the heavenly bodies.

In like manner, it appears to me highly probable that "The Christmas Tree," with its lighted tapers, which is introduced at that sacred season for the entertainment of our young people, is "a survival" of the worship of the sacred tree and of the Sun. The toys which are hung on the twigs of the tree may also be "survivals" of the offerings which were usually made to the Sun and the heavenly bodies. If I am correct in my conjecture on this subject, it throws a very interesting light on what is considered as a mere agent for the amusement of children.

Next, with respect to the Cuneiform Character. When I first went to reside in London, in 1829, I often visited the British Museum. It was the most instructive and interesting of all the public institutions which I had yet seen. I eagerly seized every opportunity I could spare to spend as many hours as possible in wandering through its extensive galleries, especially those which contained the Assyrian, Egyptian, and Greek antiquities. By careful and repeated examination of the objects arranged in them, I acquired many ideas that afforded me subjects for thought and reflection.

Amongst these objects, I was specially impressed and interested with the so-called "Arrow-head" or "Cuneiform Inscriptions" in

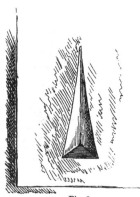

Fig. 1.

the Assyrian Department. These remarkable inscriptions were on large tablets of burnt clay. They formed the chief portion of the then comparatively limited collection of Assyrian antiquities in the British Museum. I was particularly impressed with the precision and simple beauty of these cuneiform inscriptions,—especially with the strikingly distinctive nature of what I may term the fundamental or elementary *wedge-like* form, of which the vast variety of letters or words of these inscriptions were composed. The triangular or three-sided indentation will be observed in the annexed engraving (Fig. 1).

This elementary form, placed in various positions with respect to each other, appeared to be capable of yielding an infinite variety of letters and words, as seen in Fig. 2. I may here mention that I entered upon this interesting subject with no pretensions as a linguist, nor with any idea of investigating the *meaning* of these remarkable inscriptions ; but only as a Mechanic, to ascertain the manner in which the striking characters were produced, so as to convey words and ideas through their variety of combinations.

I soon perceived that the simple but distinctive characters shown in the above representations were essentially connected with the employment of plastic clay ; this being the material most suitable for their impression, by means of a three-sided instrument or stylus. The angular extremity of this instrument, when depressed into the surface of a tablet of plastic clay in different positions and directions, would leave these cunei-form impressions in all their beautifully dis-tinct and char-acteristic forms. And thus, after the tablets had been subjected to fire and made into hard brick, the impressions have come down to us, after the lapse of thou-

Fig. 2.

sands of years, as fresh and distinct as if they had been produced but yesterday!

I was so fortunate as to have my conjectures confirmed with respect to the exact form of the instrument by which these remark-able characters are produced, by observing, in what appeared to be a hastily-formed inscription on the edge of a large brick, that the inscriber had apparently used rather more pressure on his stylus than was requisite. In consequence of which, the end of it had been so deeply depressed into the soft clay as to leave an exact counterpart of its size and form. I secured a cast of this over deep impression of the stylus, from which Fig. 3 is taken, after a photograph.

Fig. 3.

In order further to illustrate the simple mode of producing inscriptions on tablets of clay, I give in Fig. 4 a tablet inscription produced by means of the stylus which is seen laid over the tablet.

The next illustration (Fig. 5) is intended to convey an idea of the manner in which the stylus was held and applied to the surface of the clay when a cuneiform inscription was being produced. The upper, flat, or third side of the stylus enabled the inscriber to keep it in correct relative position in respect to the tablet, yielding at the same time a convenient flat surface upon which to rest the end of his finger when indenting the angular end into the clay.

Fig. 4.

Refer back to Fig. 2, and it will be found that any variety in the size of the cuneiform inscriptions may be produced by *the same* stylus, by simply depressing the angular end of it to a greater or less depth into the surface of the clay. In many of the most elaborate inscriptions, a certain lob-sidedness of the cuneiform character may be observed. This is due to the inscriber having held his stylus somewhat askew, as we do a pen in ordinary writing.

Fig. 5.

Referring to my remark that the distinctive shape of the cuneiform character was essentially due to the use of plastic clay as the most suitable material for its production, I think it highly probable that the *origin* of these inscriptions took its rise not only from the facility with which the characters could be indented on the material, but from the abundance of plastic mud which forms the natural soil

of the lands adjoining the great Assyrian rivers. This, when made into bricks, became the chief building material of the energetic

people of Babylon and the other great cities of the Tigris and Euphrates valleys. The laborious work of brickmaking was generally assigned to captives as task-work, and it appears to me highly probable that "the tale" of the brick-maker or his task-master might be most readily marked by simply indenting the side of the soft tale brick with the corner angle of a dry one; and that thus the strikingly peculiar character of the cunei-

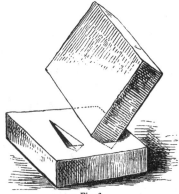

Fig. 6.

form character was produced (see Fig. 6). In course of time the elementary form was expanded into this most beautifully simple mode of communicating ideas through the agency of conventional signs or letters; being also especially suited for making historical or other records on tablets of moist clay, which, when "fired," became absolutely indestructible, so far as time is concerned.

This is abundantly proved by those marvellously perfect burnt clay tablets, covered with exquisitely minute and perfect inscriptions, which, after having remained hid in mounds of rubbish for thousands of years, among the ruins of the Assyrian cities, are

brought to light as fresh and perfect as on the day on which they were executed. These tablets now excite the wonder and admiration of all who are able to appreciate the beauty of the inscriptions, as well as of those who are speculatively curious as to the origin of written language.

Fig. 7.

This attempt to explain the probable origin of the cuneiform character may to some appear fanciful. But whether or not, it is certain that this simple and impressive character can be readily produced by the primitive means

which I have ventured to suggest. I give a cuneiform inscription (Fig. 7), which I have produced by simply employing the corner angle of an ordinary brick as the stylus for indenting the inscription on the tablet of soft clay. This might have been extended to any length, in longer as well as minuter impressions.

As soon as the capability of the cuneiform impression was adopted as the Assyrian character, it was in due time employed for inscriptions on stone or other materials, such as marble or alabaster. The chisel was then substituted for the stylus; but the characters remained in a great measure the same. In some cases a slight modification was observable, being naturally due to the change of material and the method of carving it; but in most respects the departure from the clay prototype is very slight, and the original is adhered to with remarkable integrity.

When examining some early Greek inscriptions in marble, in the British Museum, in the year 1837, I was much interested to observe the appearance of a cuneiform element in the limbs of

Fig. 8.

several Greek letters, especially in the terminals, as illustrated in Fig. 8, each limb of the letters being in itself a perfect cuneiform; and as such the terminal of each limb is at right angles to the axis, and not as now (in our modern capital letters) parallel to the line of inscription.

This apparent presence of the cuneiform element in these early Greek inscriptions suggests some very interesting historic causes which led to their introduction, and so passed from the Greek into the Roman, and eventually into the capital letters of our own

GREEK.   ROMAN.   MODERN.

Fig. 9.

alphabet. To give one instance, — though many might be cited,—take the capital letter T, and it will be found that it went from the Cuneiform into the Greek, then into the Roman, and lastly into our own letter, thus presenting a remarkable instance of the *survival of a form* from remote antiquity down to the present day.

The letters A K H I K M N Y X have the distinct remains

of their Babylonian origin in the top and bottom stroke, which is nothing more nor less than a corruption of the original or primitive arrow-headed impression of the stylus in the moist clay, begun thousands of years ago.

In a lecture which I gave at the Royal Institution in London, in 1839, and in another at the British Association at Cheltenham, in 1856, I referred to this presence of the cuneiform element in the Greek letters, illustrating the subject by actual casts from the inscriptions themselves. At Cheltenham the question gave rise to a most animated and interesting discussion, in which Dr. Whewell and Sir Thomas Phillips (the great antiquarian) took a prominent part. I understood that Sir Thomas Phillips assigned that the intermixture of cuneiform with the Greek alphabet proceeded from the Samaritans, who were originally an Assyrian colony. I find that many Greek inscriptions exhibit the cuneiform element in nearly all the letters composing them. This is a subject well worthy of the attention of our antiquarian Greek scholars, as pointing to an intimate intercourse with the Assyrians at some remote age. The distinctive character of the cuneiform in the Greek inscriptional letters could not have arisen *from chance.* Some intercommunication with the Assyrians must have taken place.

This subject is all the more interesting, as the cuneiform element appears to have passed from the Greek inscriptional letters into those of the Roman, and from thence into our own capital letters. This affords a very remarkable instance of the "survival" of a form, which, however naturally due to the plastic material in connection with which it originated, nevertheless led to its use for ages after the circumstances which led to its adoption had passed away. This tendency in mankind to cling to shapes and forms through mere traditional influences is widely observable, especially in connection with architectural forms, arrangements, and decorative details. It offers a subject of great interest to those who have a natural aptitude to investigate what I may term *the etymology* of form, a subject of the most attractive nature, especially to those who enjoy thinking and reflecting upon what they have specially observed.

Before concluding this subject I may mention that the Assyrians employed a cylindrical roller-seal in order to produce impressions in a wholesale way. This is exemplified in the annexed engraving. The mechanical principles inherent in this beautifully simple form of roller-seal, indicate a high order of ingenuity, well worthy of the originators of the arrow-headed character. In fact it is the prototype not only of the modern system of calico printing but

ASSYRIAN ROLLER SEAL.

of the Walter Printing Press, by which the *Times* and many other newspapers are now printed—a remarkable instance of the survival or restoration of a very old method of impression.

James Nasmyth

the   marks

" *an Impression before*
Letters "

INDEX.

456   INDEX.

FINIS.

Printed in the United States
By Bookmasters